Technologies, Strategies and Applications of Plant Proteomics

Technologies, Strategies and Applications of Plant Proteomics

Editor: Kaden Scott

MURPHY & MOORE
www.murphy-moorepublishing.com

www.murphy-moorepublishing.com

MURPHY & MOORE

Cataloging-in-Publication Data

Technologies, strategies and applications of plant proteomics / edited by Kaden Scott.
p. cm.
Includes bibliographical references and index.
ISBN 978-1-63987-754-6
1. Plant proteomics. 2. Plant proteins. 3. Plant molecular biology. I. Scott, Kaden.
QK898.P8 T43 2023
572.62--dc23

Murphy & Moore Publishing
1 Rockefeller Plaza,
New York City,
NY 10020, USA

ISBN 978-1-63987-754-6

Contents

Preface

Plant proteomics refers to the broad study of proteins focused on their structures and functions as governed by the physiological metabolism of plant cells. It is used in researches for studying gene products and validating their predicted biological functions. It also plays an integral role in understanding the major biochemical signaling pathways involved in the response of plants to their environment. Biotechnological applications in plant breeding rely on the data gathered from proteomic investigations. Proteomics has made it possible to identify a large variety of proteins that are essential for enhancing product quality, preserving crop health and developing resistance to stressors. This book explores the technologies, strategies and applications of plant proteomics. It consists of contributions made by international experts. A number of latest researches have been included herein to keep the readers updated with the latest concepts in this area of study.

The researches compiled throughout the book are authentic and of high quality, combining several disciplines and from very diverse regions from around the world. Drawing on the contributions of many researchers from diverse countries, the book's objective is to provide the readers with the latest achievements in the area of research. This book will surely be a source of knowledge to all interested and researching the field.

In the end, I would like to express my deep sense of gratitude to all the authors for meeting the set deadlines in completing and submitting their research chapters. I would also like to thank the publisher for the support offered to us throughout the course of the book. Finally, I extend my sincere thanks to my family for being a constant source of inspiration and encouragement.

Editor

1

Proteome Characterization of Leaves in Common Bean

Faith M. Robison [1,2], Adam L. Heuberger [3], Mark A Brick [2] and Jessica E. Prenni [1,4,*]

1 Proteomics and Metabolomics Facility, Colorado State University, Fort Collins, CO 80523, USA; E-Mail: fmr33@cornell.edu
2 Department of Soil and Crop Sciences, Colorado State University, Fort Collins, CO 80523, USA; E-Mail: Mark.Brick@ColoState.EDU
3 Department of Horticulture and Landscape Architecture, Colorado State University, Fort Collins, CO 80523, USA; E-Mail: Adam.Heuberger@colostate.edu
4 Department of Biochemistry, Colorado State University, Fort Collins, CO 80523, USA

* Author to whom correspondence should be addressed; E-Mail: jprenni@colostate.edu

---

Abstract: Dry edible bean (*Phaseolus vulgaris* L.) is a globally relevant food crop. The bean genome was recently sequenced and annotated allowing for proteomics investigations aimed at characterization of leaf phenotypes important to agriculture. The objective of this study was to utilize a shotgun proteomics approach to characterize the leaf proteome and to identify protein abundance differences between two bean lines with known variation in their physiological resistance to biotic stresses. Overall, 640 proteins were confidently identified. Among these are proteins known to be involved in a variety of molecular functions including oxidoreductase activity, binding peroxidase activity, and hydrolase activity. Twenty nine proteins were found to significantly vary in abundance (p-value < 0.05) between the two bean lines, including proteins associated with biotic stress. To our knowledge, this work represents the first large scale shotgun proteomic analysis of beans and our results lay the groundwork for future studies designed to investigate the molecular mechanisms involved in pathogen resistance.

Keywords: shotgun proteomics; LC-MS; *Phaseolus vulgaris*; plant defense; pathogen resistance; beans; leaves

---

1. Introduction

Dry edible bean (*Phaseolus vulgaris* L., "bean") is an important food crop that is grown and consumed worldwide [1]. Molecular techniques are critical to bean research to facilitate the development of high-yielding, disease resistant bean lines adapted to a wide range of environments. Current efforts in bean breeding are focused on the identification of genetic markers for important plant phenotypes in agriculture, including yield and resistance to abiotic and biotic stresses [2]. However, the functional effect of genetic markers on a plant phenotype is often not well understood and therefore additional molecular techniques focused on the analysis of RNA, proteins or metabolites, are necessary to characterize the genetic basis of plant phenotypes [3,4].

To date, the majority of plant proteomic studies have utilized 2-dimensional (2-D) gel electrophoresis for protein separation and quantitation, followed by either matrix-assisted laser desorption/ionization-mass spectrometry coupled to time of flight (MALDI-TOF) or liquid chromatography coupled to tandem mass spectrometry (LC-MS/MS) [5–12]. For example, Katam *et al.* conducted studies on peanut leaves utilizing 2-D gels in conjunction with MALDI-TOF in which they were able to visualize 300 protein spots and confidently identify 174 proteins [13]. Salavati *et al.* used a similar approach and reported visualization of 483 protein spots for the common bean proteome. Based on quantitative differences observed in the 2-D gels, 32 identified proteins were involved in the early stage of symbiosis between bean and the nitrogen fixing bacterium *Rhizobium et al.* [14]. Zadraznik *et al.* used an approach combining 2-D gels and LC-MS/MS to characterize proteome changes resulting from drought stress in bean leaves [6]. A recent study utilized a shotgun proteomics approach in which they report the detection and identification of 2600 proteins from cassava root, 300 of which showed significant changes during postharvest physiological deterioration [15].

In the study presented here, the bean leaf proteome was characterized using a shotgun LC-MS/MS approach and a comparison was performed between two bean lines that are genetically similar but contain differences in susceptibility to leaf pathogens. To our knowledge, this study represents the first comprehensive proteome characterization of bean leaves and illustrates the utility of this approach for investigations of disease response mechanisms in plants.

2. Experimental Section

2.1. Plant Material and Growth Conditions

Bean lines are derived from two major centers of domestication: Mesoamerican (small seeded types, e.g., navy, pinto) or Andean (large-seed types, e.g., kidney, alubia). Two common bean lines of Andean origin, A195 and Sacramento, were included in the analysis. A195 is large-seeded dry bean line developed at the International center for Tropical Agriculture in Colombia [16]. Sacramento is a commercial large-seeded light red kidney bean cultivar released by Sacramento Valley Milling in 1975 [17]. The two lines differ in their physiological response to white mold (caused by *Sclerotinia sclerotiorum*, angular leaf spot (caused by *Phaeoisariopsis griseola (Sacc.) Ferraris*), and bean common mosaic virus (BCMV) a non-persistent, stylet-borne, aphid-transmitted virus, with A195 being resistant and Sacramento being susceptible to all three diseases [16,18,19].

Beans ($n = 3$ for each line) were planted in 4.5 L pots with high porosity potting media that was treated with Root Shield (*Trichoderma harzianum*, strain T-22). A pre-plant control for fungus gnats (*Bradysia* species) and 10 g osmocote 19-5-8 time release fertilizer with micronutrients was incorporated pre-planting. After planting, the pots were watered to field capacity (water draining out the bottom of the pots). Plants were grown in a greenhouse with a photoperiod of a 16/8 h light/dark and were watered once a day using a timed irrigation system. The greenhouse heating set-points were 20 °C during the day and 17 °C at night, and the cooling set points were 26 °C during the day and 24 °C at night. The relative humidity was maintained at 55% during the day and 75% at night.

2.2. Protein Extraction

Healthy leaves were detached at the petiole and tissue was acquired using a 1 cm diameter leaf hole punch (approximately 12 mg dry weight per punch). A total of five punches per leaf were acquired, transferred to a 1.5 mL microcentrifuge tube and flash frozen in liquid nitrogen. Frozen tissue was ground to a fine powder using 1 cm diameter steal beads with a TissueLyser II (Qiagen, Germantown, MD, USA). Protein was extracted by the addition of 1 mL of methanol:water (80:20, *v*:*v*), vortexing for 2 h at room temperature, and incubation at −80 °C for 20 min. The sample was centrifuged at 16,000× *g* for 20 min at 4 °C, the supernatant was removed and the pellet was dried using a Savant speedvac (Thermo Scientific, Waltham, MA, USA). Three hundred microliters of 8 M urea was added to the pellet to solubilize protein and the sample was vortexed at room temperature for 15 min. Samples were centrifuged again at 10,000× *g* for 10 min and the supernatant containing the proteins was used for further analysis.

2.3. Protein Digestion

Total protein concentration was determined using a Bradford assay (Thermo Scientific) following manufacture protocols. From each sample, 30 µg of total protein was aliquoted and then precipitated by adding four volumes of −20 °C acetone, followed by a short vortex and incubation at −80 °C for 20 min. Samples were centrifuges at 10,000× *g* for 10 min at 4 °C. The protein pellet was washed with 300 µL of (−20 °C) acetone and centrifuged at 10,000× *g* for 5 min at 4 °C. The supernatant was discarded and the pellet was air dried in a dust free environment.

The dry protein pellet was solubilized using 15 µL of 8 M urea and bath sonicated for 5 min. Then, 20 µL of 0.2% (by volume) ProteaseMAX™ Surfactant (Promega, Madison, WI, USA) was added and mixed by shaking on vortexer for 5–10 min. Afterwards, 58.5 µL of 50 mM ammonium bicarbonate was added to achieve a final volume of 93.5 µL. One microliter of 0.5 M DTT (1,4-dithiothreitol) was added and incubated at 50 °C for 20 min. Then, 2.7 µL of 0.55 M IAA (iodoacetic acid) was added and incubated at room temperature in the dark for 15 min. One microliter of 1% (by volume) ProteaseMAX and 18 µL of 0.1 µg/µL trypsin (Promega) were added. Samples were incubated at 37 °C for 3 h. After the digestion was complete, samples were briefly centrifuged and 27 µL of 2.5% (by volume) TFA (trifluoroacetic acid) was added to stop enzymatic digestion. Samples were then mixed, and incubated at room temperature for 5 min. The peptides were desalted using Pierce C18 spin columns (per manufactures protocol) and dried in a speedvac. Samples were then prepared for mass spectrometry by resuspension in 5% acetonitrile (ACN)/1% formic acid (by volume) to a concentration of 1.0 µg/µL.

2.4. Mass Spectrometry

Peptides were further purified and concentrated using an on-line enrichment column (5 μm, 100 μm ID × 2cm C18 column, Thermo Scientific, Cat#: 89870). One microliter of sample was injected in 100% buffer A (0.1% formic acid, by volume) at a flow rate of 20 μL/min (20 μL injection loop) with enrichment trapping at a flow rate of 4 μL/min. Subsequent chromatographic separation was performed on a reverse phase nanospray column (EASYnano-LC, 3 μm, 75 μm ID × 100 mm C18 column, Thermo Scientific) using a 90 min linear gradient from 10%–30% buffer B (100% ACN, 0.1% formic acid by volume) at a flow rate of 400 nL/min. Peptides were eluted directly into the mass spectrometer (Orbitrap Velos, Thermo Scientific) equipped with a Nanospray Flex ion source (Thermo Scientific) and spectra were collected over a *m/z* range of 400–2000, positive mode ionization, using a dynamic exclusion limit of 2 MS/MS spectra of a given *m/z* value for 30 s (exclusion duration of 90 s). The instrument was operated in FT mode for MS detection (resolution of 60,000) and ion trap mode for MS/MS detection with a normalized collision energy set to 35%. Compound lists of the resulting spectra were generated using Xcalibur software (Thermo Scientific, version 2.2) with a S/N threshold of 1.5 and 1 scan/group.

2.5. Data Analysis

MS/MS spectra were searched against the *Phaseolus vulgaris* protein sequence database obtained from Phytozome.net and concatenated to a reverse database (63,276 sequence entries, version 16 May 2013) using the Mascot database search engine (version 2.3, Matrix Science, Boston, MA, USA). The following search parameters were included: monoisotopic mass, parent ion mass tolerance of 20 ppm, fragment ion mass tolerance of 0.8 Da, fully tryptic peptides with one missed cleavage, variable modification of oxidation of methionine and fixed modification of carbamidomethylation of cysteine.

Search results for each sample were imported and combined using probabilistic protein identification algorithms [20] implemented in Scaffold software [21] (version 4, Proteome Software, Portland, OR, USA). A peptide threshold of 99% and protein probability threshold of 90% were applied and a minimum of two unique peptides was required. A peptide false discovery rate was calculated by Scaffold based on hits to the reverse database [22]. Proteins species that contained similar peptides and could not be differentiated based on MS/MS analysis alone were grouped to satisfy the principles of parsimony.

Instrument functionality and stability was monitored using the MassQC software (Proteome Software). This software uses a set of quantitative metrics developed by the National Institute of Science and Technology (NIST) in collaboration with the National Cancer Institute's Clinical Proteomic Technologies for Cancer (CPTC) that monitor technical variability in mass spectrometry-based proteomics instrumentation [23]. Quality control samples containing a mixture of 6 trypsin digested bovine proteins were injected at least once every 24 h throughout the analysis, and the data from this run was analyzed using the MassQC software. Values for all metrics were within normal limits throughout the duration of the experiment indicating instrument stability and data robustness.

Relative quantitation was determined using both spectral counting (SpC) and average MS/MS total ion current (MS^2 TIC). [24,25]. Data was normalized using the default method in the Scaffold software. A student's t-test was applied to determine protein species that were significantly different in abundance between groups (p-value < 0.05). The resulting list of significantly different proteins was further filtered by the following criteria: proteins must be present in a minimum of 2 out of 3 biological replicates for a given group and the total normalized spectral counts for a given group must be >10. Pseudo values were added (+1 for SpC and +1000 for MS^2 TIC) prior to fold change calculations to eliminate zero values.

Gene ontology (GO) term enrichment analysis was performed in Pathway Studio Plant Web (Elsevier, Atlanta, GA, USA) and was based on *Arabidopsis thaliana* GO libraries (https://www.arabidopsis.org). Gene Set Enrichment Analysis (GSEA) [26] was conducted in Pathway Studio Plant Web using the Mann-Whitney-U Test, and significantly enriched pathways were determined using a threshold of $p < 0.05$.

3. Results and Discussion

In total, 640 protein species were confidently identified (Supplementary Table S1). Of these, 184 protein species were detected in all six leaf samples. The 640 identified protein species were mapped to genetic loci within the bean genome to evaluate the potential for shotgun proteomics to facilitate investigations into quantitative genetics, specifically protein quantitative trait loci (pQTL). Chromosome location and length was determined by aligning the detected protein annotations with transcript annotations and corresponding genomic loci provided by Phytozome.net [27] (Supplementary Table S2). Gene sequences related to the detected protein species spanned across all 11 of the *Phaseolus vulgaris* chromosomes (Figure 1). This analysis revealed a wide genomic distribution of the 640 detected leaf protein species with chromosome 9 being most saturated. Further, molecular processes related to photosynthesis, response to biotic stress and redox status were also widely distributed across the genome. These results support shotgun proteomics of plant leaves as a valuable tool to investigate genetic resistance to biotic stress.

The data was subsequently interrogated for co-localization of genes that encode proteins associated with a physiological or molecular response to biotic stress. For example, two basic chitinases (Phvul.009G116600.1 and Phvul.009G116700.1) co-localized on chromosome 9 with starting positions at 17,449,427 bp and 17,485,614 bp, respectively. Two of the three detected osmotin 34 protein genes, (Phvul.002G286500.1 and Phvul.002G286600.1) are located on chromosome 2 with starting positions at 45,043,859 bp and 45,053,480, respectively. Conversely, some of the detected protein species contained genes that do not co-localize, such as polygalacturonase inhibiting protein 1s which is found on chromosome 2 and chromosome 9 with starting positions at 36,117,844 bp on chromosome 2 and 26,512,436 bp on chromosome 9, respectively. This is consistent with several protein homologues that were detected, such as photosystem II subunit O-2 (chromosome 2 and 3) and photosystem I subunit E-2 (chromosome 7 and 8). Overall, the shotgun proteomics approach resulted in the detection of leaf protein species representative of the bean genome. Taken together, these results illustrate the potential to integrate shotgun proteomic data with traditional quantitative genetics in the context of plant based molecular investigations.

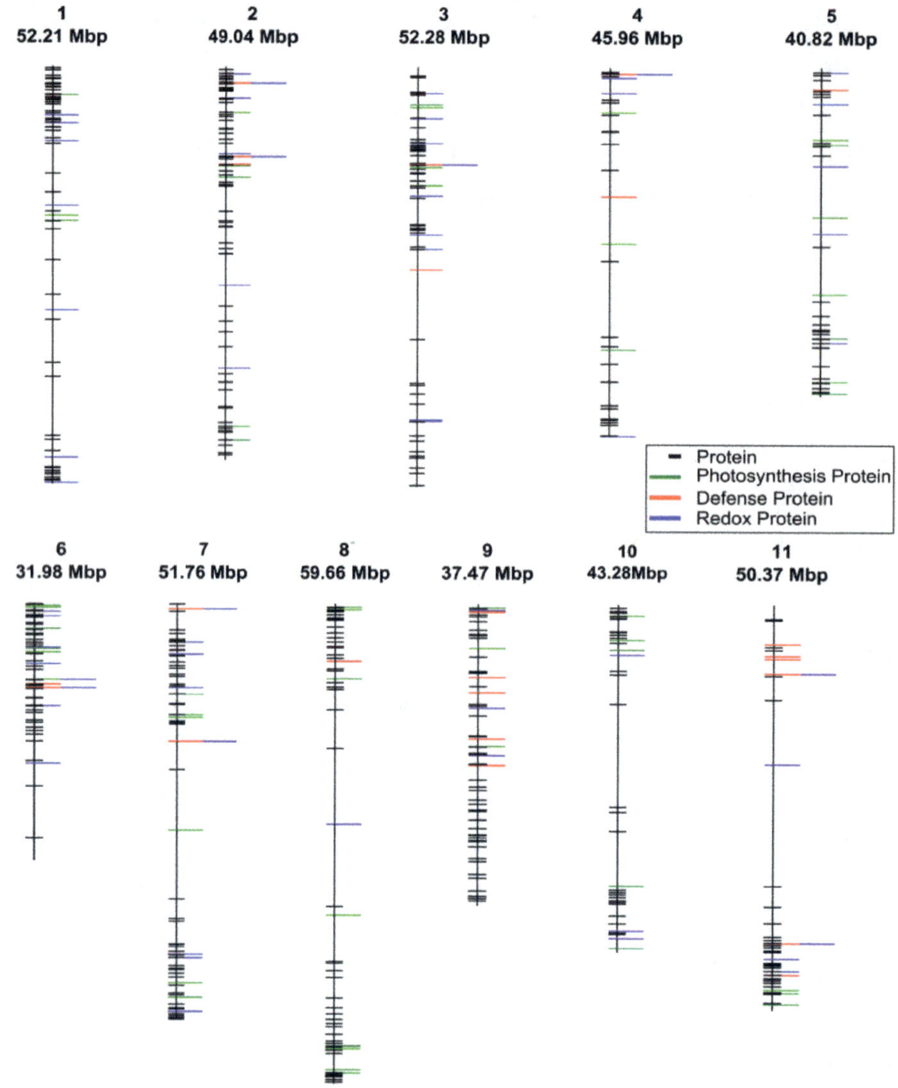

Figure 1. Genomic location of detected protein species using shotgun LC-MS/MS. Genomic locations were determined by aligning detected protein annotations with gene locations. Colors indicate proteins involved in photosynthesis (**green**), plant defense (**red**), or redox (**blue**).

3.1. Characterization of Bean Proteins

The 640 detected bean protein species were categorized based on annotated gene ontology (GO) terms (Figure 2, Supplemental Table S3). The results indicated that the detected protein species are involved in many metabolic processes, for example proteins associated with response to stimuli and that contain binding activities. For example, 43 protein species were found to be involved in defense responses to bacterium, 48 protein species were involved in photosynthesis, 81 protein species identified in ATP binding, and 63 protein species involved in the response to salt stress (Supplementary Table S3).

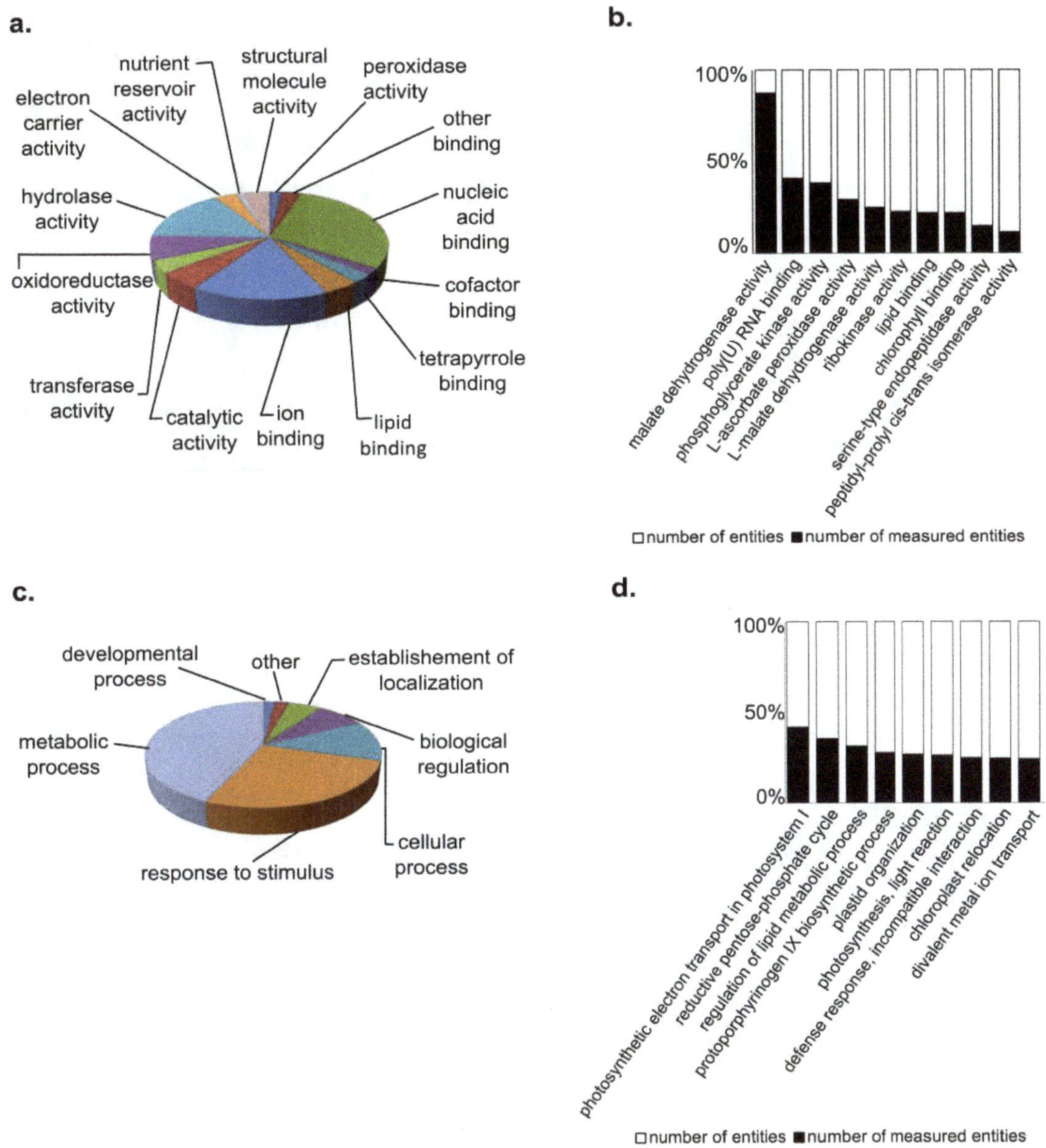

Figure 2. Gene Ontology terms for protein species detected in bean leaves. Gene Ontology (GO) terms were assigned based on a library for *Arabidopsis thaliana.* (**a**) GO terms for molecular function represented as the total number of proteins with each GO term; (**b**) The 10-most enriched molecular functions as determined based on the percent of the biochemical pathway represented (number of entities detected/number of entities in the library × 100%); GO terms were similarly evaluated for (**c**) biological process and (**d**) biological process-enriched pathways.

While the use of GO terms is instructive for global characterization of proteomes, our results illustrate the limitation of this approach when working with non-model systems. Specifically, there were eight protein species (sedoheptulose bisphosphatase, fructose-bisphosphate aldolase 2, ATP synthase δ-subunit gene, phosphoglucomutase, ascorbate peroxidase 4, Mog1/PsbP/DUF1795-like photosystem II reaction center PsbP family protein, starch branching enzyme 2.2, and NDH-dependent cyclic electron flow 1) for which GO terms indicated involvement in glucosinolate metabolic process. All eight protein species were also found to be involved in other molecular functions. To our knowledge, glucosinolates have not been detected in common bean, and the molecular function of bean proteins may be incorrectly classified due to the reliance on homology to the *Arabidopsis thaliana* GO term library.

Gene set enrichment analysis (GSEA) was also performed to identify biochemical pathways highly represented in the protein dataset (Supplementary Table S4). GSEA characterizes categorical bias within the list of protein species to identify shared functions or properties in systems biology [26]. GSEA analysis of the 640 detected protein species identified 74 pathways as enriched (Mann-Whitney-U test, $p < 0.05$). The most enriched pathways included photosynthesis and defense response pathways. For example, 13 of 15 protein species (86.67%) in the photosynthesis pathway were detected (Figure 3). An analysis of total identified protein species revealed 41 protein species involved of hormone signaling and biosynthesis with overlap of shared proteins. For example, 20 detected protein species are listed as involved in both jasmonic acid and salicylic acid pathways. Thus, future studies could utilize this approach to study the crosstalk between jasmonic and salicylic acid pathways by monitoring proteins involved in the defense network.

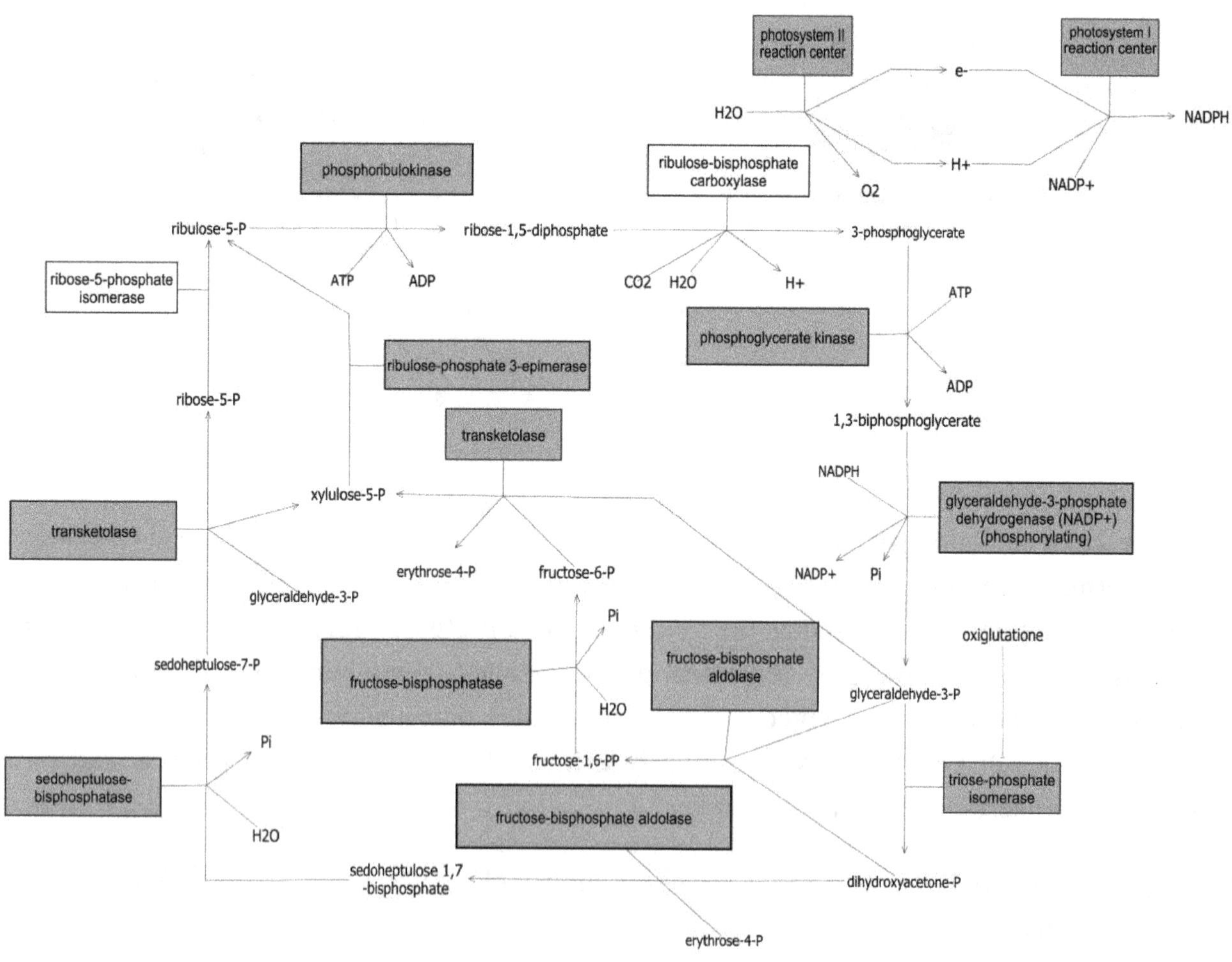

Figure 3. Detected protein species associated with photosynthesis. Proteins detected and identified in this study are shaded in grey.

3.2. Quantitative Variation in Proteins between Bean Lines

Label free quantitation by spectral counting and total MS/MS ion current revealed 29 protein species that varied between line A195 and the Sacramento (Student's *t*-test, $p \leq 0.05$). (Table 1, Supplementary Table S5). Not surprisingly, this list includes proteins involved in photosynthesis and response to biotic stress. Specifically, five protein species are involved in plant defense and five are involved in photosynthesis (Table 1). For example, Metacaspase 5 is a proteolytic enzyme involved in hydrogen peroxide-induced programmed cell death and was observed to be in higher abundance in Sacramento as compared to A195 [28,29].

Table 1. Protein species significantly different in abundance between A195 and Sacramento.

Go Term	Protein Name	Accession Number	*p*-Value	Fold Change *
Photosynthesis	light harvesting complex photosystem II subunit 6	Phvul.006G207300.1|PACid:27166485	0.0031	145.40
Photosynthesis	light-harvesting chlorophyll B-binding protein 3	Phvul.005G085800.1|PACid:27148435	<0.00010	63.33
Photosynthesis	light harvesting complex photosystem II subunit 6	Phvul.002G207500.1|PACid:27170403	<0.00010	53.33
Oxidation-Reduction, Defense	plastid transcriptionally active 16	Phvul.011G061600.1|PACid:27152979	0.019	5.97
ATP synthesis coupled proton transport	ATP synthase δ-subunit gene	Phvul.003G211100.1|PACid:27143051	0.049	2.15
--	uridylyltransferase-related	Phvul.005G181700.1|PACid:27149150	0.0052	1.93
Oxidation-Reduction, Defense	2-cysteine peroxiredoxin B	Phvul.001G194100.1|PACid:27163227	0.038	1.62
RNA Binding	elongation factor Ts family protein	Phvul.004G168100.1|PACid:27157476	0.0095	1.62
Ion Binding	copper ion binding;cobalt ion binding;zinc ion binding	Phvul.002G265700.1|PACid:27169903	0.035	1.44
Oxidation-Reduction, Defense	Oxidoreductase, zinc-binding dehydrogenase family protein	Phvul.003G060300.1|PACid:27143379	0.018	0.83
Oxidation-Reduction	thioredoxin M-type 4	Phvul.005G071200.1|PACid:27149735	0.043	0.77
--	RNA-binding (RRM/RBD/RNP motifs) family protein	Phvul.005G039000.1|PACid:27150205	0.053	0.69
--	CP12 domain-containing protein 2	Phvul.001G212900.1|PACid:27164757	0.0072	0.67
--	protochlorophyllide oxidoreductase A	Phvul.005G083700.1|PACid:27149633	0.048	0.67
Translation	Ribosomal L5P family protein	Phvul.008G059200.1|PACid:27153603	0.022	0.60
Photosynthesis	cold, circadian rhythm, and rna binding 2	Phvul.009G023700.1|PACid:27146551	0.027	0.53

Table 1. *Cont.*

Go Term	Protein Name	Accession Number	*p*-Value	Fold Change *
--	FASCICLIN-like arabinogalactan 2	Phvul.008G288800.1|PACid:27153618	0.036	0.51
Defense	metacaspase 5	Phvul.011G180300.1|PACid:27150868	0.015	0.45
Photosynthesis	2Fe-2S ferredoxin-like superfamily protein	Phvul.002G045400.1|PACid:27168785	0.017	0.33
Protein Binding	DnaJ/Hsp4Unknown cysteine-rich domain superfamily protein	Phvul.003G164500.1|PACid:27144762	0.03	0.30
--	Ribosomal protein S19e family protein	Phvul.005G018600.1|PACid:27149117	0.089	0.30
--	FASCICLIN-like arabinogalactan-protein 1 Unknown	Phvul.008G075000.1|PACid:27155369	0.079	0.22
Lipid Metabolic Process	GDSL-like Lipase/Acylhydrolase superfamily protein	Phvul.006G169100.1|PACid:27166530	0.045	0.20
--	FKBP-like peptidyl-prolyl cis-trans isomerase family protein	Phvul.011G051700.1|PACid:27151439	0.01	0.15
Translation	plastid ribosomal protein l11	Phvul.004G090600.1|PACid:27158071	0.0011	0.13
Protein Binding	Embryo-specific protein 3, (ATS3)	Phvul.005G177100.1|PACid:27149941	0.0023	0.03
Response to Stress	Unknown	Phvul.003G074600.1|PACid:27143731	0.0001	0.01
Oxidation-Reduction, Defense	Peroxidase superfamily protein	Phvul.007G082600.1|PACid:27161387	0.00015	0.01
Ion Binding	allantoinase	Phvul.006G186800.1|PACid:27164938	0.0039	0.01

* Calculated as A195/Sac.

Taken together, the variation in leaf protein abundances between A195 and Sacramento support the utility of the shotgun proteomic approach for future studies to investigating pathways associated with photosynthesis, redox, and pathogen resistance.

4. Conclusions

The objective of this study was to demonstrate the utility of shotgun proteomics as a molecular tool for the characterization of beans. Here, we report the detection and confident identification of 640 protein species from bean leaf extracts. Among these, 29 protein species were found to be significantly different in abundance ($p < 0.05$) between A195 and Sacramento, two bean lines that vary in their physiological responses to biotic stress. Characterization of the identified protein species revealed their involvement in important pathways including oxidation/reduction, photosynthesis, and defense against pathogens.

Mapping of the protein species against the bean genome demonstrates the broad coverage of detected protein species across the genome and the utility of this approach for the identification of protein QTL in the context of pathogen infection. Overall, the results presented here demonstrate the effectiveness of the shotgun proteomics approach for the analysis of plant tissue and the potential for this technique to elucidate proteome changes indicative of plant defense mechanisms. While a shotgun proteomics approach will be biased towards the most abundant proteins, our results illustrate that a sufficient breadth of proteome coverage is achieved to inform on important biological pathways and mechanisms. Importantly, this proteome characterization lays the groundwork for future experiments to focus on host proteome changes in response to pathogen inoculations which could improve our understanding of plant resistance mechanisms.

Acknowledgments

We would like to thank Barry Ogg help in growing plants, and David Goodstein and Joe Carlson for facilitating the resources at Phytozome.net.

Author Contributions

FMR, ALH, JEP, and MAB conceived and designed research. FMR conducted experiments. FMR and JEP analyzed data. FMR, ALH, JEP, and MAB wrote manuscript. All authors read and approved the manuscript.

References

1. Jones, A.L. *Phaseolus Bean: Post-Harvest Operations*; Centro Internacional de Agricultura Tropical, Cali, Colombia, 1999; pp. 1–24. Available online: http://www.fao.org/3/a-av015e.pdf (accessed on 12 August 2015).
2. Beaver, J.S.; Osorno, J.M. Achievements and limitations of contemporary common bean breeding using conventional and molecular approaches. *Euphytica* **2009**, *168*, 145–175.
3. Yang, D.F.; Du, X.H.; Yang, Z.Q.; Liang, Z.S.; Guo, Z.X.; Liu, Y. Transcriptomics, proteomics, and metabolomics to reveal mechanisms underlying plant secondary metabolism. *Eng. Life Sci.* **2014**, *14*, 456–466.
4. Kushalappa, A.C.; Gunnaiah, R. Metabolo-proteomics to discover plant biotic stress resistance genes. *Trends Plant Sci.* **2013**, *18*, 522–531.
5. Badowiec, A.; Weidner, S. Proteomic changes in the roots of germinating *Phaseolus vulgaris* seeds in response to chilling stress and post-stress recovery. *J. Plant Physiol.* **2014**, *171*, 389–398.
6. Zadraznik, T.; Hollung, K.; Egge-Jacobsen, W.; Meglic, V.; Sustar-Vozlic, J. Differential proteomic analysis of drought stress response in leaves of common bean (*Phaseolus vulgaris* L.). *J. Proteom.* **2013**, *78*, 254–272.
7. Yang, Z.B.; Eticha, D.; Fuhrs, H.; Heintz, D.; Ayoub, D.; van Dorsselaer, A.; Schlingmann, B.; Rao, I.M.; Braun, H.P.; Horst, W.J. Proteomic and phosphoproteomic analysis of polyethylene glycol-induced osmotic stress in root tips of common bean (*Phaseolus vulgaris* L.). *J. Exp. Bot.* **2013**, *64*, 5569–5586.

8. De La Fuente, M.; Borrajo, A.; Bermudez, J.; Lores, M.; Alonso, J.; Lopez, M.; Santalla, M.; de Ron, A.M.; Zapata, C.; Alvarez, G. 2-DE-based proteomic analysis of common bean (*Phaseolus vulgaris* L.) seeds. *J. Proteom.* **2011**, *74*, 262–267.
9. Yan, S.; Du, X.Q.; Wu, F.; Li, L.; Li, C.Y.; Meng, Z. Proteomics insights into the basis of interspecific facilitation for maize (*Zea mays*) in faba bean (*Vicia faba*)/maize intercropping. *J. Proteom.* **2014**, *109*, 111–124.
10. Demartini, D.R.; Carlini, C.R.; Thelen, J.J. Global and targeted proteomics in developing jack bean (*Canavalia ensiformis*) seedlings: An investigation of urease isoforms mobilization in early stages of development. *Plant Mol. Biol.* **2011**, *75*, 53–65.
11. Torres, N.L.; Cho, K.; Shibato, J.; Hirano, M.; Kubo, A.; Masuo, Y.; Iwahashi, H.; Jwa, N.S.; Agarwal, G.K.; Rakwal, R. Gel-based proteomics reveals potential novel protein markers of ozone stress in leaves of cultivated bean and maize species of panama. *Electrophoresis* **2007**, *28*, 4369–4381.
12. Huang, B.; Chu, C.H.; Chen, S.L.; Juan, H.F.; Chen, Y.M. A proteomics study of the mung bean epicotyl regulated by brassinosteroids under conditions of chilling stress. *Cell. Mol. Biol. Lett.* **2006**, *11*, 264–278.
13. Katam, R.; Basha, S.M.; Suravajhala, P.; Pechan, T. Analysis of peanut leaf proteome. *J. Proteome Res.* **2010**, *9*, 2236–2254.
14. Salavati, A.; Taleei, A.; Bushehri, A.A.; Komatsu, S. Analysis of the proteome of common bean (*Phaseolus vulgaris* L.) roots after inoculation with rhizobium etli. *Protein Pept. Lett.* **2012**, *19*, 880–889.
15. Vanderschuren, H.; Nyaboga, E.; Poon, J.S.; Baerenfaller, K.; Grossmann, J.; Hirsch-Hoffmann, M.; Kirchgessner, N.; Nanni, P.; Gruissem, W. Large-scale proteomics of the cassava storage root and identification of a target gene to reduce postharvest deterioration. *Plant cell* **2014**, *26*, 1913–1924.
16. Singh, S.R.; Teran, H.; Lema, M.; Schwartz, H.F.; Miklas, P.N. Registration of white mold resistant dry bean germplasm line a 195. *J. Plant Regist.* **2007**, *1*, 62–63.
17. Shanahan, J.F.; Schwartz, H.F.; Brick, M.A.; Schweissing, F.C.; Meyer, R.F.; Pearson, C.H.; Hain, J.P.; Johnson, C.L.; Barry Ogg, J. *Colorado Dry Bean Variety Performance Trials*; Technical Report TR93-8; Colorado Agricultural Experiment Station, Fort Collins, CO, USA, December 1993. Available online: http://cospl.coalliance.org/fedora/repository/co%3A9650/ucsu20211tr938internet.pdf/ (accessed on 12 August 2015).
18. Hart, L.P.; Saettler, A.W.; Correa, F. *Angular leaf spot of dry beans*. Extension Bulletin, E-1753; Michigan State University, East Lansing, MI, USA, February 1984. Available online: http://archive.lib.msu.edu/DMC/Ag.%20Ext.%202007-Chelsie/PDF/e1753-1984.pdf (accessed on 12 August 2015).
19. Bello, M.H.; Moghaddam, S.M.; Massoudi, M.; McClean, P.E.; Cregan, P.B.; Miklas, P.N. Application of in silico bulked segregant analysis for rapid development of markers linked to bean common mosaic virus resistance in common bean. *BMC Genom.* **2014**, *15*, 13, doi: 10.1186/1471-2164-15-903
20. Keller, A.; Nesvizhskii, A.I.; Kolker, E.; Aebersold, R. Empirical statistical model to estimate the accuracy of peptide identifications made by ms/ms and database search. *Anal. Chem.* **2002**, *74*, 5383 5392.

21. Searle, B.C.; Turner, M.; Nesvizhskii, A.I. Improving sensitivity by probabilistically combining results from multiple ms/ms search methodologies. *J. Proteome Res.* **2008**, *7*, 245–253.
22. Kall, L.; Storey, J.D.; MacCoss, M.J.; Noble, W.S. Assigning significance to peptides identified by tandem mass spectrometry using decoy databases. *J. Proteome Res.* **2008**, *7*, 29–34.
23. Rudnick, P.A.; Clauser, K.R.; Kilpatrick, L.E.; Tchekhovskoi, D.V.; Neta, P.; Blonder, N.; Billheimer, D.D.; Blackman, R.K.; Bunk, D.M.; Cardasis, H.L.; *et al.* Performance metrics for liquid chromatography-tandem mass spectrometry systems in proteomics analyses. *Mol. Cell. Proteom.* **2010**, *9*, 225–241.
24. Liu, H.; Sadygov, R.G.; Yates, J.R., III. A model for random sampling and estimation of relative protein abundance in shotgun proteomics. *Anal. Chem.* **2004**, *76*, 4193–4201.
25. Freund, D.M.; Prenni, J.E. Improved detection of quantitative differences using a combination of spectral counting and MS/MS total ion current. *J. Proteome Res.* **2013**, *12*, 1996–2004.
26. Subramanian, A.; Tamayo, P.; Mootha, V.K.; Mukherjee, S.; Ebert, B.L.; Gillette, M.A.; Paulovich, A.; Pomeroy, S.L.; Golub, T.R.; Lander, E.S.; *et al.* Gene set enrichment analysis: A knowledge-based approach for interpreting genome-wide expression profiles. *Proc. Natl. Acad. Sci. USA* **2005**, *102*, 15545–15550.
27. Goodstein, D.M.; Shu, S.Q.; Howson, R.; Neupane, R.; Hayes, R.D.; Fazo, J.; Mitros, T.; Dirks, W.; Hellsten, U.; Putnam, N.; *et al.* Phytozome: A comparative platform for green plant genomics. *Nucleic Acids Res.* **2012**, *40*, D1178–D1186.
28. Van der Hoorn, R.A.L.; Jones, J.D. The plant proteolytic machinery and its role in defence. *Curr. Opin. Plant Biol.* **2004**, *7*, 400–407.
29. Watanabe, N.; Lam, E. Recent advance in the study of caspase-like proteases and bax inhibitor-1 in plants: Their possible roles as regulator of programmed cell death. *Mol. Plant Pathol.* **2004**, *5*, 65–70.

2

Comparative Proteomic Analysis of *Puccinellia tenuiflora* Leaves under Na_2CO_3 Stress

Juanjuan Yu [1,2], Sixue Chen [3], Tai Wang [4], Guorong Sun [5] and Shaojun Dai [1,*]

1 College of Life and Environmental Sciences, Shanghai Normal University, Shanghai 200234, China; E-Mail: yujuan8186@gmail.com

2 Sanquan Medical College, Xinxiang Medical University, Xinxiang 453003, He'nan, China

3 Department of Biology, Genetics Institute, Plant Molecular and Cellular Biology Program, Gainesville, FL 32610, USA; E-Mail: schen@ufl.edu

4 Institute of Botany, Chinese Academy of Sciences, Beijing 100093, China; E-Mail: twang@ibcas.ac.cn

5 Binzhou Polytechnic College, Binzhou 256603, Shandong, China; E-Mail: grsun@live.cn

* Author to whom correspondence should be addressed; E-Mail: daishaojun@hotmail.com

---

Abstract: Soil salt-alkalinization is a widespread environmental stress that limits crop growth and agricultural productivity. The influence of soil alkalization caused by Na_2CO_3 on plants is more severe than that of soil salinization. Plants have evolved some unique mechanisms to cope with alkali stress; however, the plant alkaline-responsive signaling and molecular pathways are still unknown. In the present study, Na_2CO_3 responsive characteristics in leaves from 50-day-old seedlings of halophyte *Puccinellia tenuiflora* were investigated using physiological and proteomic approaches. Comparative proteomics revealed 43 differentially expressed proteins in *P. tenuiflora* leaves in response to Na_2CO_3 treatment for seven days. These proteins were mainly involved in photosynthesis, stress and defense, carbohydrate/energy metabolism, protein metabolism, signaling, membrane and transport. By integrating the changes of photosynthesis, ion contents, and stress-related enzyme activities, some unique Na_2CO_3 responsive mechanisms have been discovered in *P. tenuiflora*. This study provides new molecular information toward improving the alkali tolerance of cereals.

Keywords: proteomics; halophyte; *Puccinellia tenuiflora*; Na_2CO_3 response

---

Abbreviation: 6PGDH, 6-phosphogluconate dehydrogenase; AAIR, acetohydroxy acid isomeroreductase; AIPLP, aluminum-induced protein-like protein; ALD, aldolase; AST, aspartate aminotransferase; ATP ase CF1, ATP synthase CF1 alpha subunit; ATPase 1, ATP synthase alpha subunit; CA, carbonic anhydrase; CAT, catalase; CPS, counts per second; DREPP, developmentally regulated plasma membrane polypeptide; FBA, fructose-bisphosphate aldolase; FRKs, fructokinases; FTR, ferredoxin-thioredoxin reductase; Fv/Fm, maximum quantum efficiency of PSII photochemistry; Fv'/Fm', PSII maximum efficiency; GAPDH, glyceraldehyde-3-phosphate dehydrogenase; GLO, glyoxalase; GLP, germin-like protein 1; Gs, Stomatal conductance; IDs, identities; IF4A, eukaryotic initiation factor 4A; M30, membrane-associated 30 kDa protein; MDA, malondialdehyde; MDH, malate dehydrogenase; MS, methionine synthase; PDI, protein disulfide-isomerase; PGK, phosphoglycerate kinase; Pn, photosynthetic rate; POD, peroxidase; PPIase, peptidyl-prolyl cis-trans isomerase; PRK, phosphoribulokinase; Proteasome α3, proteasome alpha type 3; PSII, photosystem II; qNP, non-photochemical quenching coefficient; ROS, reactive oxygen species; RP, Ribosomal protein S1-like RNA-binding domain; RRM, RNA recognition motif; Rubisco LSU, Rubisco large subunit; Rubisco SSU, Rubisco small subunit; SOD, superoxide dismutase; TIL, temperature stress-induced lipocalin; TIM, triosephosphate isomerase; Tr, transpiration rate; V-ATPase, vacuolar-type H^+-ATPase.

1. Introduction

Salt-alkali soil is one of the major abiotic constraints limiting crop distribution and yield worldwide [1–3]. Plant salinity tolerance has been extensively studied, but the understanding of plant alkalinity tolerance is lacking. The alkali salt affects plant growth and development through sodium toxicity and high pH, which are more likely to cause serious damage than neutral salt to the plant [4]. The high soil pH (mainly attributed to carbonate salts, e.g., Na_2CO_3 and $NaHCO_3$) could directly affect nutrient uptake, organic acid balance, and ion homeostasis, especially the pH stability at cellular and whole plant levels [4,5]. Meanwhile, some halophyte species can naturally survive in high alkaline soil, and have evolved various regulatory and metabolic mechanisms for alkali tolerance.

Alkali grass (*Puccinellia tenuiflora*) is an alkali tolerant halophyte species that can survive in highly alkaline soil (e.g., pH 10). Thus, it is considered as an outstanding pasture for soil improvement. Although some salt tolerant mechanisms in *P. tenuiflora* have been studied before [6], few studies were focused on the specific molecular mechanisms underlying alkali tolerance. To cope with saline or alkaline stress, *P. tenuiflora* has developed various strategies, such as ion balance [7–9], osmotic adjustment [9–11], and reactive oxygen species (ROS) scavenging [12]. Previous studies have revealed that *P. tenuiflora* can remarkably accumulate citric acid in leaves and roots when exposed to alkaline stress. This is different from salt stress under which citric acid levels remain unchanged [9,11]. The accumulation of citric acid in *P. tenuiflora* may play an important role in pH adjustments used to cope with alkaline stress [9,11]. Besides, several ion salt-responsive genes encoding antiporters/channel proteins in *P. tenuiflora* have been isolated and transformed into yeast, rice and Arabidopsis to test their biological functions. These genes include *PutPMP3-1/2* [13], *PutHKT2;1* [14], *PutAKT1* [15], *KPutB1* [16], *PutCAX1* [17], and *PtNHA1* [18]. The specific functions of these genes have been summarized in our previous article [6]. Furthermore, some candidate salt/alkali-responsive genes/proteins in *P. tenuiflora* have been found using high-throughput transcriptomic and proteomic approaches. A cDNA library was constructed for *P. tenuiflora* under 450 mM $NaHCO_3$ stress for 48 h. It contained a total of 95 differentially regulated transcripts [19]. Our previous comparative proteomic analysis revealed 93 unique NaCl-responsive proteins in *P. tenuiflora* leaves [6]. These studies have provided important information for understanding salt-tolerance mechanisms and candidate gene functions. However, the alkali responsive molecular processes remain elusive.

In the present study, we analyzed the characteristics of *P. tenuiflora* leaves in response to Na_2CO_3 using physiological and comparative proteomic approaches. By integrating the changes of photosynthesis, ROS scavenging enzymes activities, ion contents, and alkali-responsive proteins, some unique mechanisms of *P. tenuiflora* in response to Na_2CO_3 have been revealed, leading to better understanding of the underlying molecular mechanisms of alkali tolerance in cereals.

2. Results

2.1. Effects of Na_2CO_3 Stress on the Growth and Photosynthesis of P. tenuiflora

To evaluate the effects of alkaline stress on the growth of *P. tenuiflora*, shoot length, leaf fresh weight, dry weight, and water content were determined (Figure 1). The shoot length declined gradually with the increase in Na_2CO_3 concentration (Figure 1A). The fresh weight and water content decreased when seedlings were treated with 95 mM Na_2CO_3 (Figure 1B,D).

Figure 1. Biomass of *P. tenuiflora* seedlings grown under Na_2CO_3 conditions. (**A**) shoot length of seedlings; (**B**) fresh weight of leaves; (**C**) dry weight of leaves; (**D**) water content in leaves. The values were determined after plants were treated with 0 mM, 38 mM, and 95 mM Na_2CO_3 for seven days, and were presented as means ± SE (n = 9). The different small letters indicate significant differences ($p < 0.05$).

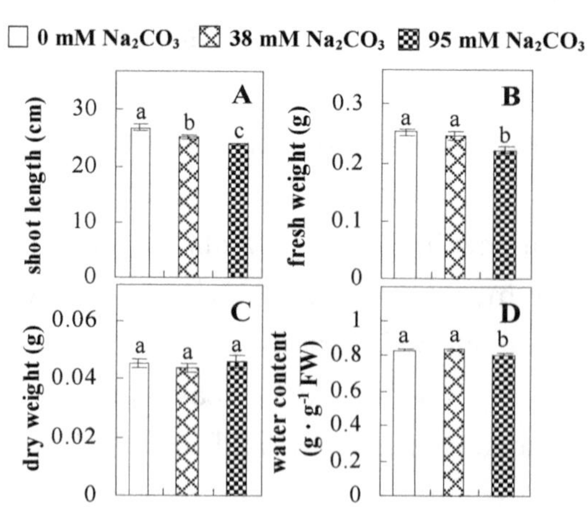

The photosynthesis indexes of *P. tenuiflora* under Na_2CO_3 treatment were analyzed. After seven days of 38 mM and 95 mM Na_2CO_3 treatments, the *P. tenuiflora* seedlings did not show obvious damage to leaf morphology (data not shown), implying the high capacity of *P. tenuiflora* seedlings to tolerate Na_2CO_3. However, photosynthesis was affected by Na_2CO_3 stress. Stomatal conductance (Gs) (Figure 2A), photosynthetic rate (Pn) (Figure 2B), and transpiration rate (Tr) (Figure 2C) exhibited little changes under 38 mM Na_2CO_3 treatment, but showed marked decreases under 95 mM Na_2CO_3. In addition, chlorophyll fluorescence parameters were monitored to determine the performance of photosystem II (PSII) photochemistry. The maximum quantum efficiency of PSII photochemistry (Fv/Fm) (Figure 2D) and the PSII maximum efficiency (Fv'/Fm') (Figure 2E) were not significantly altered under 38 mM Na_2CO_3, but were reduced remarkably under 95 mM Na_2CO_3. The non-photochemical quenching coefficient (qNP) (Figure 2F) remained constant under 38 mM and then increased under 95 mM Na_2CO_3 treatment.

Figure 2. Photosynthetic characteristics (A, B, C) and chlorophyll fluorescence parameters (D, E, F) of *P. tenuiflora* leaves under Na_2CO_3 treatment. (**A**) stomata conductance (Gs); (**B**) photosynthesis rate (Pn); (**C**) transpiration rate (Tr); (**D**) Fv/Fm; (**E**) Fv'/Fm'; (**F**) qNP. The values were determined after plants were treated with 0 mM, 38 mM, and 95 mM Na_2CO_3 for seven days, and were presented as means ± SE ($n = 9$). The different small letters indicate significant differences ($p < 0.05$).

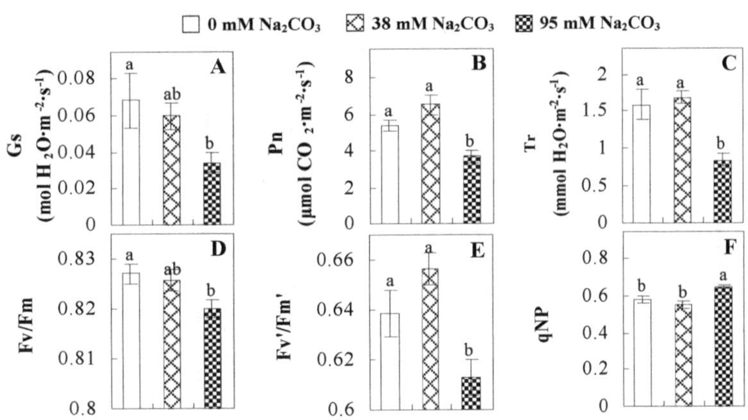

2.2. Changes to Leaf Osmotic Potential, Plasma Membrane Integrity and Antioxidant Enzyme Activities

Leaf osmotic potential showed a significant decrease under Na_2CO_3 treatments (Figure 3A), indicating the seedlings suffered from osmotic stress. The electrolyte leakage ratio (Figure 3B) and malondialdehyde (MDA) contents (Figure 3C) were increased significantly under Na_2CO_3. This indicates that the plasma membrane integrity was damaged by Na_2CO_3 treatment, probably resulting from ROS generated under high pH and ion stress conditions. The activities of representative antioxidative enzymes were altered with different patterns under Na_2CO_3 stress. The superoxide dismutase (SOD) activity decreased under Na_2CO_3 stress (Figure 3D), but the peroxidase (POD) activity increased under 38 mM Na_2CO_3 (Figure 3E), and the catalase (CAT) activity increased obviously under both Na_2CO_3 concentrations (Figure 3F).

2.3. Ion Content Changes in Leaves under Na_2CO_3 Stress

Ion homeostasis is important in plant response to salt stress. Leaf Na content increased with increases in Na_2CO_3 concentration (Figure 4A). Leaf K contents also increased under 95 mM Na_2CO_3 treatment (Figure 4B). This led to declined K/Na ratios under the Na_2CO_3 treatment (Figure 4C). Furthermore, Na and K contents on the leaf surface increased gradually with the increases in Na_2CO_3 concentration, and K/Na ratio was reduced on the leaf surface (Figure 4D,E,F). Moreover, as an important signaling messenger, calcium contents in different subcellular compartments of epidermal cells and mesophyll cells were affected by Na_2CO_3. In epidermal cells, calcium content in the cell wall and cytoplasm increased gradually with the increases in Na_2CO_3 concentration, but reduced in the vacuoles under 38 mM Na_2CO_3 treatment (Figure 4G). Similarly, in mesophyll cells, calcium contents in the cell wall and cytoplasm increased under 95 mM Na_2CO_3 treatments, while the calcium content in vacuoles reduced sharply in the Na_2CO_3 treated samples (Figure 4H).

Figure 3. Changes of some antioxidant-related indexes in leaves of *P. tenuiflora* under Na_2CO_3 treatment. (**A**) osmotic potential; (**B**) electrolyte leakage ratio; (**C**) MDA contents; (**D**) SOD activity; (**E**) POD activity; (**F**) CAT activity. The values were determined after plants were treated with 0 mM, 38 mM, and 95 mM Na_2CO_3 for seven days, and were presented as means ± SE ($n = 6$). The small letters indicate significant difference ($p < 0.05$).

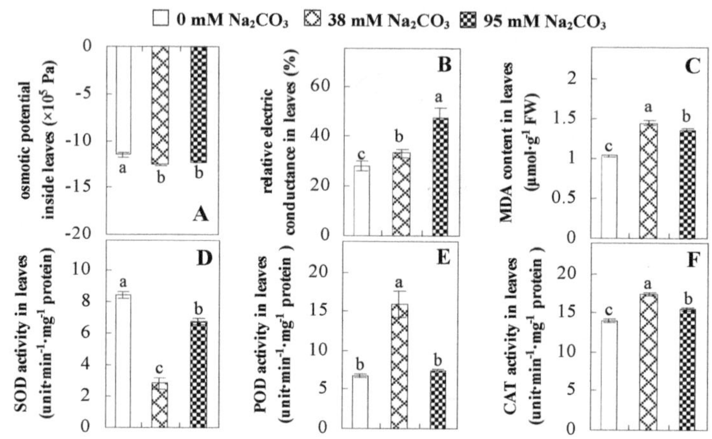

Figure 4. Changes in ionic contents in leaves of *P. tenuiflora* under Na_2CO_3 stress. (**A**) Na content in leaves; (**B**) K content in leaves; (**C**) K/Na ratio in leaves; (**D**) Na content outside leaves; (**E**) K content outside leaves; (**F**) K/Na ratio outside leaves; (**G**) Ca content in epidermal cells; (**H**) Ca content in mesophyll cells. The values were determined after plants were treated with 0 mM, 38 mM, and 95 mM Na_2CO_3 for seven days, and were presented as means ± SE ($n = 6$). The different letters indicate significant differences ($p < 0.05$). CPS, counts per second.

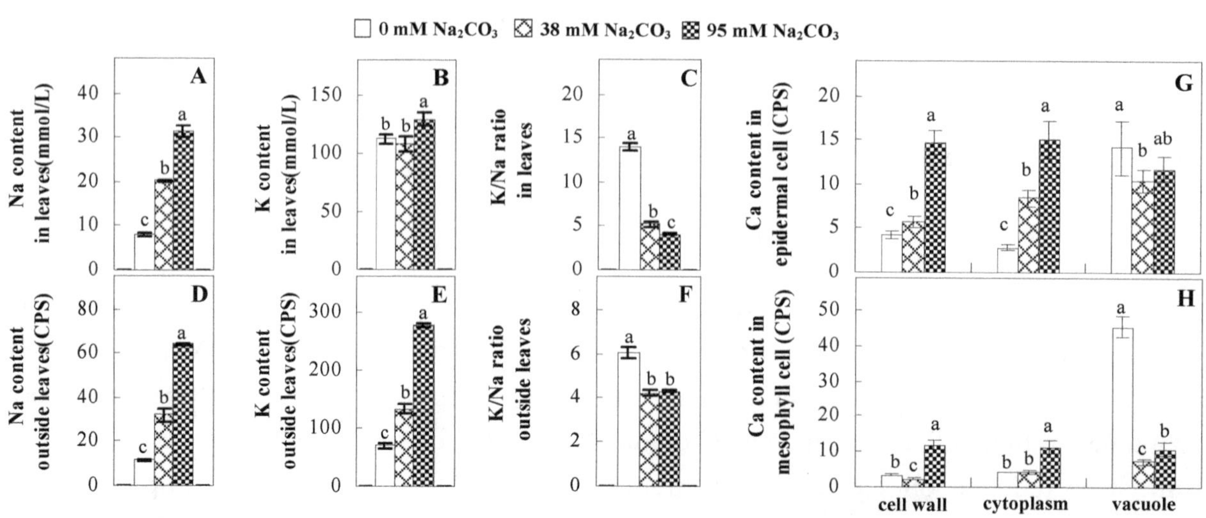

2.4. Identification and Functional Categorization of Na_2CO_3 Responsive Proteins

The alkali-responsive protein profiles were obtained using 2-DE analysis of leaf samples under 0 mM, 38 mM, and 95 mM Na_2CO_3 treatments (Figure 5, Supplemental Figure S1). Approximately 1000 Coomassie Brilliant Blue-stained protein spots were detected on the pI 4–7 gels. A total of 43 protein spots were identified using LC ESI Q-TOF MS/MS and Mascot database searching (Figure 5, Table 1). Based on Gene Ontology, BLAST alignments, and information from literature, the protein identities (IDs) were classified into 10 functional categories including photosynthesis, stress and defense, membrane and transport, carbohydrate and energy metabolism, amino acid metabolism, transcription, protein synthesis, protein folding and transporting, protein degradation, and signaling (Table 1). Among these categories, carbohydrate and energy metabolism (41%), transcription and protein metabolism (27%), photosynthesis (14%), as well as stress and defense (9%), were over-represented.

Figure 5. 2-DE profiling of proteins extracted from *P. tenuiflora* leaves treated with different concentrations of Na_2CO_3. Gels A, B, and C are protein samples from leaves treated with 0, 38, and 95 mM Na_2CO_3 for seven days, respectively. Molecular weight (MW) in kilodaltons (kDa) and pI of proteins are indicated on the right and top of gel A and B, respectively. The 43 gel spots with protein IDs were marked with spot numbers. Detailed information can be found in Table 1.

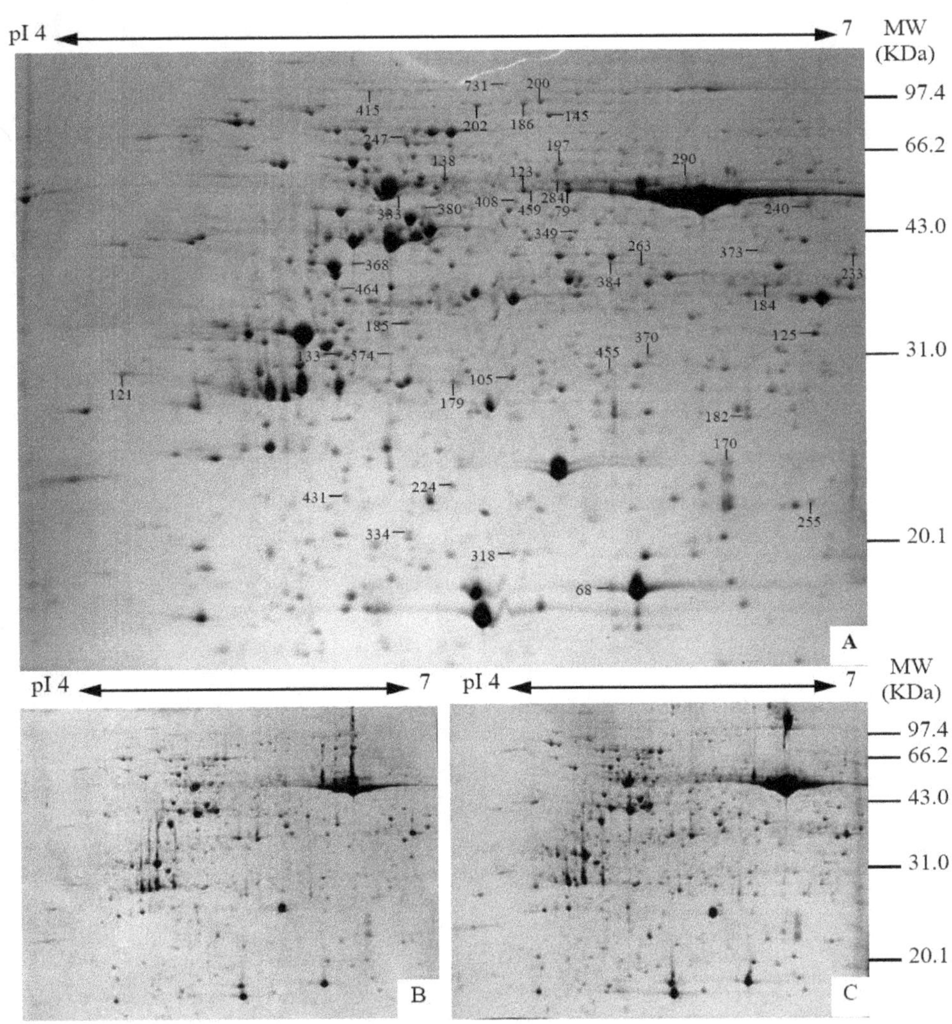

Table 1. Proteins and their relative changes in leaves from *P.tenuiflora* under Na_2CO_3 treatment.

Spot No. [a]	Protein name [b]	Plant species [c]	gi Number [d]	Thr. MW (Da)/pI [e]	Exp. MW (Da)/pI [f]	Sco [g]	Cov (%) [h]	QM [i]	V% ± SE [j] 0, 38, 95 mM Na_2CO_3
	Photosynthesis (6)								
182	carbonic anhydrase, chloroplast precursor (CA)	*Hordeum vulgare*	729003	35,736/8.93	30,029/6.47	125	6	2	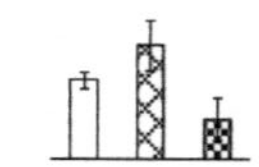
459	Rubisco large subunit (Rubisco LSU)	*Tristachya leucothrix*	125991685	51,294/6.23	64,412/5.71	86	19	8	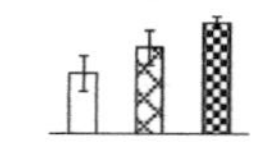
240	Rubisco large subunit (Rubisco LSU)	*Orobanche coerulescens*	46410750	47,410/6.45	62,047/6.68	176	10	4	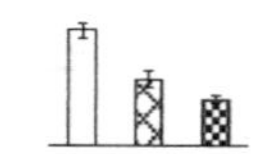
68	Rubisco small subunit (Rubisco SSU)	*Avena sterilis subsp. ludoviciana*	3790104	19,030/8.29	4088/5.98	78	9	2	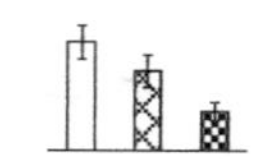
368	phosphoribulokinase (PRK), chloroplast precursor	*Oryza sativa*	125578	44,486	53,645/5.08	210	14	6	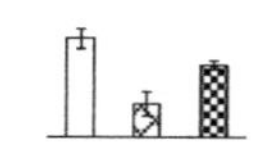
318	ferredoxin-thioredoxin reductase, variable chain (FTR)	*Zea mays*	2498397	10,937/5.69	9350/5.63	49	14	2	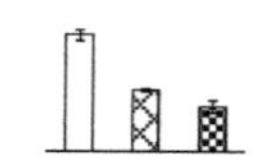
	Stress and defense (4)								
224	temperature stress-induced lipocalin (TIL)	*Triticum aestivum*	18650668	21,809/5.50	19,791/5.42	114	7	1	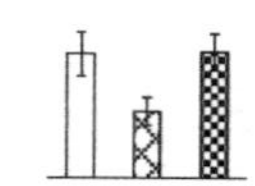
185	glyoxalase I (GLO)	*O. sativa (japonica cultivar-group)*	16580747	32,861/5.51	44,631/5.27	202	15	6	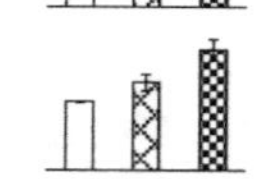
170	germin-like protein 1 (GLP)	*O. sativa*	4239821	22,017/6.01	22,973/6.40	84	10	1	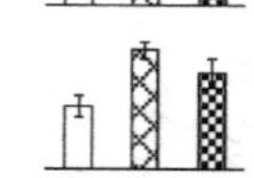
574	aluminum-induced protein-like protein (AIPLP)	*Setaria italica*	124263781	27,004/6.05	39,899/5.21	114	9	2	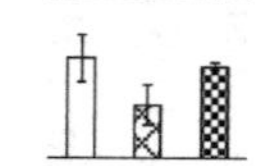
	Membrane and transport (3)								
133	Os01g0233000,containing pfam05558 DREPP plasma membrane polypeptide domain (DREPP)	*O.sativa (japonica cultivar-group)*	115435500	21,788/4.92	40,022/4.03	89	7	2	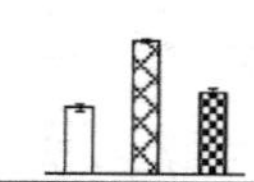

Table 1. *Cont.*

Spot No. [a]	Protein name [b]	Plant species [c]	gi Number [d]	Thr. MW (Da)/pI [e]	Exp. MW (Da)/pI [f]	Sco [g]	Cov (%) [h]	QM [i]	V% ± SE [j] 0, 38, 95 mM Na_2CO_3
125	Vacuolar-type H^+-ATPase (V-ATPase)	*H. vulgare*	2493132	53,806/5.12	42,714/6.71	70	3	2	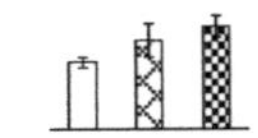
370	membrane-associated 30 kDa protein, chloroplast precursor (M30)	*Pisum sativum*	729842	35,709/9.30	39,410/6.12	80	6	2	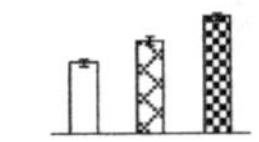
	Carbohydrate and energy metabolism (18)								
464	Os08g0113100, containing cd01167 fructokinases (FRKs) domain	*O. sativa (japonica cultivar-group)*	115474481	35,893/5.02	49,892/5.04	127	5	1	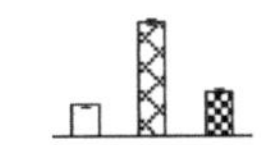
233	putative fructose-bisphosphate aldolase (FBA)	*Phleum pratense*	5419990	25,028/7.79	54,664/6.85	132	17	2	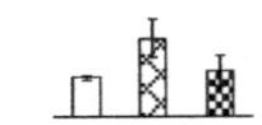
373	cytoplasmic aldolase (ALD)	*O. sativa*	218157	39,151/6.56	55,317/6.53	226	10	7	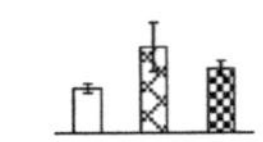
179	triosephosphate isomerase, cytosolic (TIM)	*Secale cereale*	1174749	27,138/5.24	35,657/5.43	79	13	3	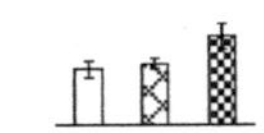
384	Os03g0129300, containing pfam00044 Gp_dh_N (GAPDH) domain	*O. sativa (japonica cultivar-group)*	115450493	47,537/6.22	53,685/5.98	239	12	8	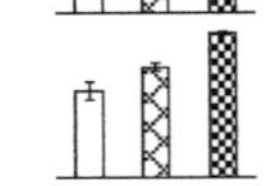
380	phosphoglycerate kinase (PGK)	*Vitis vinifera*	147843754	42,510/6.29	62,210/5.33	137	9	3	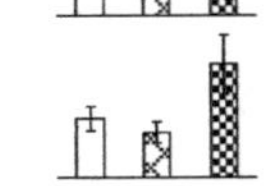
138	enolase2	*Z. mays*	162460735	48,418/5.70	66,697/5.4	143	10	3	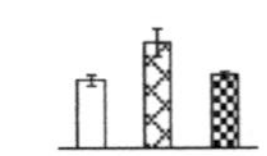
184	malate dehydrogenase, mitochondrial precursor (MDH)	*Citrullus lanatus var. lanatus*	126896	36,406/9.68	50,177/6.54	78	3	2	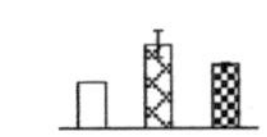
408	putative cytosolic 6-phosphogluconate dehydrogenase (6PGDH)	*Z. mays*	3342802	53,204/6.24	63,107/5.65	64	4	2	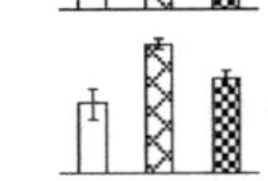
79	ATP synthase alpha (ATPase 1)	*T. aestivum*	81176509	55,557/5.70	64,698/5.84	361	12	4	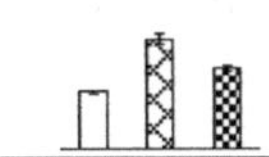

Table 1. *Cont.*

Spot No. [a]	Protein name [b]	Plant species [c]	gi Number [d]	Thr. MW (Da)/pI [e]	Exp. MW (Da)/pI [f]	Sco [g]	Cov (%) [h]	QM [i]	V% ± SE [j] 0, 38, 95 mM Na_2CO_3
123	ATP synthase alpha subunit (ATPase 1)	*T. aestivum*	81176509	55,557/5.70	64,820/5.68	207	39	19	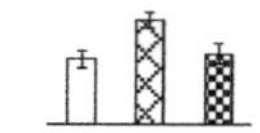
290	ATP synthase alpha subunit (ATPase 1)	*Elymus sibiricus*	51556908	55,549/6.03	66,615/6.26	77	20	8	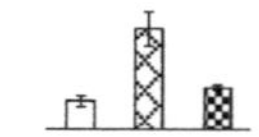
284	ATP synthase CF1 alpha subunit (ATP ase CF1)	*Agrostis stolonifera*	118430299	55,491/6.11	66,044/5.8	88	24	9	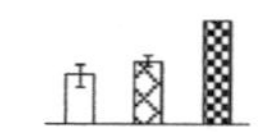
105	mitochondrial ATP synthase precursor (ATPase 1)	*T. aestivum*	47607439	27,090/7.71	36,228/5.63	62	6	2	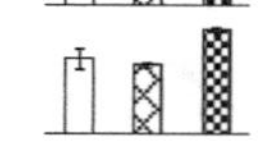
200	Os12g0230100, containing two AAA ATPase family protein domain	*O. sativa (japonica cultivar-group)*	115487910	102,068/6.62	78,280/5.74	87	16	13	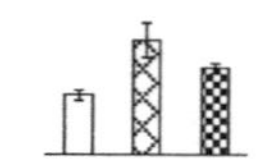
202	hypothetical protein OsI_036614, containing two AAA ATPase family protein domain	*O. sativa (indica cultivar-group)*	125536167	156,826/7.37	77,750/5.52	73	4	10	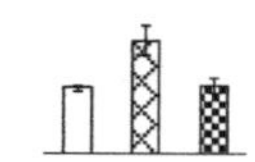
186	unnamed protein product, containing cd00009 the AAA, ATPases domain	*Vitis vinifera*	157343871	50,950/5.8	78,076/5.68	146	37	16	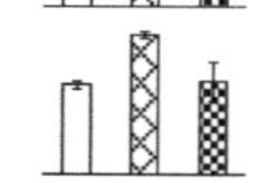
247	hypothetical protein OsI_023646, containing AAA, ATPases domain	*O. sativa (indica cultivar-group)*	125556808	64,173/5.65	72,936/5.27	74	22	12	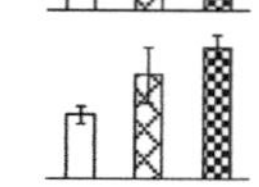
	Amino acid metabolism (2)								
349	aspartate aminotransferase (AST)	*O. sativa*	29468084	46,016/5.90	58,376/5.86	375	16	7	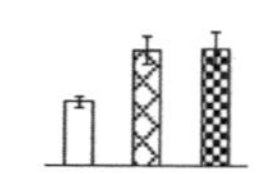
145	methionine synthase (MS)	*H. vulgare subsp. vulgare*	50897038	84,794/5.68	76,241/5.77	112	4	3	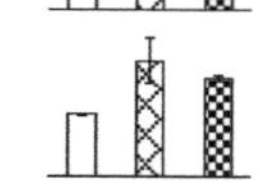
	Transcription related (1)								
121	cp31BHv, containing cd00590 RNA recognition motif (RRM) domain	*H. vulgare subsp. vulgare*	3550483	30,662/4.76	37,167/4.27	115	8	2	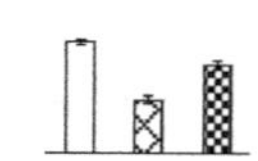
	Protein synthesis (2)								
333	eukaryotic initiation factor 4A (IF4A)	*O. sativa (japonica cultivar-group)*	303844	47,187/5.29	64,045/5.24	166	12	3	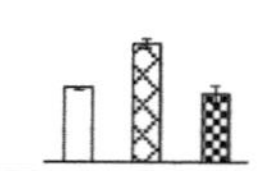

Table 1. *Cont.*

Spot No. [a]	Protein name [b]	Plant species [c]	gi Number [d]	Thr. MW (Da)/pI [e]	Exp. MW (Da)/pI [f]	Sco [g]	Cov (%) [h]	QM [i]	V% ± SE [j] 0, 38, 95 mM Na_2CO_3
415	Os07g0168000, containing cd00164 Ribosomal protein S1-like RNA-binding domain(RP)	*O. sativa (japonica cultivar-group)*	115470767	98,023/5.64	79,993/5.15	84	3	2	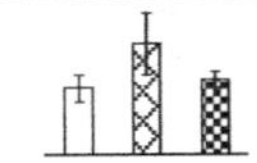
	Protein folding and transporting (3)								
255	peptidyl-prolyl cis-trans isomerase, chloroplast precursor (PPIase)	*Glycine max*	9899901	18,841/8.49	17,018/6.69	146	19	2	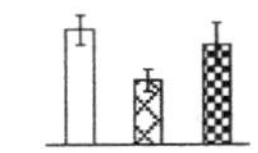
263	protein disulfide-isomerase precursor (PDI)	*Nicotiana tabacum*	1848212	40,082/5.99	53,400/6.09	92	6	3	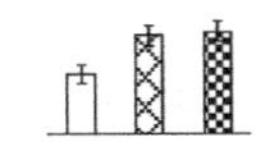
731	putative SecA	*O. sativa (japonica cultivar-group)*	52075758	114,899/5.78	80,931/5.62	88	11	10	
	Protein degradation (3)								
455	Os01g0811100, containing cd03751 proteasome alpha type 3 (Proteasome α3) domain	*O. sativa (japonica cultivar-group)*	115440617	27,506/5.75	37,044/5.97	107	6	2	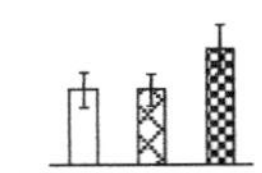
431	unknown, containing cd03751 proteasome alpha type 3 (Proteasome α3) domain	*H. vulgare*	117670154	27,448/5.82	18,201/5.05	167	15	4	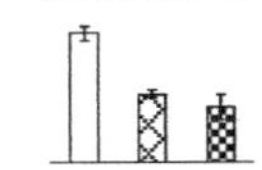
197	Os05g0573700, containing pfam07991 acetohydroxy acid isomeroreductase, catalytic domain (AAIR)	*O. sativa (japonica cultivar-group)*	115465569	62,680/6.01	68,899/5.81	254	11	6	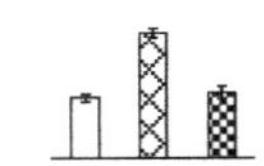
	Signaling (1)								
334	plant adhesion molecule PAM1	*Arabidopsis thaliana*	22531279	34,306/8.80	12,776/5.27	74	17	6	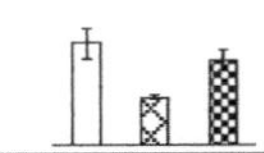

[a] Assigned spot number as indicated in Figure 5; [b] The name and functional categories of the proteins identified using LC ESI Q-TOF MS/MS; [c] The plant species that the peptides matched to; [d] Database accession numbers from NCBInr; [e, f] Theoretical (e) and experimental (f) mass (kDa) and pI of identified proteins. Experimental values were calculated using Image Master 2D Platinum Software. Theoretical values were retrieved from the protein database; [g] The amino acid sequence coverage for the identified proteins; [h] The Mascot score obtained after searching against the NCBInr database; [i] The number of unique peptides identified for each protein; [j] The mean values of protein spot volumes relative to total volume of all the spots. Three Na_2CO_3 treatments (0 mM, 38 mM, 95 mM) were performed. Error bars indicate ± standard error (SE).

Figure 6. Hierarchical clustering analysis of the expression profiles of the identified 43 proteins. The three columns represent different treatments, *i.e.*, 0 mM, 38 mM, and 95 mM. The rows represent individual proteins. The protein cluster is on the left, and the treatment cluster is on the top. The increased and decreased protein spots were indicated in red and green, respectively. The intensities of the colors increase with the increase of expression differences, as shown in the bar on the top. The protein spot numbers are listed on the right, and the letters before the spot numbers represent various functional categories of the proteins. A, photosynthesis; B, stress and defense; C, membrane and transport; D, carbohydrate and energy metabolism; E, amino acid metabolism; F, transcription related; G, protein synthesis; H, protein folding and transporting; I, protein degradation; J, signaling.

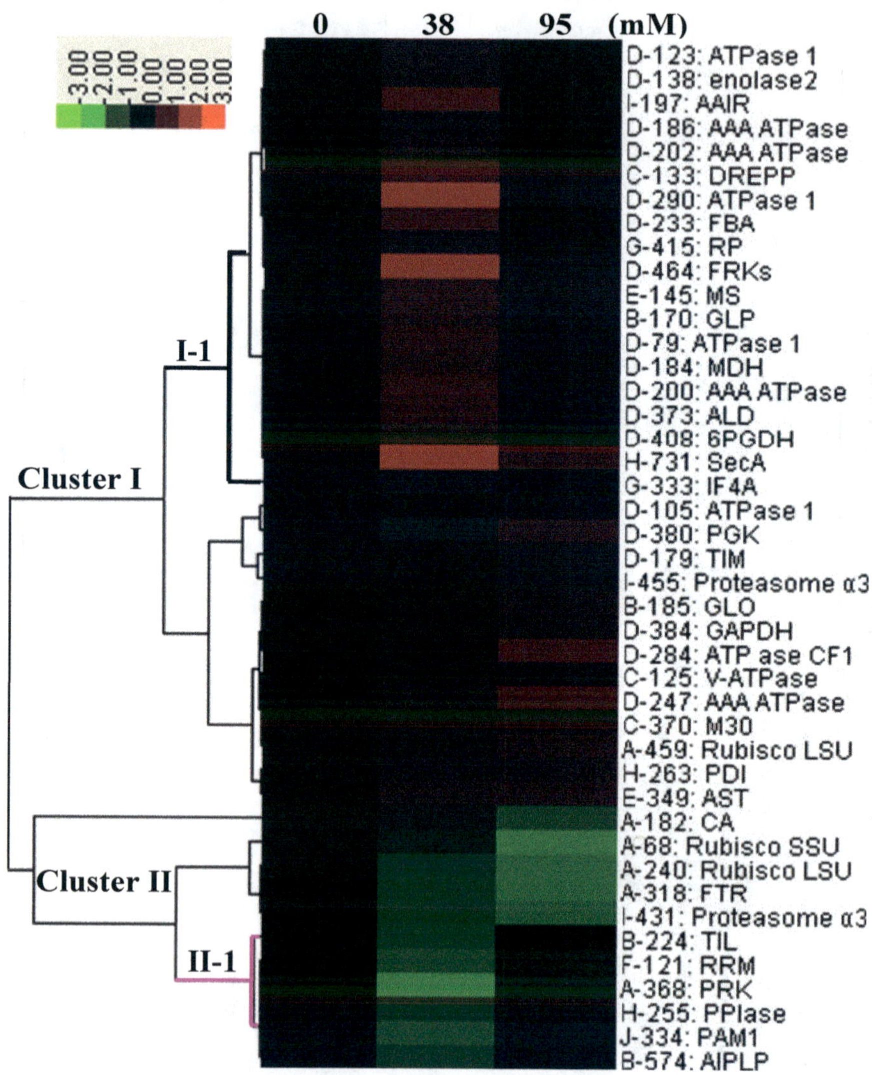

2.5. Protein Clustering and the Dynamics of Protein Networks

One important goal of system biology is to understand the interdependence of proteins and their expression profiles in a certain tissue or other biological samples [6,20]. An effective method to determine the regulatory mechanisms for protein interactions is the application of hierarchical clustering algorithms used in DNA microarray experiments. With this method, the proteins appearing on the same branches are assumed to be involved in related molecular functions [6,20]. Thus, to analyze the expression characteristics of proteins involved in each functional category, we performed hierarchical clustering analysis of the 43 IDs, which revealed two main clusters. Cluster I included 32 alkali-induced IDs and cluster II contained 11 alkali-reduced IDs (Figure 6). The number of alkali-induced proteins was obviously larger than that of reduced proteins. The analysis of protein functional categories showed a heterogeneous distribution between the two clusters (Figure 6). For example, most of the IDs involved in carbohydrate and energy metabolism were grouped in cluster I, whereas photosynthesis-related proteins were mainly in cluster II (Figure 6). This suggests that a switch of biological processes occurred in the course of alkaline treatment. Interestingly, the clustering result supports the previous notion that 38 mM is the turning point concentration of Na_2CO_3 for *P. tenuiflora* [12,21,22], because there were 28 out of 43 protein IDs whose protein abundances changed. Among them, 22 proteins (sub-cluster I-1) reached the maximum levels, and six proteins (subcluster II-1) got to the minimum levels under 38 mM Na_2CO_3 treatment (Figure 6). Such protein patterns correlated well with our aforementioned physiological results, e.g., the photosynthetic capability (e.g., Pn, and qP) and antioxidant-related indexes.

3. Discussion

3.1. Photosynthesis Is Inhibited by Na_2CO_3

The effects of salinity and alkalinity on plant growth and development vary amongst plants. For glycophytes, salt stress generally reduces plant growth and development, but for most halophytes, moderate salt accumulation promotes the plant growth [23]. However, the less-tolerant dicotyledonous halophytes and monocotyledons halophytes, especially grasses, grow better in non-salinized conditions [23,24]. Our results showed that the growth of *P. tenuiflora*, a monocot halophyte grass, was not obviously affected by low Na_2CO_3 concentration, but was inhibited by higher Na_2CO_3 concentration (Figure 1). This correlates well with the biomass changes of *P. tenuiflora* under NaCl stress [6]. In this study, Gs, Pn, and Tr of *P. tenuiflora* seedlings exhibited little changes at 38 mM Na_2CO_3, but decreased significantly at 95 mM Na_2CO_3 (Figure 2A,B,C), indicating photosynthesis was reduced at the higher Na_2CO_3 concentration. Moreover, Fv/Fm and Fv'/Fm' were stable at 38 mM Na_2CO_3, but reduced at 95 mM Na_2CO_3 (Figure 2D,E). This implied that the efficiency of PSII photochemistry was not affected by the low Na_2CO_3 concentration, but inhibited by the high Na_2CO_3 concentration. The reduced Fv/Fm of seedlings under 95 mM Na_2CO_3 implied the occurrence of photoinhibition in *P. tenuiflora* under the higher concentration of Na_2CO_3. The decrease of Fv/Fm was usually accompanied by increase of thermal dissipation, which was evaluated by nonphotochemical quenching of Chl fluorescence (qNP) [25]. Here qNP was maintained almost constant at 38 mM Na_2CO_3, but increased significantly at 95 mM Na_2CO_3 (Figure 2F), which corresponded with the change of Fv/Fm. This result implied that thermal dissipation remained steady at 38 mM Na_2CO_3, but increased at 95 mM Na_2CO_3.

Previous studies have found that salt and alkali stresses affected photosynthetic carbon fixation [1,26]. Our present proteomics data revealed that some of the enzymes in Calvin cycle were reduced by Na_2CO_3 stress, which implied that the decrease in Pn was due to the less efficient carbon fixation under Na_2CO_3 stress. These enzymes included carbonic anhydrase (CA), RuBisCO, phosphoribulokinase (PRK), and ferredoxin-thioredoxin reductase (FTR). CA can help increase the concentration of CO_2 within the chloroplast in order to increase the carboxylation rate of RuBisCO [27]. The change tendency of CA is similar to those of Gs, Pn, and Tr in *P. tenuiflora*. These results indicate that the decrease of Pn is mainly resulted from the declined carbon fixation. RuBisCO catalyzes the first major step of carbon fixation in C3 plants, and PRK is the key enzyme that functions in phosphorylating RuP into ribulose-1,5-bisphosphate (RuBP) in the Calvin cycle [28]. FTR is an iron-sulfur enzyme, which links light to enzyme regulation in oxygenic photosynthesis, catalyzing the activation of fructose 1,6-bisphosphatase [29]. The decline of these enzymes in the Calvin cycle under Na_2CO_3 treatment could lead to the decrease of carbon fixation.

3.2. Antioxidant Mechanisms in Leaves to Cope with Na_2CO_3

Salt and alkali stresses enhance the production of ROS, resulting in various ROS-associated perturbations in the seedlings [2]. Chloroplasts are key intracellular ROS generators. In chloroplasts, the production of O_2^- is mainly determined by the balance between absorption and utilization of light energy [25]. The energy consumption for CO_2 assimilation suppressed by Na_2CO_3 stress led to ROS imbalance, which would cause oxidative damage to enzymes and thus the photosynthetic apparatus [2,25,30]. In the present study, the significantly increased electrolyte leakage ratio and MDA contents indicate that plasma membrane was damaged by lipid peroxidation under Na_2CO_3 stress [30].

Our results have revealed several mechanisms of light energy balance and antioxidation used to cope with Na_2CO_3 stress in *P. tenuiflora* seedlings. Salt accumulation on the leaf surface would prevent excessive light absorption in Na_2CO_3 stressed plants. The contents of Na and K on the surface of *P. tenuiflora* leaves increased gradually with increasing Na_2CO_3 concentrations (Figure 4A,B,D,E), which suggested that *P. tenuiflora* possesses the ability to secrete salts under Na_2CO_3 stress. Other ions such as Ca, Mg, and Si have also been found to increase in concentration on the *P. tenuiflora* leaf surface with the increasing Na_2CO_3 concentrations [10]. The accumulation of salts on leaf surfaces developed greater surface reflectance, contributing to reduce light absorption [31]. At the same time, thermal dissipation was enhanced to remove excess energy at high Na_2CO_3 concentration. The increase of qNP demonstrates that the thermal dissipation was enhanced to protect the photochemical apparatus under Na_2CO_3 stress (Figure 2F). In addition, the enhancement of the xanthophyll cycle would increase the thylakoid pH, which is helpful for induced thermal dissipation [25]. The increase of temperature-induced lipocalin (TIL) in our proteomics results supported this notion. Lipocalin was found to be a key enzyme in the xanthophyll cycle responsible for protection against photo-oxidative damage [32]. It was also found to be increased in *Solanum lycopersicum* under salt stress [33].

The ROS scavenging system was activated in seedlings to cope with Na_2CO_3 stress. The activities of POD (Figure 3E) and CAT (Figure 3F), the two enzymes involved in the removal of H_2O_2, were increased under Na_2CO_3 treatments, especially under 38 mM Na_2CO_3. However, SOD activity was reduced dramatically under Na_2CO_3 stress (Figure 3D). SOD is an enzyme for dismutation of O_2^- to produce H_2O_2. Thus, in *P. tenuiflora* seedlings under Na_2CO_3, H_2O_2 might be mainly produced in peroxisome from oxidation of glycolate during photorespiration rather than from dismutation of O_2^- [6]. In addition, our proteomics data showed that other antioxidant and detoxification mechanisms were enhanced in seedlings to cope with Na_2CO_3 stress. Glyoxalase (GLO) was found to be induced in seedlings with increasing concentrations of Na_2CO_3. GLO is a member of the glyoxalase system that carries out the detoxification of methylglyoxal and other reactive aldehydes produced in plant metabolism. Previous studies have documented that salinity stress induced high accumulation of methylglyoxal, a potent cytotoxin in various plant species [34]. In transgenic tobacco plants, overexpression of glyoxalase I can tolerate an increase in methylglyoxal and maintain high levels of reduced glutathione under salinity stress [35]. Our results implied that GLO might be an important candidate for conferring high alkali tolerance in *P. tenuiflora*. Furthermore, germin-like protein (GLP) was also induced under Na_2CO_3 stress. Germin was first detected in germinating wheat seeds, but its homologs have now become ubiquitous in the plant kingdom and have various functions, not only during embryogenesis, but also in biotic or abiotic stress conditions [36]. The increased GLPs were detected in *Nicotiana tabacum* leaf apoplast [37] and *Arabidopsis thaliana* roots [38] after exposure to salt stress. Wheat germin has been shown to display oxalate oxidase activity; this activity is shared among most plants [36]. Germin-like oxalate oxidase is involved in degrading the oxalic acid, a highly toxic chemical, through production of H_2O_2 [36]. Thus, the Na_2CO_3 increased GLP might provide an explanation for the decrease of oxalic acids in shoots and roots of *P. tenuiflora* [9]. Additionally, an aluminum-induced protein-like protein (AIPLP) was induced under 95 mM Na_2CO_3 treatment. Previous studies have shown that the AIPLP was not specific to aluminum stress, but also involved in other metal, wounding [39], and drought stresses [40]. This protein might also contribute to the tolerance to Na_2CO_3 stress in *P. tenuiflora* when at a high concentration.

3.3. Ion Homeostasis and Transport under Na_2CO_3 Stress

The intracellular ion homeostasis is fundamental to living cells. Under salinity conditions, high apoplastic levels of Na^+ would alter the aqueous and ionic thermodynamic equilibrium, resulting in hyperosmotic stress, ionic imbalance, and toxicity [26]. Thus, proper regulation of ion flux is necessary for cells to keep the concentrations of toxic ions low and to accumulate essential ions. Our results implied that *P. tenuiflora* developed some protective mechanisms to reestablish cellular ion homeostasis through selective salt accumulation or exclusion, *in vivo* compartmentalization, and Ca^{2+} signaling.

Maintaining a high cytosolic K^+/Na^+ ratio is one of the most important mechanisms for plant salt tolerance [8]. In our study, the Na content in leaves increased remarkably in all alkaline treatments. However, the K content did not increase under 38 mM Na_2CO_3, but increased slightly under 95 mM Na_2CO_3 (Figure 4A,B). This led to the decreased intracellular K/Na ratio in *P. tenuiflora* leaves (Figure 4C), although some Na ions were secreted to the leaf surface (Figure 4D). It is obvious that the ion homeostasis in *P. tenuiflora* leaves was affected under Na_2CO_3 stress. In previous studies, *P. tenuiflora* had lower net Na^+ uptake rates than wheat (less than 50% under 150 mM NaCl) [8], which indicates that *P. tenuiflora* has a greater capacity than wheat to restrict unidirectional Na^+ influx to maintain low net Na^+ uptake[9]. Besides, the increased contents of K and Na on the leaves surface with the increase of external Na_2CO_3 concentrations (Figure 4D,E) support the hypothesis that *P. tenuiflora* leaves could exude salts through stomata or together with wax secretion [10].

Ion compartmentalization in different tissues can facilitate their metabolic functions [2]. The salt-inducible Na^+/H^+ antiporter is in charge of Na removal from the cytoplasm or compartmentalization in the vacuoles [2]. The vacuolar Na^+/H^+ antiporters were induced by $NaHCO_3$ in *P. tenuiflora*, suggesting its key role in pH regulation under alkaline conditions [41]. Vacuolar-type Na^+/H^+ antiporter was mainly driven by the proton gradient across the vacuolar membrane generated by vacuolar type H^+-ATPases (V-ATPases) [2,42]. The V-ATPase is indispensable for plant growth under normal conditions due to its roles in energizing secondary transport, maintaining solute homeostasis, and facilitating vesicle fusion. Under stress conditions (e.g., salinity, drought, cold, acid, anoxia, and heavy metals), the survival of the cells depends strongly on maintaining or adjusting the activities of the V-ATPases [2,42]. In the present study, a subunit of V-ATPase was induced under Na_2CO_3 stress. The corresponding increase of Na content in vacuoles under Na_2CO_3 stress was much higher than in cytoplasm of epidermal cells and mesophyll cells [43]. These findings suggest that the Na_2CO_3-induced V-ATPase was required to energize the tonoplast for ion uptake into the vacuoles.

Importantly, calcium content was changed with diverse cellular structures (e.g., cell wall, cytoplasm, and vacuole) in epidermal cells and mesophyll cells in leaves of *P. tenuiflora* when exposed to Na_2CO_3 stress (Figure 4G,H). Calcium is a principal signaling molecule for salinity tolerance [2,3]. High salinity leads to increased cytosolic Ca^{2+}, which initiates the stress signal transduction pathways [3]. In this study, the calcium content in cytoplasm of epidermal cells (Figure 4G) and mesophyll cells (Figure 4H) increased significantly under 95 mM Na_2CO_3. In contrast, the calcium content in vacuoles of epidermis cells (Figure 4G) and mesophyll cells (Figure 4H) decreased. This indicates that the increased cytosolic Ca^{2+} might be transported from the apoplast and intracellular compartments [2]. In addition, our proteomics data revealed that a developmentally regulated plasma membrane polypeptide (DREPP PM)-like protein increased under Na_2CO_3 stress. DREPP-like protein contains a possible Glu-rich site at the C terminus responsible for calcium binding. DREPP-like protein has been found to be increased temporarily in rice under cold acclimation [44] and salt stress [45]. This result suggests that DREPP-like protein may be associated with the Ca^{2+} signal transduction pathway in the seedlings of *P. tenuiflora* under Na_2CO_3 stress.

3.4. Enhancement of Energy Supply and Other Specialized Metabolism

In this study, nine protein IDs were carbohydrate metabolism-related enzymes, and nine were involved in energy production (Table 1). Among them, seven IDs (representing six unique proteins) were enzymes in glycolysis, including fructokinases (FRK), fructose-bisphosphate aldolase (FBA), triosephosphate isomerase (TIM), glyceraldehyde 3-phosphate dehydrogenase (GAPDH), phosphoglycerate kinase (PGK), and enolase. They were all induced under Na_2CO_3 stress. In fact, glycolysis shared a series of reversible reactions with gluconeogenesis. Interestingly, except for FRK, the remaining five enzymes catalyzing the reversible reactions were shared between glycolysis and gluconeogenesis. FRK, an enzyme that irreversibly catalyzes the transfer of a phosphate group from ATP to fructose in glycolysis, is the most important gateway in the control of sugar influx into glycolysis. Thus, the increase of FRK under Na_2CO_3 along with the other five enzymes would contribute to glucose breakdown for energy generation to cope with Na_2CO_3 stress. In addition, malate dehydrogenase (an enzyme in citric acid cycle) and cytosolic 6-phosphogluconate dehydrogenase (an enzyme involved in pentose phosphate pathway) were induced by Na_2CO_3 stress. Moreover, five homologs of ATP synthases and four proteins containing the AAA ATPase family protein domain were induced with similar patterns to aforementioned carbohydrate metabolism-related enzymes in leaves under Na_2CO_3 stress. This implies that the enhancement of the metabolism provides ATP for plant adaptation to Na_2CO_3 stress. Similar results have also been found in salt-stressed *Salicornia europaea* [20] and *P. tenuiflora* [6].

Two amino acid metabolism-related proteins, aspartate aminotransferase (AST) and methionine synthase (MS), increased in levels under Na_2CO_3 treatment (Table 1). AST catalyzes the interconversion of aspartate and α-ketoglutarate to oxaloacetate and glutamate, and plays an important role in nitrogen assimilation and biosynthesis of amino acids [20]. MS is an enzyme that catalyzes the final step in the regeneration of methionine from homocysteine [46]. L-methionine is the substrate for the synthesis of *S*-adenosyl-L-methionine (AdoMet), which is the major methyl-group donor for numerous transmethylations important in plants secondary metabolism. Lignin biosynthesis has been suggested to represent a major sink for AdoMet consumption in vascular plants. Thus, the increase of MS in Na_2CO_3-stressed *P. tenuiflora* could reflect an enhanced demand of AdoMet in the synthesis of lignin. In addition, AST and MS have been thought to be relevant to cell wall lignification [20], and a previous study has reported the increased deposition of lignin in the vascular tissues of plants under salinity stress [46]. Therefore, the increase of these two proteins might imply that Na_2CO_3 treatment can increase the vessel development in *P. tenuiflora.* The enhanced cell wall lignification increased the mechanical rigidity of cell wall, strengthening the vascular tissue and subsequently permitting water to be conducted through the xylem under negative pressure without collapse of the vessels. This structural alteration may enhance the cell-to-cell pathway for water transport, impart greater selectivity and reduced ion uptake, and compensate for diminished bulk flow of water and solutes along the apoplastic pathway [6,20,46].

Additionally, a number of proteins involved in transcription, protein synthesis, protein folding and transport, as well as protein degradation showed Na_2CO_3 response (Table 1). Among them, two Na_2CO_3-induced proteins were involved in protein folding. One was protein disulfide-isomerase (PDI), a member of a thioredoxin superfamily inserting disulfide bonds into folding proteins. PDI is specialized to accommodate the structural features of the membrane and secreted proteins. The other was a putative SecA, the mobile subunit of an integral membrane transporter in *Escherichia coli*, consuming ATP during the insertion and deinsertion phases of its catalytic cycle while guiding preprotein segments across the membrane [47]. While the exact function of SecA in plants exposed to akalinity stress is not known, its function is likely to affect membrane protein dynamics in *P. tenuiflora* under alkaline stress.

4. Experimental Section

4.1. Plant Cultivation and Treatment

Seeds of *Puccinellia tenuiflora* (Turcz.) scribn. et Merr. were sowed on pearlite and cultured in a controlled environment chamber under white fluorescent light (300 μM m^{-2} s^{-1}; 13 h light/11 h dark) at 25 °C and about 75% relative humidity for 50 days, as previously described [6,12]. The plants were divided into three groups. In order to correlate with a number of our previous studies that used 0, 0.4%, 1.0% Na_2CO_3 treatments, these plants were treated with 0 mM, 38 mM and 95 mM Na_2CO_3, respectively. Three biological replicates were grown and treated with different concentrations. To maintain stable Na concentrations in Hoagland medium, we changed the culture medium daily and monitored the solution ion content and the osmotic potential using a SevenMulti Neutral meter (Mettler Toledo, London, UK) and a vapor pressure osmometer (5520, Wescor Inc., Logan, UT, USA), respectively. Seven days later, the treated and untreated leaves were harvested and used fresh or immediately frozen in liquid nitrogen and stored at −80 °C for physiological and proteomic analysis. Shoot length and fresh weight were measured right after harvesting. Dry weight was determined after dehydration at 60 °C until a constant weight was maintained. Leaf water content was estimated as the difference between the fresh weight and dry weight divided by the fresh weight [6].

4.2. Photosynthesis and Chlorophyll Fluorescence Analysis

Net photosynthetic rate (Pn), stomatal conductance (Gs), and transpiration rate (Tr) were determined at 10:00 am using a portable photosynthesis system LI-COR 6400 (LI-COR Inc., Lincoln, NE, USA). The induction kinetics of chlorophyll fluorescence was recorded at room temperature using a pulse modulation chlorophyll fluorometer (FMS-2, Hansatech, King's Lynn, UK). After dark adaptation for 30 min, the initial fluorescence yield (Fo) in weak modulated light (0.12 μmol photons m^{-1} s^{-1}, 600 Hz), and maximum fluorescence yield (Fm) emitted during a saturating light pulse (5000 μmol photons m^{-2} s^{-1}, 0.7 s) were measured. Variable fluorescence (Fv) was calculated by the formula: Fv = Fm−Fo. Then Fm' was measured using an irradiance of 400 μmol photons m^{-2} s^{-1} as the actinic light and 5000 μmol photons m^{-2} s^{-1} as the saturating flashes. Fo' was measured using an irradiance of far-red light (1.67 μmol photons m^{-2} s^{-1}, 3 s). The variable fluorescence (Fv') was also calculated by Fv' = Fm'−Fo'. Maximal photochemical efficiency of PSII (Fv/Fm), photochemical efficiency of PSII in the light (Fv'/Fm'), non-photochemical quenching coefficient (qNP; 1−(Fm'−Fo')/(Fm−Fo)), could be acquired directly from the instrument [12].

4.3. Analysis of Osmotic Potential, Electrolyte Leakage, MDA Content, and Antioxidant Enzymes

Osmotic potential was determined by a vapor pressure osmometer (5520, Wescor Inc., Logan, UT, USA). The MDA content was determined using the thiobarbituric acid (TBA) reaction as described by Wang *et al.* [48]. The electrolyte leakage ratio was obtained according to Akram *et al.* [49]. The three antioxidant enzymes (SOD, POD, and CAT) were extracted and detected according to the method of Yu *et al.* [6].

4.4. Determination of Ion Content

For measuring the relative contents of K and Na on the leaf surface, fresh leaves (1 × 2 cm) were cemented to a sample plate, then observed and analyzed with XL-30 surrounding scanning electron microscope (The Netherlands Phillips Company, Eindhoven, Netherlands) and Kevex energy peatrometer (Thermo Company, Schaumburg, IL, USA). The working voltage was 20 kV, and the time difference for the measurement of samples was 60 s. The relative contents of elements were represented as CPS (counts per second), which were computed by subtracting the background value from peak values of varied elements.

The sample preparation and microscopic analysis for Ca determination was done according to Qi *et al.* [50]. The relative Ca contents in diverse cellular structures (including cell wall, cytoplasm, vacuole) were detected by transmission electron microscope (TEM)-X ray microanalysis of 1 μM sections after freeze-drying and water-free embedding in plastic. The leaves (2 mm × 2 mm) were placed in aluminum mesh boxes, and then frozen quickly in the pre-cooling isopentane and propane mixture (the volume ratio is 1:3). Once frozen, the material was placed in a freeze-dryer for removing all of the water. Dry materials were then loaded into the T-type vacuum permeable tube, and infiltrated with diethyl ether in vacuum at 27 °C for 24 h. After being infiltrated with phenylethylene-butyl methacrylate under normal temperature, the materials were transferred into capsules for polymerization. The embedding materials were cut into 1 μm thickness sections on an ultramicrotome (Reichert Ultracut E, Vienna, Austria). For X-ray microanalysis, dry sections were transferred to folding copper grids, coated with carbon and stored over silica gel. Sections were analyzed in a Hitachi 800 transmission electron microscope (fitted with an energy-dispersive X-ray analyzer EDAX 9100). The accelerating voltage was 150 kV, the angle between sample and probe was 33°, and the counting time was 60 s. The relative Ca contents in diverse cellular structures (including cell wall, cytoplasm, vacuole) were also represented by CPS.

For measuring the concentrations of K^+ and Na^+ in the seedling leaves, 0.5 g fresh leaves of each treatment were cut to 2 cm, and put into a small beaker, then immersed in 30 mL deionized water and shaken for 30 min. The leaves were taken out and put into another small beaker with about 30 mL deionized water, then cooked for 15 min in the autoclave with high temperature (121 °C) in order to kill the cells and release all the ions. Finally, after the temperature of the solution was cooled to room temperature, the leaves were filtered from the solution, and the extracted solution was added to the volume of 50 mL. We used the SevenMulti tester to determine the concentrations of Na^+ and K^+.

4.5. Protein Sample Preparation, 2DE, and Image Analysis

The proteins from leaves under different treatment conditions were extracted according to the method of Wang *et al.* [48]. Protein samples were prepared independently from three different batches of plants. Protein concentration was determined using a Quant-kit according to manufacture's instructions (GE Healthcare, Piscataway, NJ, USA). The protein samples were separated and visualized using 2DE approaches according to Yu *et al.* [6]. Protein samples were separated on 24 cm IPG strips (pH 4–7 linear gradient) through isoelectric focusing (IEF) in the first dimension, followed by 12.5% SDS-PAGE gels in the second dimension. About 1.6 mg protein was loaded per gel. Gels were stained by Coomassie Brilliant Blue (CBB). Three biological replicates for each sample and three technical replicates for each biological replicate were run on 2D gels. Gel image acquisition and analysis were conducted as previously described [48]. For image acquisition, the gels were scanned using an ImageScanner III (GE Healthcare) at a resolution of 300 dpi and 16-bit grayscale pixel depth. The images were analyzed with ImageMaster 2D software (version 5.0) (GE Healthcare, Piscataway, NJ, USA).The average vol% values were calculated from three technical replicates to represent the final vol% values of each biological replicate. Spots with more than 1.5-fold change among the treatments and a p value smaller than 0.05 were considered to be differentially expressed.

4.6. Protein Identification and Database Searching

The differentially expressed spots were excised from the gels and digested with trypsin [51]. MS/MS spectra were acquired on a LC ESI Q-TOF MS/MS (AB Sciex, Framingham, MA, USA) according to the method of Yu *et al.* [6]. The MS/MS spectra were searched against the NCBInr protein databases [52] using Mascot software (Matrix Sciences, London, UK). The taxonomic category was green plants. The searching criteria were according to Yu *et al.* [6].

4.7. Protein Classification and Hierarchical Cluster Analysis

The identified proteins were searched against the NCBI database [52] and UniProt database [53] to determine if their functions were known. Combined with the result of BLAST alignments, these proteins were classified into different categories based on biochemical functions. Self-organizing tree algorithm hierarchical clustering of the expression profiles was performed on the log transformed fold change values of protein spots [20,54].

4.8. Statistical Analysis

All results were presented as means ± standard error (SE) of at least three replicates. Data were analyzed by one-way ANOVA using the statistical software SPSS 17.0 (SPSS Inc., Chicago, IL, USA). The treatment mean values were compared by post hoc least significant difference (LSD) test. A p value less than 0.05 was considered statistically significant.

5. Conclusion

Current proteomics allows detailed inspection of individual proteins on a large scale and their dynamic changes underlying different alkaline-responsive cellular processes. In the present study, we discovered several Na_2CO_3-induced enzymes, such as xanthophyll cycle-related TIL, ROS scavenging enzymes (e.g., GLO and GLP), enzymes involved in carbohydrate metabolism, and amino acid metabolism-related proteins (e.g., AST and MS). A number of proteins involved in carbon assimilation, transcription, as well as proteins synthesis and fate were found to be reduced under Na_2CO_3 treatment. In addition, several proteins were proposed to be involved in cross-tolerance (e.g., AIPLP and DREPP-like protein). More than half of the Na_2CO_3-responsive proteins reached the maximum or minimum abundances under 38 mM Na_2CO_3 treatment. This correlates to the performance of plant growth and physiological indexes (e.g., photosynthetic capability and antioxidant enzyme activities), which supports the previous notion that 38 mM Na_2CO_3 is the turning point concentration for *P. tenuiflora.* Based on the integration of proteomics and physiological results, our study revealed several Na_2CO_3-responsive mechanisms in *P. tenuiflora,* including declined photosynthesis (e.g., light absorption and carbon assimilation) and activation of multiple antioxidant mechanisms (e.g., salt ion exclusion and compartmentalization; POD pathway, CAT pathway, and glyoxalase system), as well as enhanced energy supply and other specialized metabolisms. All these findings provide useful molecular information toward improving alkali tolerance of cereals.

Acknowledgments

The project was supported by grants from National Natural Science Foundation of China (No. 31071194, 31270310), Fundamental Research Funds for the Central Universities (No. DL11EA01), China, and Funds for Distinguished Young Scientists of Heilongjiang Province, China (No. JC201011) to S.D.

References

1. Munns, R.; Tester, M. Mechanisms of salinity tolerance. *Annu. Rev. Plant Biol.* **2008**, *59*, 651–681.
2. Parida, A.K.; Das, A.B. Salt tolerance and salinity effects on plants: A review. *Ecotoxicol. Environ. Saf.* **2005**, *60*, 324–349.
3. Tuteja, N. Mechanisms of high salinity tolerance in plants. *Methods Enzymol.* **2007**, *428*, 419–438.
4. Yang, C.; Chong, J.; Li, C.; Kim, C.; Shi, D.; Wang, D. Osmotic adjustment and ion balance traits of an alkali resistant halophyte *Kochia sieversiana* during adaptation to salt and alkali conditions. *Plant Soil* **2007**, *294*, 263–276.
5. Yang, C.; Shi, D.; Wang, D. Comparative effects of salt and alkali stresses on growth, osmotic adjustment and ionic balance of an alkali-resistant halophyte *Suaeda glauca* (Bge.). *Plant Growth Regul.* **2008**, *56*, 179–190.
6. Yu, J.; Chen, S.; Zhao, Q.; Wang, T.; Yang, C.; Diaz, C.; Sun, G.; Dai, S. Physiological and proteomic analysis of salinity tolerance in *Puccinellia tenuiflora*. *J. Proteome Res.* **2011**, *10*, 3852–3870.
7. Peng, Y.H.; Zhu, Y.F.; Mao, Y.Q.; Wang, S.M.; Su, W.A.; Tang, Z.C. Alkali grass resists salt stress through high K^+ and an endodermis barrier to Na^+. *J. Exp. Bot.* **2004**, *55*, 939–949.
8. Wang, C.M.; Zhang, J.L.; Liu, X.S.; Li, Z.; Wu, G.Q.; Cai, J.Y.; Flowers, T.J.; Wang, S.M. *Puccinellia tenuiflora* maintains a low Na^+ level under salinity by limiting unidirectional Na^+ influx resulting in a high selectivity for K^+ over Na^+. *Plant Cell Environ.* **2009**, *32*, 486–496.
9. Guo, L.; Shi, D.; Wang, D. The key physiological response to alkali stress by the alkali-resistant halophyte *Puccinellia tenuiflora* is the accumulation of large quantities of organic acids and into the rhyzosphere. *J. Agron. Crop Sci.* **2010**, *196*, 123–135.
10. Sun, G.R.; Peng, Y.Z.; Shao, H.B.; Chu, L.Y.; Zhao, X.N.; Min, H.Y.; Cao, W.Z.; Wei, C.X. Does *Puccinelia tenuiflora* have the ability of salt exudation? *Colloid. Surf. B* **2005**, *46*, 197–203.
11. Shi, D.C.; Yin, S.J.; Yang, G.H.; Zhao, K.F. Citric acid accumulation in an alkali-tolerant plant *Puccinellia tenuiflora* under alkaline stress. *Acta Bot. Sin.* **2002**, *44*, 537–540.
12. Wang, Y.X.; Sun, G.R.; Suo, B.; Chen, G.; Wang, J.B.; Yan, Y. Effects of Na_2CO_3 and NaCl stresses on the antioxidant enzymes of chloroplasts and chlorophyll fluorescence parameters of leaves of *Puccinellia tenuiflora* (Turcz.) scribn.et Merr. *Acta Physiol. Plant* **2008**, *30*, 143–150.
13. Zhang, C.Q.; Nishiuchi, S.; Liu, S.K.; Takano, T. Characterization of two plasma membrane protein 3 genes (*PutPMP3*) from the alkali grass, *Puccinellia tenuiflora*, and functional comparison of the rice homologues, *OsLti6a/b* from rice. *BMB Rep.* **2008**, *41*, 448–454.
14. Ardie, S.W.; Xie, L.; Takahashi, R.; Liu, S.; Takano, T. Cloning of a high-affinity K^+ transporter gene *PutHKT2;1* from *Puccinellia tenuiflora* and its functional comparison with *OsHKT2;1* from rice in yeast and *Arabidopsis*. *J. Exp. Bot.* **2009**, *60*, 3491–3502.
15. Ardie, S.W.; Liu, S.; Takano, T. Expression of the AKT1-type K^+ channel gene from *Puccinellia tenuiflora*, *PutAKT1*, enhances salt tolerance in *Arabidopsis*. *Plant Cell Rep.* **2010**, *29*, 865–874.
16. Ardie, S.W.; Nishiuchi, S.; Liu, S.; Takano, T. Ectopic expression of the K^+ channel beta subunits from *Puccinellia tenuiflora* (*KPutB1*) and rice (*KOB1*) alters K^+ homeostasis of yeast and *Arabidopsis*. *Mol. Biotechnol.* **2011**, *48*, 76–86.

17. Liu, H.; Zhang, X.; Takano, T.; Liu, S. Characterization of a *PutCAX1* gene from *Puccinellia tenuiflora* that confers Ca^{2+} and Ba^{2+} tolerance in yeast. *Biochem. Biophys. Res. Commun.* **2009**, *383*, 392–396.
18. Wang, X.; Yang, R.; Wang, B.; Liu, G.; Yang, C.; Cheng, Y. Functional characterization of a plasma membrane Na^+/H^+ antiporter from alkali grass (*Puccinellia tenuiflora*). *Mol. Biol. Rep.* **2011**, *38*, 4813–4822.
19. Wang, Y.; Yang, C.; Liu, G.; Jiang, J. Development of a cDNA microarray to identify gene expression of *Puccinellia tenuiflora* under saline-alkali stress. *Plant Physiol. Biochem.* **2007**, *45*, 567–576.
20. Wang, X.; Fan, P.; Song, H.; Chen, X.; Li, X.; Li, Y. Comparative proteomic analysis of differentially expressed proteins in shoots of *Salicornia europaea* under different salinity. *J. Proteome Res.* **2009**, *8*, 3331–3345.
21. Wang, Y.X.; Sun, G.R.; Wang, J.B.; Chen, G.; Cao, W.Z.; Liu, A.P.; Wang, C. Comparison of the effects of Na_2CO_3 and NaCl stresses on antioxidant enzymes of chloroplasts of *Puccinellia tenuiflora* seedlings. *Acta Prataculturae Sin.* **2007**, *16*, 81–86.
22. Yan, X.F.; Xiao, W.; Sun, G.R. Physiological reaction of seedlings of *Puccinellia tenuiflora* under the condition of salt stress I. Effect of salt stress on the growth of *Puccinellia tenuiflora* seedlings. *Heilongjiang J. Anim. Sci. Vet. Med.* **1994**, *3*, 1–3.
23. Flowers, T.J.; Colmer, T.D. Salinity tolerance in halophytes. *New Phytol.* **2008**, *179*, 945–963.
24. Glenn, E.P.; Brown, J.J.; Blumwald, E. Salt tolerance and crop potential of halophytes. *Crit. Rev. Plant Sci.* **1999**, *18*, 227–255.
25. Niyogi, K.K. Photoprotection Revisited: Genetic and molecular approaches. *Annu. Rev. Plant Physiol. Plant Mol. Biol.* **1999**, *50*, 333–359.
26. Zhang, H.; Han, B.; Wang, T.; Chen, S.; Li, H.; Zhang, Y.; Dai, S. Mechanisms of plant salt response: Insights from proteomics. *J. Proteome Res.* **2012**, *11*, 49–67.
27. Badger, M.R.; Price, G.D. The role of carbonic anhydrase in photosynthesis. *Annu. Rev. Plant Biol.* **1994**, *45*, 369–392.
28. Raines, C.A. The Calvin cycle revisited. *Photosynth. Res.* **2003**, *75*, 1–10.
29. Droux, M.; Jacquot, J.P.; Miginac-Maslow, M.; Gadal, P.; Huet, J.; Crawford, N.; Yee, B.; Buchanan, B. Ferredoxin-thioredoxin reductase, an iron-sulfur enzyme linking light to enzyme regulation in oxygenic photosynthesis: Purification and properties of the enzyme from C3, C4, and cyanobacterial species. *Arch. Biochem. Biophys.* **1987**, *252*, 426–439.
30. Miller, G.; Suzuki, N.; Ciftci-Yilmaz, S.; Mittler, R. Reactive oxygen species homeostasis and signalling during drought and salinity stresses. *Plant Cell Environ.* **2010**, *33*, 453–467.
31. Demmig-Adams, B.; Adams Iii, W.W. Photoprotection and other responses of plants to high light stress. *Annu. Rev. Plant Biol.* **1992**, *43*, 599–626.
32. Bugos, R.C.; Hieber, A.D.; Yamamoto, H.Y. Xanthophyll cycle enzymes are members of the lipocalin family, the first identified from plants. *J. Biol. Chem.* **1998**, *273*, 15321–15324.
33. Chen, S.; Gollop, N.; Heuer, B. Proteomic analysis of salt-stressed tomato (*Solanum lycopersicum*) seedlings: Effect of genotype and exogenous application of glycinebetaine. *J. Exp. Bot.* **2009**, *60*, 2005–2019.

34. Yadav, S.K.; Singla-Pareek, S.L.; Ray, M.; Reddy, M.; Sopory, S. Methylglyoxal levels in plants under salinity stress are dependent on glyoxalase I and glutathione. *Biochem. Biophys. Res. Commun.* **2005**, *337*, 61–67.
35. Yadav, S.K.; Singla-Pareek, S.L.; Reddy, M.; Sopory, S. Transgenic tobacco plants overexpressing glyoxalase enzymes resist an increase in methylglyoxal and maintain higher reduced glutathione levels under salinity stress. *FEBS Lett.* **2005**, *579*, 6265–6271.
36. Bernier, F.; Berna, A. Germins and germin-like proteins: Plant do-all proteins. But what do they do exactly? *Plant Physiol. Biochem.* **2001**, *39*, 545–554.
37. Dani, V.; Simon, W.J.; Duranti, M.; Croy, R.R. Changes in the tobacco leaf apoplast proteome in response to salt stress. *Proteomics* **2005**, *5*, 737–745.
38. Jiang, Y.; Yang, B.; Harris, N.S.; Deyholos, M.K. Comparative proteomic analysis of NaCl stress-responsive proteins in *Arabidopsis* roots. *J. Exp. Bot.* **2007**, *58*, 3591–3607.
39. Snowden, K.C.; Richards, K.D.; Gardner, R.C. Aluminum-induced genes (induction by toxic metals, low calcium, and wounding and pattern of expression in root tips). *Plant Physiol.* **1995**, *107*, 341–348.
40. Ozturk, Z.N.; Talamé, V.; Deyholos, M.; Michalowski, C.B.; Galbraith, D.W.; Gozukirmizi, N.; Tuberosa, R.; Bohnert, H.J. Monitoring large-scale changes in transcript abundance in drought-and salt-stressed barley. *Plant Mol. Biol.* **2002**, *48*, 551–573.
41. Kobayashi, S.; Abe, N.; Yoshida, K.T.; Liu, S.; Takano, T. Molecular cloning and characterization of plasma membrane- and vacuolar-type Na^+/H^+ antiporters of an alkaline-salt-tolerant monocot, *Puccinellia tenuiflora*. *J. Plant Res.* **2012**, *125*, 587–594.
42. Dietz, K.J.; Tavakoli, N.; Kluge, C.; Mimura, T.; Sharma, S.; Harris, G.; Chardonnens, A.; Golldack, D. Significance of the V-type ATPase for the adaptation to stressful growth conditions and its regulation on the molecular and biochemical level. *J. Exp. Bot.* **2001**, *52*, 1969–1980.
43. Chen, G.; Yang, W.; Sun, G.R.; Peng, Y.Z.; Zhou, X.Y.; Zhou, W.D.; Du, K.; Zhang, B. The content of K and Na in epidermis and mesophyll cells of *Puccinellia tenuiflora* seedings leaves under Na_2CO_3 stress. *Acta Ecol. Sin.* **2007**, *27*, 5408–5413.
44. Kawamura, Y.; Uemura, M. Mass spectrometric approach for identifying putative plasma membrane proteins of *Arabidopsis* leaves associated with cold acclimation. *Plant J.* **2003**, *36*, 141–154.
45. Cheng, Y.; Qi, Y.; Zhu, Q.; Chen, X.; Wang, N.; Zhao, X.; Chen, H.; Cui, X.; Xu, L.; Zhang, W. New changes in the plasma-membrane-associated proteome of rice roots under salt stress. *Proteomics* **2009**, *9*, 3100–3114.
46. Sanchez-Aguayo, I.; Rodriguez-Galan, J.M.; Garcia, R.; Torreblanca, J.; Pardo, J.M. Salt stress enhances xylem development and expression of *S*-adenosyl-L-methionine synthase in lignifying tissues of tomato plants. *Planta* **2004**, *220*, 278–285.
47. Economou, A.; Wickner, W. SecA promotes preprotein translocation by undergoing ATP-driven cycles of membrane insertion and deinsertion. *Cell* **1994**, *78*, 835–843.
48. Wang, X.; Chen, S.; Zhang, H.; Shi, L.; Cao, F.; Guo, L.; Xie, Y.; Wang, T.; Yan, X.; Dai, S. Desiccation tolerance mechanism in resurrection fern-ally *Selaginella tamariscina* revealed by physiological and proteomic analysis. *J. Proteome Res.* **2010**, *9*, 6561–6577.

49. Akram, M.S.; Ashraf, M.; Shahbaz, M.; Aisha Akram, N. Growth and photosynthesis of salt-stressed sunflower (*Helianthus annuus*) plants as affected by foliar-applied different potassium salts. *J. Plant Nutr. Soil Sci.* **2009**, *172*, 884–893.
50. Qi, L.; Fritz, E.; Tianqing, L.; Hüttermann, A. X-ray microanalysis of ion contents in roots of *Populus maximowiczii* grown under potassium and phosphorus deficiency. *J. Plant Physiol.* **1991**, *138*, 180–185.
51. Dai, S.; Li, L.; Chen, T.; Chong, K.; Xue, Y.; Wang, T. Proteomic analyses of *Oryza sativa* mature pollen reveal novel proteins associated with pollen germination and tube growth. *Proteomics* **2006**, *6*, 2504–2529.
52. The National Center for Biotechnology Information (NCBI). Available online: http://www.ncbi.nlm.nih.gov/ (accessed on 8 August 2012).
53. The Universal Protein Resource (UniProt). Available online: http://www.ebi.uniprot.org/ (accessed on 8 August 2012).
54. Eisen Lab. Available online: http://rana.lbl.gov/EisenSoftware (accessed on 8 August 2012).

3

Alterations in Soluble Class III Peroxidases of Maize Shoots by Flooding Stress

Claudia-Nicole Meisrimler [1], Friedrich Buck [2] and Sabine Lüthje [1,*]

[1] University of Hamburg, Biocenter Klein Flottbek and Botanical Garden, Oxidative Stress and Plant Proteomics Group, Ohnhorststraße 18, D-22609 Hamburg, Germany; E-Mail: c_m_2406@yahoo.de

[2] University Hospital Eppendorf, Institute of Clinical Chemistry, Campus Science, Martinistraße 52, D-20246 Hamburg, Germany; E-Mail: buck@uke.de

* Author to whom correspondence should be addressed; E-Mail: s.luthje@botanik.uni-hamburg.de

---

Abstract: Due to changing climate, flooding (waterlogged soils and submergence) becomes a major problem in agriculture and crop production. In the present study, the effect of waterlogging was investigated on peroxidases of maize (*Zea mays* L.) leaves. The plants showed typical adaptations to flooding stress, *i.e.*, alterations in chlorophyll *a/b* ratios and increased basal shoot diameter. Seven peroxidase bands could be detected by first dimension modified SDS-PAGE and 10 bands by first dimension high resolution Clear Native Electrophoresis that altered in dependence on plant development and time of waterlogging. Native isoelectric focusing revealed three acidic to neutral and four alkaline guaiacol peroxidases that could be further separated by high resolution Clear Native Electrophorese in the second dimension. One neutral peroxidase (pI 7.0) appeared to be down-regulated within four hours after flooding, whereas alkaline peroxidases (pI 9.2, 8.0 and 7.8) were up-regulated after 28 or 52 h. Second dimensions revealed molecular masses of 133 kDa and 85 kDa for peroxidases at pI 8.0 and 7.8, respectively. Size exclusion chromatography revealed native molecular masses of 30–58 kDa for peroxidases identified as class III peroxidases and ascorbate peroxidases by mass spectrometry. Possible functions of these peroxidases in flooding stress will be discussed.

Keywords: flooding; water logging; guaiacol peroxidases; native PAGE; soluble proteins; shoot; *Zea mays* L.

1. Introduction

Weather records documented a steady and significant increase in flooding events over the past six decades [1]. As a consequence, crop fields are more often overflooded by extreme water levels of rivers and heavy rain falls. Survival of plants under those conditions depends on physiological, morphological and metabolic adaptations [2].

Depending upon the moisture or water level on the field, flood, submergence or soil saturation, can be distinguished for waterlogging. Two types of flooding are generally discriminated in the field: (1) waterlogging, in which root and some portion of the shoot are under water or the soil appears water saturated without free-standing water; and (2) complete submergence, where the whole plant is under water [3].

Waterlogged soils provoke iron toxicity and low oxygen levels in roots. Oxygen levels are characterized by two terms: (1) Hypoxia, reduction of oxygen below optimal levels and (2) anoxia, the complete lack of oxygen, which occurs in soils that are exposed to long-term flooding and complete submerging [4].

Waterlogging resistant plants like maize (*Zea mays* L.) adapt to waterlogged conditions by developing aerenchyma in roots for ventilation and some wetland plant species form an apoplastic barrier at the outer cell layers of roots to reduce radial oxygen loss [5]. The apoplastic barrier composition is not well understood, but one potential component is suberin, which accumulates at the hypodermal/exodermal cell layers of the roots under waterlogged soil conditions. However, variation between plant species makes evaluation of the significance of suberin in prevention of radial oxygen loss rather difficult.

Depending on flooding conditions—short-term (<two weeks) or long-term submergence—plants evolved two different strategies [6]. Plants temporally flooded like maize or tufted hairgrass (*Deschampsia cespitosa* L.) show low oxygen quiescence or avoidance syndrome, whereas species like deep water rice (*Oryza sativa* L.) showed low oxygen escape syndrome [7–9]. One of the key players in rice and wetland species grown under submerged conditions is ethylene, which induces (i) aerenchyma in the root cortex by programmed cell death; (ii) adventitious root growth and (iii) elongation of internode by regulation of gibberellic acid biosynthesis and sensitivity [3]. These adaptations provide leaf contact with the atmosphere under submerged conditions and enhanced gas diffusion. The molecular mechanisms induced by flooding have been intensively investigated in deep water rice and *SNORKEL* genes that encode transcription factors of the AP2/ethylene response factor (ERF) family subgroup VII have been discovered [7]. In contrast to the low oxygen escape syndrome, plant species confronted with short-term flooding stress (partial or complete submergence) maintained steady energy conservation without shoot elongation [8]. It is known that another member of the AP2/ERF family mediates the quiescence syndrome (SUB1A) [10].

Aside, nutrient uptake and photosynthesis are affected by flooding in general and changes in chlorophyll a/b ratios in the foliage were observed in both cases [11–13]. Furthermore, the role of reactive oxygen species (ROS) has been discussed recently for both stresses [5,14]. Low photon utilisation of flooded plants could result in the production of ROS like superoxide anion radicals, singlet oxygen, hydrogen peroxides and hydroxyl radicals [15]. These ROS are very reactive and provoke damage to lipid membranes and proteins. To manage the level of ROS plants have antioxidants (e.g., ascorbate, glutathione and tocopherols) and ROS scavenging enzymes like superoxide dismutase or peroxidases [16]. Peroxidase activity is used as a general stress marker. Class III peroxidases (secretory pathway) are antioxidative systems involved in several physiological functions including plant development, cell wall related processes and oxidative stress [17–20]. Due to their reactive cycles, heme-containing peroxidases are involved in both production and detoxification of ROS and are affected under several stress conditions [19–21]. Cytosolic ascorbate peroxidase of soybean (*Glycine max* L.), a flooding sensitive plant species, decreased under submerged conditions [22,23]. In contrast to this observation, peroxidase activity increased in flooding tolerant clover and an additional isoperoxidase was induced [24]. Although total peroxidase activity of plant extracts is a stress marker, results are not clear, because of the high amount of isoenzymes that may be differentially regulated [25]. To distinguish between several isoenzymes and to identify peroxidases involved in a specific stress response, proteomic approaches are state of the art [25,26]. Additionally, advantages of proteomic approaches in studying flooding have been summarised [27]. Protocols for separation of class III peroxidases by native and in-native 2D-PAGE and detection by peroxidase specific in-gel stains have been published by our team [26,28].

Besides rice, grasses like barley (*Hordeum vulgare* L.), wheat (*Triticum aestivum* L.) and maize belong to the flooding tolerant plants. Although maize is one of the most important crop plants in agriculture and biochemical studies indicate its flooding tolerance [27], proteomic approaches have not been presented for leaves of waterlogged maize. In the present study, profiles of soluble proteins were analyzed from leaves of control and waterlogged maize plants that showed typical stress symptoms. Alterations in profiles of class III peroxidases were investigated by modified SDS-PAGE and native isoelectric focusing (IEF) combined with guaiacol staining. IEF-gels were transferred to second dimension modified SDS-PAGE or high resolution Clear Native Electrophoresis (hrCNE) for further separation of isoperoxidases. Possible functions of identified peroxidases in flooding stress will be further discussed in the results and discussion section.

2. Experimental

2.1. Plant Material

Maize plants (*Zea mays* L. cv. Gelber Badischer Landmais, Saatenunion, Hannover, Germany) were grown in the green house (28 °C at day; 16–18 °C at night; 1000 $\mu mol/m^2 \cdot s \pm 50 \mu mol/m^2 \cdot s$) for 28 days on potting soil. At day 29, plants were flooded continuously. Flooding conditions were done without additional oxygen supply. Oxygen concentration, pH and water temperature were checked constantly. Water temperature was steadily 20 °C ± 0.5 °C and pH was 5.6 for all three time points. The water level was held at 15 cm above the soil surface. The control plants were kept in soil without flooding and were continuously watered indirectly from the bottom and once per day from the top. Water content was held between 20% and 30%. All leaves of the shoot were harvested from control and flooded plants 4 h, 28 h and 52 h after induction of flooding. Samples were taken always at the same time point of the day and for each condition four pools containing five biological replicates were collected. Shoot length was determined for each time point using the same 20 plants.

Statistics and diagrams were calculated using OriginPro 8.5.1.G (Additive GmbH, Friedrichsdorf, Germany). For all measurements, standard deviation was calculated and student's *t*-test was used to determine the significance of changes (control *versus* stressed sample).

2.2. Determination of Chlorophyll Concentrations

Leaves were grinded with liquid nitrogen before chlorophyll was extracted using a 90% acetone solution. After extraction for 30 min in the dark, extract was filtered and volume was made up to 50 mL. Chlorophyll *a* and chlorophyll *b* concentrations were determined spectrophotometrically, using the absorption maximum at 663 nm for chlorophyll a and the maximum at 646 nm for chlorophyll b [29]. Based on the absorption, chlorophyll concentration per g fresh weight was calculated for control and stressed plants.

2.3. Protein Extraction

Soluble proteins of shoots were separated from the microsomal proteins by differential centrifugation as described elsewhere [30]. Soluble proteins were concentrated and desalted using spin columns (Millipore, MWC 10,000, Schwalbach, Germany) and protein amounts were quantified as described by Bradford [31] using bovine serum albumin as standard. Samples were stored at −76 °C until further use.

2.4. Size Exclusion Chromatography

Proteins were separated by size exclusion using an HPLC-System (ÄKTA, Amersham Pharmacia Biotech, Freiburg, Germany) with a 2-mL loop. All steps were performed at 4 °C. Samples were concentrated (Centricon YM-10 concentrators; Millipore, Bedford, MA, USA). Concentrated fractions (40–60 μL) or calibration proteins (thyroglobulin (669 kDa), ferritin (440 kDa), catalase (232 kDa), aldolase (158 kDa), bovine serum albumin (68 kDa), horseradish peroxidase (44 kDa), and ribonuclease A (13.7 kDa), Amersham Pharmacia Biotech) were applied on a Superdex 200 column (HR 10/30, GE Healthcare) equilibrated with four column volumes of phosphate buffer (50 mM Na_3PO_4 (pH 7.0), 150 mM NaCl, 1 mM CHAPS, 1 mM EDTA and 1 mM ascorbate). Proteins were eluted by 1.5 column volumes of buffer. The flow rate was 0.5 mL min^{-1}. The fraction size was 0.5 mL. Peroxidase containing fractions were identified by a microassay in 96 well plates. The assay contained 20 μL protein fraction, 180 μL 50 mM Na-acetate buffer, pH 5.5, 25 μL guaiacol (826 mM) and H_2O_2 (8.8 mM) each. Estimates of the molecular masses of peroxidases were calculated using a semi-logarithmic plot of the molecular mass values for the calibration proteins against the elution volumes. For each sample, three biological replicates have been separated.

2.5. Gel Electrophoresis

One dimensional modified SDS-PAGE (12% acrylamide, no reducing agents, no heating of the samples), native IEF-PAGE and hrCNE were used for separation of soluble proteins. Electrophoresis of modified SDS-PAGE was done at 200 V and 4 °C. First dimensions IEF was accomplished in a mini gel cell (Biorad, Munich, Germany). Gels (0.075 × 7 × 8 cm) contained 4 M urea, 2% 3-[(3-cholamidopropyl)dimethylammonio]-1-propanesulfonate (CHAPS), 2% carrier ampholytes pH 3–10 (Serva, Heidelberg, Germany) and 5% acrylamide. Electrophoresis was carried out for 120 min at 100 V, 90 min at 250 V and 30 min at 350 V at 4 °C with 10 mM phosphoric acid and 20 mM NaOH as respectively anode and cathode buffer [28]. Isoelectric points were calculated in comparison with the pH of gel segments derived from control lanes. IEF-PAGE in the first dimension was followed by activity in-gel staining or by the second dimension modified SDS-PAGE and hrCNE). First dimension hrCNE was casted as continues gradient gel (6%–15% acrylamide concentration), electrophoresis was conducted for 45 min at 100 V, followed by 500 V and restriction to 10 mA per gel until the ponceau S reached the bottom of the gel. Gel lanes of the first dimension were equilibrated in loading buffer for modified SDS-PAGE (125 mM Tris-HCl, 0.2% (*w/v*) SDS, 20% (*w/v*) glycerol, and 0.004% (*w/v*) bromo-phenol, pH 6.8) or hrCNE (75 mM imidazole, 1.5 M 6-aminohexanoic acid (ACA), 0.03% Na-deoxycholate and 0.004% ponceau S pH 7.0) for 20 min at room temperature and applied to the second dimension modified SDS-PAGE or hrCNE [21,23]. Second dimension modified SDS-PAGE and hrCNE were performed similar to the first dimension described above. Gels were stained directly with guaiacol/H_2O_2 after the electrophoresis was finished. After 5 min of activity staining, gels were scanned as TIF-file for documentation (400 DPI, Perfection V700 Photo, EPSON GmbH, Meerbusch, Germany). Prestained marker (Fermentas, St. Leon-Rot, Germany) was used as a standard for all SDS-PAGEs. Gels used for guaiacol staining were generally loaded with 40 µg soluble proteins, except for proteins loaded on first dimension hrCNE (25 µg) used for the calculation of native molecular mass of peroxidases. Fractions, resulting from the separation by SEC, tested positive for guaiacol activity in the micro assay were also analysed by one dimensional PAGE. For each active fraction, 25 µL were mixed with the PAGE corresponding sample buffer and loaded on the gel.

2.6. Peroxidase Detection

In-gel peroxidase staining was accomplished with guaiacol/H_2O_2 (1:1) in 50 mM Na-acetate buffer pH 5.0, containing 10 mM $CaCl_2$ [28].

2.7. Protein Digestion

The gel bands were cut out, the proteins reduced with DTT (10 mM, 56 °C, 30 min.), the cysteine residues modified with iodoacetamide (55 mM, ambient temperature, 20 min. in the dark) and the protein in-gel digested with trypsin (conditions: 5 ng trypsin/µL (sequencing grade modified trypsin, Promega, Madison, WI, USA) in 50 mM NH_4HCO_3, 37 °C, 16 h).

After digestion, the gel pieces were repeatedly extracted (50% acetonitrile/5% formic acid) and the combined extracts dried down in a vacuum concentrator.

2.8. Mass Spectrometry

For QTOF, Premier tandem MS analysis peptide extracts were dried down in a vacuum concentrator and resuspended in 20 mL 0.1% formic acid. The samples were centrifuged at maximal speed in an Eppendorf centrifuge and 2–4 µL of the digest used for an LC-MS run. LC-MS runs were done on a QTOF Premier tandem mass spectrometer (Waters-Micromass, Eschborn, Germany) equipped with an Aquity UPLC (Waters, Germany). Samples were applied onto a trapping column (Waters nanoAquity UPLC column, C18, 180 µm × 20 mm), washed for 10 min with 5% acetonitrile, 0.1% formic acid (5 µL/min) and then eluted onto the separation column (Waters nanoAquity UPLC column, C18, 1.7 µm BEH130, 75 µm × 200 mm, 200 nL/min) with a gradient (A, 0.1% formic acid; B, 0.1% formic acid in acetonitrile, 5%–50% B in either 60 or 120 min). The spray was done from a silica emitter with a 10 µm tip (PicoTip FS360-20-10, New Objective) at a capillary voltage of 1.5 kV. For data acquisition the MS^Etechnique was applied: alternating scans (0.95 s, 0.05 s interscan delay) with low (4 eV) and high (ramp from 20–35 eV) collision energy was recorded [32,33]. The data were evaluated with the software package Protein Lynx Global Server version 2.3 (Waters) searching the Uniprot database (Jan 2014 update) and the Peroxibase. At intervals of 10 s, a lockspray spectrum (1 pmol/µL [Glu1] Fibrinopeptide B (Sigma)) was recorded. Using lockspray correction a mass accuracy of <7 ppm was achieved in the MS mode.

3. Results and Discussion

Additional information on the results can be found in the Supplemental data. The change of specific physiological parameters, morphological and anatomical changes have been shown to be typical for flooding stress. Chlorophyll a/b ratio and changes in the morphology were used to prove that plants showed typical changes in these parameters. These parameters can be often seen in context with the level of ROS, like superoxide anion radical, singlet oxygen, hydrogen peroxide and hydroxyl radical, which are highly reactive. They can provoke damage to various molecules, therefore they are tightly managed to protect cells against oxidative stress. It is known that peroxidases are part of this ROS scavenging mechanism; therefore, it is important to understand their regulations under flooding stress. Additionally, they play a role in cell wall loosening and reorganisation, such as needed for the formation of aerenchyma. Peroxidases were studied on two levels: (1) The abundance was studied by modified SDS-PAGE and (2) changes in activity were studied by native PAGE. Both methods were combined with guaiacol peroxidase specific in-gel staining and identification was done by LC-MS.

3.1. Physiological Parameters

After flooding, the plants oxygen concentrations in the water were measured continuously because, as with waterlogging stress in nature, no adjustments were done to change oxygen concentration in one direction or another. After four hours of waterlogging, oxygen concentration was 84%, after 28 h it decreased to 22% and on the last measurement point, after 52 h, it was down to 2%. Typical symptoms of flooding stress were found for maize after three days of waterlogging. The roots of flooded plants tended to become negatively gravitropic (Figure 1A). Shoot length was compared for control and flooded plants at four, 28 and 52 h. The ratio of the average shoot growth (4 h: 1.0, 28 h: 1.0 and 52 h: 1.1) of control *versus* submerged plants did not change significantly (Figure 1B), but showed already a tendency of a decreased growth of the stressed plants. Furthermore, after 52 h of flooding stress, the shoot stem diameter was increased about 24% within stressed plants in comparison to the control plants (Figure 1C,D). Additional to these parameters, chlorophyll a and chlorophyll b content was determined by spectrophotometric measurements (Figure 1E). At the first day of flooding stress, the chlorophyll b concentration decreased about 17.2% after four hours, if compared to chlorophyll a. At the second day, chlorophyll b content was decreased about 24.3% compared to chlorophyll a. At the third day, chlorophyll b content was lowered about 17.8% in comparison to the chlorophyll a content.

Figure 1. Morphological adaption of maize to submergence. (**a**) Phenotype of control plants and plants stressed by submergence at the end of the growing period; (**b**) Shoot length was measured for control and stressed plants over the three days of flooding; (**c**) Shoot basis of control and submerged plants; (**d**) Comparison of basal shoots of control and stressed samples at the end of the experiment; (**e**) Chlorophyll a/b ratio. All measurements were done for $n \geq 20$ biological replicates. Except for (**e**) measurements were done for $n \geq 5$ biological replicates. Control, light grey, flooded plants, dark grey. Significant changes were marked with an asterisk.

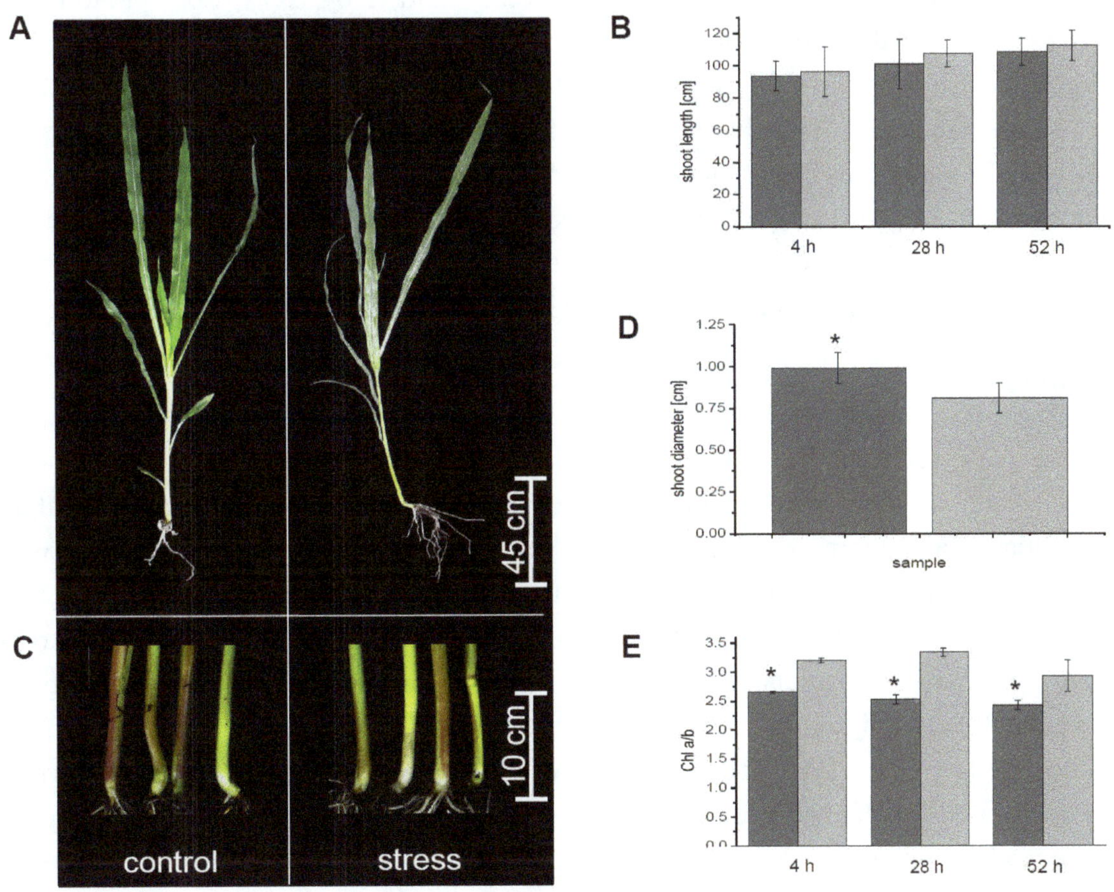

Maize plants showed typical phenotypes after three days of waterlogging (Figure 1). Growth of the shoots was compared for control and flooded plants for 4, 28 and 52 h (three days of flooding). The ratio of the average shoot growth of stressed and control plants did not change significantly. Shoot growth was only slightly decreased for water logged plants after 52 h (Figure 1B). These observations confirm published data for water logged maize [27]. Reduced elongation growth is based on the negative effect of flooding on photosynthesis and is in accordance with the low oxygen quiescence syndrome of maize [9,10,27].

The decrease of chlorophyll a/b ratio (Figure 2) was shown to be a typical reaction to flooding stress in the past [6,27]. The decrease of the chlorophyll a/b ratio seems to be a good marker as its change appears shortly after the plant is exposed to flooding stress, but validations are usually needed, because of variation in reaction to flooding stress in different species. Also, the thickening of the shoot stem diameter after a few days of flooding is in accordance with published data for maize [27]. The thickening of the basal shoot was shown to be based on the aerenchyma formation in the root cortex, which is the most studied morphological response to flooding stress [27,34]. Aerenchyma provides a continuous system of interconnected aerial spaces with a lower resistance for oxygen transport. Aerenchyma formation allows root growth and soil exploration under anaerobic conditions by oxygen transport from aerial shoots to submerged roots [27,34].

In the past, it was shown that continuous flooding over time causes a decrease in photosynthetic capacity of mesophyll cells and finally to an overall reduction of photosynthetic activity [27]. The lower photosynthetic activity is based on the lower chlorophyll content, reduced activity of carboxylation enzymes and oxidative damage of photosystem II by ROS. Low photon utilisation of flooded plants results mostly in ROS production [35]. The level of ROS, like superoxide anion radical, singlet oxygen, hydrogen peroxide and hydroxyl radical, which are highly reactive and provoke damage to various molecules, is tightly managed to protect the cells against oxidative stress. Plants contain antioxidants like ascorbate, glutathione and membrane embedded quinones (e.g., tocopherols and ubiquinone) and enzymes with ability to scavenge ROS and regenerate the antioxidants [28,30]. Peroxidases are part of this ROS scavenging mechanism, but they also play a role in cell wall loosening and reorganisation, such as needed for the formation of aerenchyma.

3.2. Differential Regulation of Soluble Peroxidases—1D and 2D PAGE Analysis

Peroxidases play roles in the ROS scavenging mechanism and in cell wall loosening and reorganisation. Data at hand showed alterations of peroxidase profiles, increases in abundance of specific peroxidases and increase or decrease of guaiacol peroxidase activities under waterlogging conditions. These observations were in accordance to published data [36,37].

Molecular mass of peroxidases and over all profiles were observed for control and stressed plants for all three time points. Four guaiacol peroxidase bands were detected in all samples after separation by modified SDS-PAGE (Figure 2A; band A, B, D, E). The peroxidase profiles were different for the observed time points (Figure 2A). From the relative stress to control (s/c) ratios, relative abundance change was calculated for the bands of the modified SDS-PAGE (Figure 2B). The strongest regulated band of peroxidase abundance was band B, with 133 kDa (Figure 2B). This band was significantly decreased after four hours of waterlogging, whereas it was increased after 28 and 52 h of waterlogging. Furthermore, peroxidase profile of four hours waterlogging exhibited a decreased band C, independent of the flooding stress. Overall, bands A–D were significantly decreased in number after four hours of waterlogging (Figure 2B). After 28 h of waterlogging, stress intensity of bands B–D were significantly increased. At 52 h of waterlogging, only band B was significantly changed. Band E was not significantly changed at any of the observed time points. The overall amount of peroxidase activity per time point, calculated from all bands per lane from all technical/biological replicates, result in the following order: stress day 2 ≥ stress day 3 ≥ control 3 ≥ control 2 ≥ control 1 ≥ stress day 1.

Figure 2. First dimension gel electrophoresis and guaiacol/H_2O_2 staining (**a**) Guaiacol/H_2O_2 staining after separation by modified SDS-PAGE. The pre-stained marker is shown on the left, indicated with M at the top of the gel. Significantly detected guaiacol bands were amounted with the letters of A–E, referring to their mass; (**b**) Relative activity of the significantly detected bands A–E in the modified SDS-PAGE ($n \geq 3$). The corresponding bands are indicated on the x-axis. Dark grey, s1/c1, light grey, s2/c2, middle grey, s3/c3 (s, stress, c, control, 1–3, day after stress induction); (**c**) Guaiacol/H_2O_2 staining after separation by native isoelectric focusing polyacrylamide gel electrophoresis (IEF-PAGE). The picture was inverted to enhance the visibility. Next to the pI, peroxidase identifiers are indicated on the left hand; (**d**) Relative activity of the significantly detected bands with the pI of 9.6–5.9 in the native IEF ($n \geq 3$). The corresponding bands are indicated on the x-axis. Dark grey, s1/c1, light grey, s2/c2, middle grey, s3/c3 (s, stress, c, control, 1, 4 h, 2, 28 h, 3, 52 h). For the gels, the type of sample was indicated at the top with control or stress. At the bottom of the gel, the day after stress induction was specified. Significant changes between control and the associated stressed sample were marked with an asterisk (student's t-test).

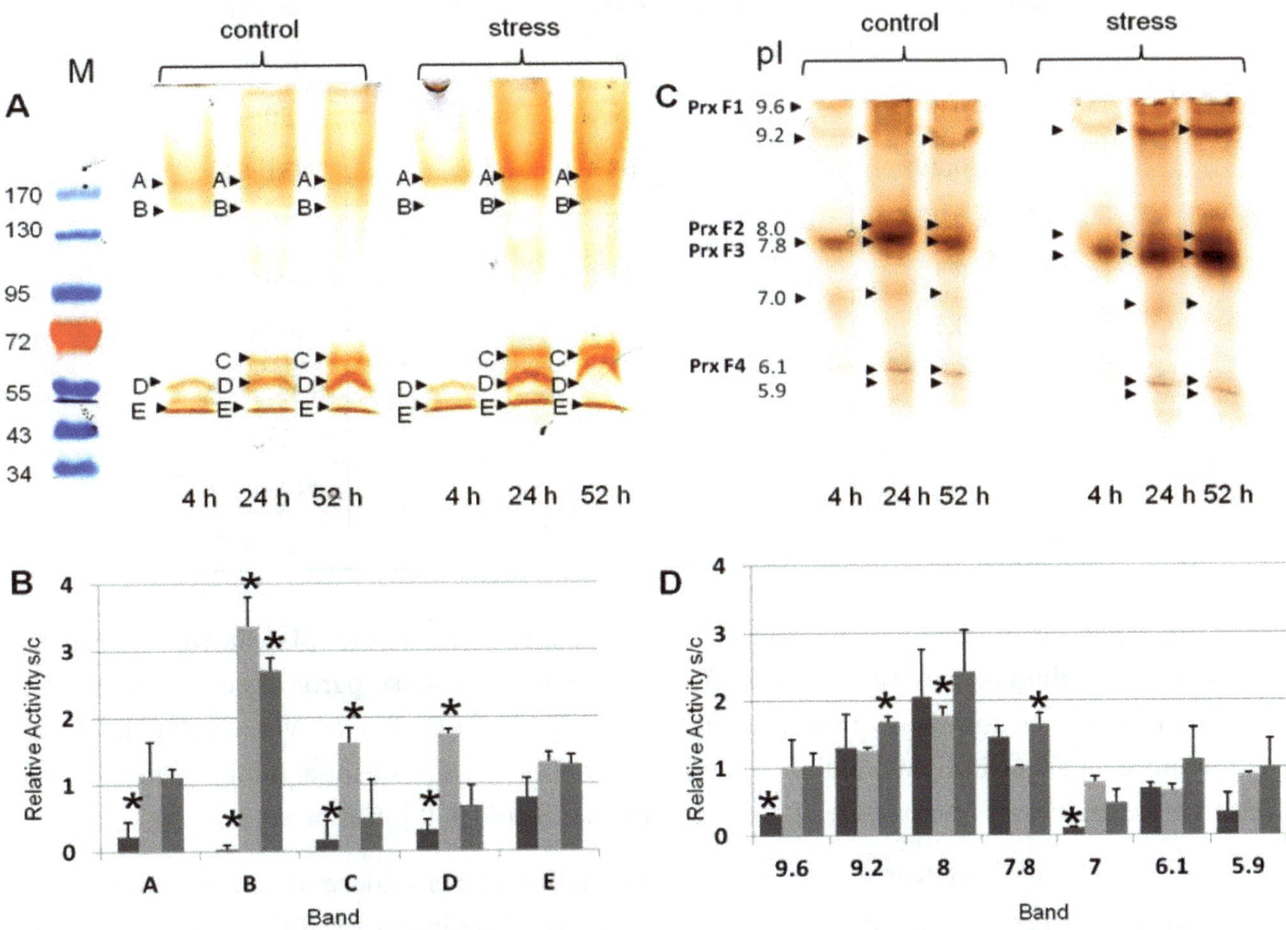

Each lane of modified SDS-PAGE was cut into four pieces, digested and used for identification of proteins by LC-MS (Supplemental Table S1), but peroxidases were not identified. Possible explanations for the lag of peroxidase identifications are the relatively high sensitivity of the guaiacol staining in comparison to standard staining, e.g., CCB, overlay of high abundant proteins with the same molecular mass and inefficient tryptic digestion based on the nature of the non-reducing SDS-PAGE.

Similar to the modified SDS-PAGE, peroxidase bands of the first dimension native IEF-PAGE were used to obtain isoelectric points (pI) and corresponding peroxidase profiles (Figure 2C; Table 1). Overall peroxidase profiles were comparable to the profiles of the modified SDS-PAGE (Figure 2B). Samples of the first time point showed a different peroxidase profile from the samples of the two following time points, independent of the sample treatment. Semi-quantitative analysis of the activity bands was performed for the native IEF-PAGE, as described for modified SDS-PAGE (Figure 2C). Significant changes in activity were observed after four hours of waterlogging at pI 9.6 (PrxF1) and 7.0, which were both decreased in the stressed sample. At 28 h of waterlogging, only the band with the pI of 8.0 (PrxF2) was significantly changed, if the ration of stress to control was compared. The band with the pI of 9.2 was only significantly changed at 52 h after induction of waterlogging. The band with the pI of 7.8 (PrxF3) was similar to the band with pI 9.2 significantly increased after 52 h in the waterlogged sample, if compared to the control (Figure 2D).

Table 1. Summary of peroxidase properties separated by one-dimensional gel-electrophoresis.

pI^{ex} *Native IEF*	kDa *Modified SDS-PAGE 1D*	kDa *hrCNE*
9.6 ± 0.3	183 ± 7	637 ± 7
9.2 ± 0.4	133 ± 5	330 ± 7
8.0 ± 0.2	68 ± 1	431 ± 8
7.8 ± 0.2	55 ± 1	219 ± 8
7.0 ± 0.1	47 ± 2	200 ± 9
6.1 ± 0.1		162 ± 8
		136 ± 0.6
		125 ± 2
		117 ± 1.7
		32 ± 1.7

Aside the specific regulations by flooding, peroxidases with acidic pIs showed a regulation independent of flooding on day two (Figure 2). It is possible that these peroxidases are differentially regulated depending on the developmental stage of the shoot [14]. Plants were grown in the glass house; therefore, changes in light conditions are also an option for this change, but daily measurements showed light intensity was comparable at all three time points (~1000 $\mu mol/m^2 \cdot s$).

In contrast to modified SDS-PAGE that allows estimation of peroxidase abundance, native PAGE allows estimation of peroxidase activities by quantification of the intensity of the guaiacol peroxidase bands [21]. As shown in Figure 2, nearly all isoenzymes increased by waterlogged conditions from time point one to three, showing an overall induction of soluble peroxidases. This was observed for both methods, modified SDS-PAGE and native IEF-PAGE, suggesting a relation between peroxidase abundance and activity. Abundance and activity are not related for all proteins, especially if proteins are activated by post-translational modifications [38].

First dimension hrCNE was used to calculate the native molecular mass of guaiacol peroxidases (Table 1). Based on the resolution and the high activity of the peroxidases (saturation of the bands), hrCNE could not be used for quantitation. Two different amounts of total protein (25 μg and 40 μg) were loaded to the gels to ensure the detection of both strong and faint bands (strong activity can cover light activity). Finally, 10 bands, ranging from 32–637 kDa, were detected (Table 1).

Aside the band at 637 kDa, bands of 133 kDa and higher were detected in the modified SDS-PAGE. In both cases, molecular mass detected is fairly high for peroxidases, indicating an association with a protein complex [39–41]. This protein band may also present peroxidase aggregates or polymers. These results were confirmed by two dimensional gels, namely native IEF/modified SDS-PAGE and native IEF/hrCNE (Figure 3, Table 2). In the native IEF/modified SDS-PAGE combination, two peroxidases with pI 8.0 (PrxF2)/133 kDa and pI 7.8 (PrxF3)/85 kDa were detected, suggesting from their masses to be a dimer and trimer, as the identified peroxidase have an theoretical molecular mass of 27–38 kDa (Table 3). The same spots were detectable in the hrCNE with a native mass of 200 kDa.

Peroxidase profiles in the second dimension hrCNE varied for the different samples and confirmed the results of the first dimension native IEF-PAGE. The smallest amount of peroxidase spots was detectable in the samples after 4 h independent of the stress, whereas the highest amount of spots was detected after 52 h (Supplemental Figure S2). The spot with the pI at pH 8.0/7.8 (PrxF2/F3) and at pI 9.2 were clearly separated in the second dimension into two spots (pI 9.2, J, N; pI 8.0, O, Q; pI 7.8, P,R) with different native molecular masses (Figure 3C, Table 2). Spots in the second dimension were only used to get a better view on the isoenzymes with similar pI that could not be separated by native IEF-PAGE. Based on gel to gel variation, these gels were not used for quantitation, and identification by MS was not successful.

Table 2. Summary of peroxidase properties separated by two-dimensional gel-electrophoresis.

Spot Name	**pI^{ex}** *Native IEF*	**kDa** *Modified SDS-PAGE*	**kDa** *hrCNE*
H	9.8 ± 0.2	n.d.	637 ± 10
J	9.2 ± 0.1	n.d.	637 ± 6
K	8.8 ± 0.2	n.d.	637 ± 5
L	9.6 ± 0.2	n.d.	440 ± 5
M	9.6 ± 0.2	n.d.	330 ± 8
N	9.2 ± 0.1	n.d.	370 ± 7
O	8.0 ± 0.3	n.d.	370 ± 7
P	7.8 ± 0.3	n.d.	330 ± 8
F$^{SDS-PAGE}$/Q^{hrCNE}	8.0 ± 0.3	133 ± 8	200 ± 4
G$^{SDS-PAGE}$/R^{hrCNE}	7.8 ± 0.3	85 ± 4	200 ± 6
S	6.1 ± 0.1	n.d.	139 ± 5
T	5.9 ± 0.2	n.d.	115 ± 2

Figure 3. Second dimension gel electrophoresis for samples exposed to three days of waterlogged soil. (**a**) Guaiacol staining of the second dimension modified SDS-PAGE after separation by IEF-PAGE in the first dimension. The pI of the guaiacol detected spots in the first dimension was indicated at the top of the gel; (**b**) CCB staining of the gel shown in (**c**); (**c**) Guaiacol staining of the second dimension hrCNE after separation by IEF-PAGE in the first dimension. The pI of the guaiacol detected spots in the first dimension is indicated at the top of the gel.

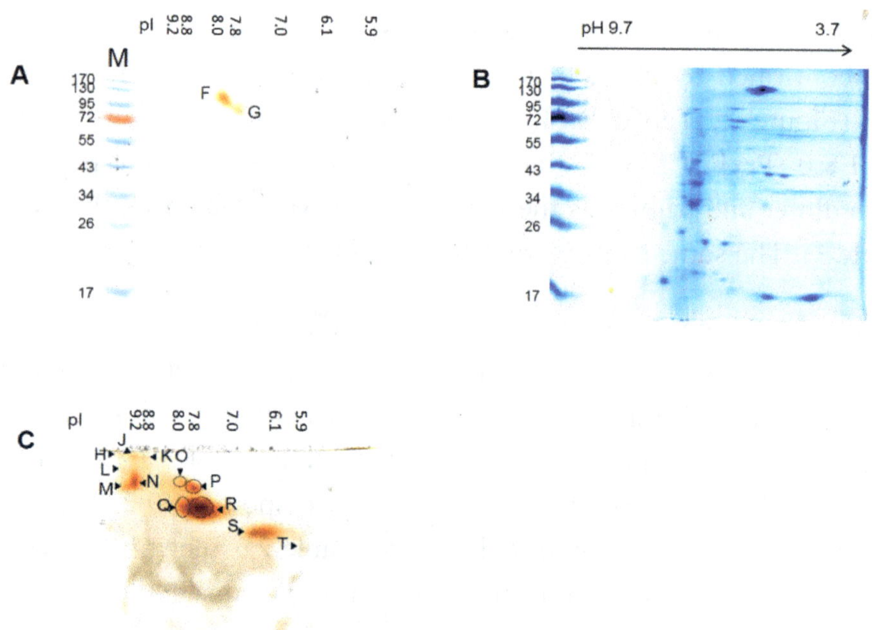

3.3. Sensitivity of Soluble Shoot Peroxidases against SDS

Second dimensions after native IEF-PAGE was separated by modified SDS-PAGE and hrCNE. Modified SDS-PAGE resulted in only two spots with a molecular mass of 133 kDa and 85 kDa and a pI of pH 8.0/7.8 (PrxF2/F3) remained active. To be sure that the protein transfer from the first to the second dimension was performed correctly, gels were stained with colloidal Coomassie Blue (CCB, Figure 3B). CCB staining proved a good transfer of proteins to the second dimension. In order to increase the potential to detect further active peroxidase spots, hrCNE was used as an alternative method for the second dimension. Using hrCNE as second dimension, it was possible to detect 12 peroxidase spots by guaiacol/H_2O_2 staining.

The two spots, 133 kDa and 85 kDa (PrxF2/F3), were the only detectable spots in the second dimension modified SDS-PAGE, showing that they were more stable than the other peroxidases. Spots with a comparable size were also detected in the electrophoretic analysis after separation by SEC, whereas others were not detectable anymore, also showing that these peroxidase multimers or complexes showed greater stability than others. Why these two peroxidase spots are more stable than others has to be further investigated. The identified ZmPrx66 in the band PrxF2 was earlier detected in root plasma membranes samples separated by modified SDS-PAGE [42]. However, most peroxidases investigated in the present study appeared to be sensitive against SDS. At least guaiacol staining did not work properly in second dimension modified SDS-PAGE, possibly due to loss of the heme group; even in the first dimension SDS showed no negative effect on the detection. An explanation for the greater sensitivity to SDS in the second dimension could be actually based on the fact that it was used as a second dimension and the stabilising factor was separated from the peroxidases in the first dimension, resulting in increased sensitivity.

In the past, modified SDS-PAGE was regularly used to study root membrane peroxidase abundance in the second dimension; these enzymes appeared to be more stable compared to the soluble isoperoxidases of the present study [15,20,21,23]. Henceforth, this case could be evidence for the lower stability of soluble peroxidase monomers in the shoot of maize. In any case, this result will need further investigation.

3.4. SEC and Identification of Peroxidases by LC-MS

To confirm the results from the gel electrophoresis, samples were separated by SEC. Peroxidase elution from the column was followed by guaiacol/H_2O_2 micro assay. Analysis of the different samples and biological replicates showed molecular mass from 40–287 kDa with significant variation between the separations. Furthermore, different peroxidase could not be clearly separated (Supplemental Figures S4 and S5). Additionally, active fractions were separated by one dimensional modified SDS-PAGE, native IEF and hrCNE and peroxidases detected by guaiacol/H_2O_2 in gel staining (Supplemental Figure S5, Table S2). Observed profiles showed strong similarities independent of the separated sample (control, stress). Molecular mass calculated for the detected bands confirmed bands from one dimensional electrophoresis separation without SEC (Table 2). Aside that, molecular mass calculated from the fraction number of the SEC varied strongly from the detected bands in the gels. Finally, only bands with a molecular mass above 120 kDa were detected after gel electrophoresis of SEC fractions, independent of the electrophoresis method. Native IEF separation was not possible for most of the samples based on a disturbed electric flow. If separation was possible, activity was detected at pI 6.1, 7.8, 8.0 and 9.6, which confirmed the primary results mentioned above. All spots detected in the native IEF were identified, after tryptic digestion, by MS as peroxidases. Experimental and theoretical properties of the identified peroxidases were summarized in Table 3, while the complete MS data set for identification of the peroxidases can be found in the Supplemental (Supplemental Table S3). The pre-separation of SEC overlaying proteins with similar pI, but different molecular mass, meant they were excluded from the separation of native IEF without diminishing the concentration of the protein. Therefore, the chance of identifying a specific protein, e.g., peroxidase, was much higher than in a first dimension modified SDS-PAGE.

In most activity spots, multiple peroxidases were identified (Table 3). With the experimental pI of the identified peroxidases (PrxF1-F4), they can be assigned to the bands found in the native IEF without pre-separation by SEC. Besides class III peroxidases, also ascorbate peroxidases (APx, class I peroxidases) were identified (ZmAPx01 and 02). These peroxidases play a major role under oxidative stress and have been shown to be regulated under stress conditions [35,43]. Even APx was identified in the bands PrxF2 and PrxF3; usually they cannot use guaiacol as substrate. In soybean, flooding stress regulated APx [22,23]. We suggest that the APxs identified do not contribute to the detected activity, which would be in accordance with earlier results. Additionally, a plasma membrane associated peroxidase (ZmPrx66) was identified in the analysed soluble fraction, which might be due to (i) a contamination or (ii) the proteins disband under specific conditions from the plasma membrane. If the second point is the case, it will have major influence on the understanding of the stress–peroxidase relation. Aside ZmPrx66, APx1 and APx2, eight more peroxidases were identified in the spots Prx F1 to Prx F4. ZmPrx06 (also named peroxidase J), ZmPrx118, ZmPrx97, ZmPrx124, ZmPrx125, ZmPrx07, ZmPrx38, ZmPrx106 were identified in the maize genome but further information on these soluble peroxidases are not known [44–46].

Based on KEGG (Kyoto Encyclopedia of Genes and Genomes) calculations related pathways for ZmPrx118 are the phenylpropanoid [47] and the lignin biosynthesis [48], as well as the phenylalanine metabolism [49]. ZmPrx07 was identified by the NCBI blast as ZmPrx66 precursor and showed 99% similarity to ZmPrx66, making it highly reasonable that the identified peroxidase in the spot Prx F2 is the plasma membrane associated ZmPrx66. ZmPrx42 identified in the band Prx F2 was predicted before as pmPOX3-1 [42]. In both cases, the functions discussed were removal of H_2O_2, oxidation of toxic reductants, biosynthesis and degradation of lignin, suberisation, auxin catabolism, response to environmental stresses such as wounding, pathogen attack and oxidative stress. These functions might be dependent on each isoenzyme/isoform in each plant tissue. Three of the identified peroxidases have been shown to be induced by biotic or abiotic stress factors (Table 3). According to the PeroxiBase, ZmPrx97, identified in band PrxF1 with the pI of 9.6, and ZmPrx66, ZmPrx42, identified in the band PrxF2 with the pI of 8.0, are induced by drought and salt stress. Our former data showed alterations of ZmPrx66 abundance at washed plasma membranes by elicitors, salicylic acid and H_2O_2 [25].

Class III peroxidases may build a complex functional network of different isoenzymes that appears tightly regulated under stress conditions. Depending on a stressor and plant stress responses, distinct isoperoxidases seem to be up-regulated and/or down-regulated. This was shown for maize under submerged conditions (Figures 2 and 3), by different signalling compounds and by oxidative stress [25]. The observed soluble peroxidases have not been able to be assigned to a specific localisation in the cell up to now. Therefore, separation of the peroxidase function in the protective cycle or in the flooding induced leaf growth has to be further investigated. Increased lipid peroxidation and guaiacol peroxidase and APX activity have been demonstrated for maize by flooding in young maize seedlings, but resulting peroxidases were not identified [50].

Thus, ROS scavenging may be one of the major functions of guaiacol peroxidases induced under waterlogging conditions. Peroxidases may also be involved in the process of adaptation. Aerenchyma formation is correlated to programmed cell death (*i.e.*, ROS production) and cell wall stiffening. In cell wall fractions of pea (*Pisum sativum* L.) roots, alkaline isoperoxidases of ionically bound fraction appeared to be involved in elongation growth, whereas covalently bound peroxidases with acidic pI were suggested to be involved in cell wall related functions [51]. Furthermore, extracellular isoperoxidases have been demonstrated to be involved in ROS production [41,52]. ROS production has been demonstrated during root hair formation [53]. Thus, functions in formation of adventive roots may also be possible. Localisation and biochemical properties of flood-induced isoperoxidases will need further studies to clarify their physiological functions in maize.

Table 3. Identified peroxidases by LC-MS. Separated samples by size exclusion chromatography were followed by native IEF and stained with guaiacol/H_2O_2. Detected spots were tryptical digested and analysed by LC-ESI-MS/MS. Identified peroxidases and their properties are listed in the table. Detailed information about the MS results can be found in the Supplemental Table S3. Name: name used in the publication; pI exp: experimental pI resulting from the calculation of the activity band in native IEF after SEC separation; MW^{exp}: experimental molecular mass (kDa), resulting from the SEC separation; Accession: accession for the peroxidase in the searched database; pI^{th}: theoretical pI given by the database; MW^{th}: theoretical molecular mass (kDa) given by the database; Peptides: amount of identified peptides; Class: classification of the peroxidase identified; Localisation: known cellular localisation; Inducers/Repressors: known inducers or repressors of the identified peroxidase based on information provided by peroxibase [54].

Band	pI^{exp}	MW^{exp}	Accession	Database	pI^{th}	MW^{th}	Peptides	Class	Localisation	Inducers/Repressors
Prx F1	9.6	34–51	ZmPrx06	Peroxibase	6.2	33	6	III	-	*induced by cyst nematode infection, pathogen interaction*
			ZmPrx118		5.5	37	10	III	-	-
			ZmPrx97		6.6	38	12	III	-	*induced by salt stress*
			ZmPrx124		4.7	37	15	III	-	-
			ZmPrx125		8.6	34	10	III	-	-
Prx F2	8.0	34–58	ZmPrx66	UniProt	8.0	33	2	III	PM	*induced by drought, elicitors, salicylic acid, wounding and H_2O_2*
			ZmPrx42	Peroxibase	5.3	33	9	III	PM	-
			ZmPrx07	Peroxibase	8.0	34	12	III	-	-
			ZmAPx01	UniProt	5.7	27	27	I	cytosolic	-
			ZmAPx02	UniProt	5.7	28	9	I	cytosolic	-
Prx F3	7.8	34–58	ZmAPx01	UniProt/Peroxibase	5.7	27	8	I	cytosolic	-
			ZmPrx38	Peroxibase	9.2	38	10	III	-	-
			ZmPrx07	Peroxibase	8.0	34	12	III	-	-
Prx F4	6.1	45	ZmPrx106	Peroxibase	8.4	34	40	III	-	-

4. Conclusions

In the present study, the effect of waterlogging on maize peroxidase profiles has been investigated. Isoperoxidases were altered in protein abundance, and increased guaiacol peroxidase abundance and activity was detected by proteomic approaches. The combination of native IEF-PAGE and hrCNE appears to be a powerful set-up to investigate protein adaptations under stress conditions. Second dimension modified SDS-PAGE appears problematic for most soluble guaiacol peroxidases, except for PrxF2 which was identified as ZmPrx66 amongst others, probably because of instability (e.g., lost heme groups). In the past, second dimension modified SDS-PAGE was regularly used for analysis of membrane-bound peroxidases.

These peroxidases appeared to be stable under these circumstances. A recent study of our group suggests a high number of membrane bound heme-peroxidases [19] that may participate in the complex network of class III peroxidases. Thus, investigation of membrane-bound isoperoxidases will be needed to fully understand the regulatory network of peroxidases involved in abiotic and biotic stresses and other cellular mechanisms.

Acknowledgments

This work was supported by the Excellence Initiative of the University of Hamburg (Postdoc Grant to C.N.M.) and Deutsche Forschungsgemeinschaft (DFG Lu-668/4-4).

Author Contributions

C.N.M performed plant experiments, sample preparation, gel analyses and data evaluation. S.L. conducted size exclusion experiments and micro assays. F.B. was responsible for mass spectrometry and data analysis. The manuscript was written by all authors. Figures and tables were prepared by C.N.M.

References

1. Number of flood events by continent and decade since 1950. Available online: http://www.grida.no/graphicslib/detail/number-of-flood-events-by-continent-and-decade-since-1950_10c2 (accessed on 28 March 2014).
2. Bailey-Serres, J.; Fukao, T.; Gibbs, D.J.; Holdsworth, M.J.; Leem, S.C.; Licausi, F.; Perata, P.; Voesenek, L.A.; van Dongen, J.T. Making sense of low oxygen sensing. *Trends Plant Sci.* **2012**, *17*, 129–138.
3. Ahmed, F.; Rafii, M.Y.; Ismail, M.R.; Juraimi, A.S.; Rahim, H.A.; Asfaliza, R.; Latif, M.A. Waterlogging tolerance of crops: Breeding, mechanism of tolerance, molecular approaches, and future prospects. *Biomed. Res. Int.* **2013**, doi:10.1155/2013/963525.
4. Sairam, R.K.; Dharmar, K.; Chinnusamy, V.; Meena, R.C. Waterlogging-induced increase in sugar mobilization, fermentation, and related gene expression in the roots of mung bean (*Vigna radiata*). *J. Plant Physiol.* **2009**, *166*, 602–616.

5. Watanabe, K.; Nishiuchi, S.; Kulichikhin, K.; Nakazono, M. Does suberin accumulation in plant roots contribute to waterlogging tolerance? *Front. Plant Sci.* **2013**, *4*, doi:10.3389/fpls.2013.00178.
6. Colmer, T.D.; Voesenek, L.A.C.J. Flooding tolerance: Suites of plants traits in variable environments. *Funct. Plant Biol.* **2009**, *36*, 665–681
7. Hattori, Y.; Nagai, K.; Furukawa, S.; Song, X.J.; Kawano, R.; Sakakibara, H.; Wu, J.; Matsumoto, T.; Yoshimura, A.; Kitano, H.; *et al.* The ethylene response factors SNORKEL1 and SNOKEL2 allow rice to adapt to deep water. *Nature* **2009**, *460*, 1026–1030.
8. Bailey-Serres, J.; Voesenek, L.A.C.J. Flooding stress: Acclimations and genetic diversity. *Annu. Rev. Plant Biol.* **2008**, *59*, 313–336.
9. Bailey-Serres, J.; Voesenek, L.A.C.J. Life in the balance: A signaling network controlling survival of flooding. *Curr. Opin. Plant Biol.* **2010**, *13*, 489–494.
10. Fukao, T.; Bailey-Serres, J. Submergence tolerance conferred by Sub1A is mediated by SLR1 and SLRL1 restriction of gibberellin responses in rice. *Proc. Natl. Acad. Sci. USA* **2008**, 105, 16814–16819.
11. Gomes, A.R.; Kozlowski, T.T. Growth responses and adaptations of *Fraxinus pennsylvanica* seedlings to flooding. *Plant Physiol.* **1980**, *66*, 267–271.
12. Gonzáles, J.A.; Gallardo, M.; Hilal, M.; Rosa, M.; Prado, F.E. Physiological responses of quinoa (*Chenopodium quinoa* Willd.) to drought and waterlogging stresses: Dry matter partitioning. *Bot. Stud.* **2009**, *50*, 35–42.
13. Pereira, E.G.; Oliva, M.A.; Rosado-Souza, L.; Mendes, G.C.; Colares, D.S.; Stopato, C.H.; Almeida, A.M. Iron excess affects rice photosynthesis through stomatal and non-stomatal limitations. *Plant Sci.* **2013**, *201–202*, 81–92.
14. Steffens, B.; Steffen-Heins, A.; Sauter, M. Reactive oxygen species mediates growth in submerged plants. *Front Plant Sci.* **2013**, *4*, doi:10.3389/fpls.2013.00179.
15. Farmer, E.E.; Müller, M.J. ROS-mediated lipid peroxidation and RES-activated signaling. *Ann. Rev. Plant Biol.* **2013**, *64*, 429–450.
16. Asada, K. Production and scavenging of reactive oxygen species in chloroplasts and their functions. *Plant Physiol.* **2006**, *141*, 391–396.
17. De Gara, L. Class III peroxidases and ascorbate metabolism in plants. *Phytochem. Rev.* **2004**, *3*, 195–205.
18. Kawano, T. Roles of the reactive oxygen species-generating peroxidase reactions in plant defense and growth induction. *Plant Cell Rep.* **2003**, *21*, 829–837.
19. Passardi, F.; Cosio, C.; Penel, C.; Dunand, C. Peroxidases have more functions than a Swiss army knife. *Plant Cell Rep.* **2005**, *24*, 255–265.
20. Lüthje, S.; Meisrimler, C.N.; Hopff, D.; Möller, B. Phylogeny, topology, structure and functions of membrane-bound class III peroxidases in vascular plants. *Phytochemistry* **2011**, *72*, 1124–1135.
21. Lüthje, S.; Möller, B.; Perrineau, F.C.; Wöltje, K. Plasma membrane electron pathways and oxidative stress. *Antioxid. Redox Signal.* **2013**, *18*, 2163–2183.
22. Shi, F.; Yamamoto, R.; Shimamura, S.; Hiraga, S.; Nakayama, N.; Nakamura, T.; Yukawa, K.; Hachinohe, M.; Matsumot, H.; Komatsu, S. Cytosolic ascorbate peroxidase 2 (cAPX2) is involved in the soybean response to flooding. *Phythochemistry* **2008**, *69*, 1295–1303.

23. Kausar, R.; Hossain, Z.; Makino, T.; Komatsu, S. Characterization of ascorbate peroxidase in soybean under flooding and drought stress. *Mol. Biol. Rep.* **2012**, *39*, 10573–10579.

24. Simova-Stoilova, L.; Demirevska, K.; Kingston-Smith, A.; Feller, U. Involvement of the leaf antioxidant system in the response to soil flooding in two *Trifolium* genotypes differing in their tolerance to waterlogging. *Plant Sci.* **2012**, *183*, 43–49.

25. Mika, A.; Boenisch, M.J.; Hopff, D.; Lüthje, S. Membrane-bound guaiacol peroxidases are regulated by methyl jasmonate, salicylic acid, and pathogen elicitors. *J. Exp. Bot.* **2010**, *61*, 831–841.

26. Meisrimler, C.N.; Planchon, S.; Renaut, J.; Sergeant, K.; Lüthje, S. Alteration of plasma membrane-bound redox systems of iron deficient pea roots by chitosan. *J. Proteomics* **2011**, *74*, 1437–1449.

27. Komatsu, S.; Hiraga, S.; Yanagawa, Y. Proteomics techniques for development of flood tolerant crops. *J. Proteome Res.* **2012**, *11*, 68–78.

28. Lüthje, S.; Meisrimler, C.N.; Hopff, D.; Schütze, T.; Köppe, J.; Heino, K. Class III peroxidases. *Methods Mol. Biol.* **2014**, *1072*, 687–706.

29. Porra, R.J.; Thompson, W.A.; Kriedemann, P.E. Determination of accurate extinction coefficients and simultaneous equation for assaying chlorophylls a and b extracted with four different solvents: Verification of the concentration of chlorophyll standards by atomic absorption spectroscopy. *Biochim. Biophys. Acta* **1989**, *975*, 384–394.

30. Lüthje, S.; van Gestelen, P.; Córdoba-Pedregosa, M.C.; González-Reyes, J.A.; Asard, H.; Villalba, J.M.; Böttger, M. Quinones in plant plasma membranes—A missing link? *Protoplasma* **1998**, *205*, 43–51.

31. Bradford, M.M. A rapid and sensitive method for the quantification of microgram quantities of protein dye binding. *Anal. Biochem.* **1976**, *72*, 248–254.

32. Silva, J.C.; Denny, R.; Dorschel, C.A.; Gorenstein, M.; Kass, I.J.; Li, G.Z.; McKenna, T.; Nold, M.J.; Richardson, K.; Young, P.; *et al.* Quantitative proteomic analysis by accurate mass retention time pairs. *Anal. Chem.* **2005**, *77*, 2187–2200.

33. Li, G.Z.; Vissers, J.P.; Silva, J.C.; Golick, D.; Gorenstein, M.V.; Geromanos, S.J. Database searching and accounting of multiplexed precursor and product ion spectra from the data independent analysis of simple and complex peptide mixtures. *Proteomics* **2009**, *9*, 1696–1719.

34. Striker, G.G. Flooding Stress on Plants: Anatomical, Morphological and Physiological Responses. *Botany* **2012**, *1*, 3–28.

35. Asada, K.; Takahashi, M. Production and scavenging of active oxygen in chloroplasts. In *Photoinhibition*; Kyle, D.J., Osmond, C.B., Arntzen, C.J., Eds.; Elsevier: Amsterdam, The Netherlands, 1987; pp. 227–287.

36. Pourabdal, L.; Heidary, R.; Farboodnia, T.T. Effects of three different flooding periods on some anatomical, morphological and biochemical changings in maize (*Zea mays* L.) seedlings. *Asian J. Plant Sci.* **2008**, *7*, 90–94.

37. Luo, F.L.; Thiele, B.; Janzik, I.; Zeng, B.; Schurr, U.; Matsubara, S. De-submergence responses of antioxidative defense systems in two wetland plants having escape and quiescence strategies. *J. Plant Physiol.* **2012**, *169*, 1680–1689.

38. Greenbaum, D.; Colangelo, C.; Williams, K.; Gerstein, M. Comparing protein abundance and mRNA expression levels on a genomic scale. *Genome Biol.* **2003**, *4*, doi:10.1186/gb-2003-4-9-117.
39. Führs, H.; Götze, S.; Specht, A.; Erban, A.; Gallien, S.; Heintz, D.; van Dorsselaer, A.; Kopka, J.; Braun, H.P.; Horst, W.J. Characterization of leaf apoplastic peroxidases and metabolites in *Vigna unguiculata* in response to toxic manganese supply and silicon. *J. Exp. Bot.* **2009**, *60*, 1663–1678.
40. Lüthje, S.; Hopff, D.; Schmitt, A.; Meisrimler, C.N.; Menckhoff, L. Hunting for low abundant redox proteins in plant plasma membranes. *J. Proteomics* **2009**, *72*, 475–483.
41. Minibayeva, F.; Kolesnikov, O.; Chasov, A.; Beckett, R.; Lüthje, S.; Vylegzhanina, N.; Buck, F.; Böttger, M. Wound-induced apoplastic peroxidase activities: Their roles in the production and detoxification of reactive oxygen species. *Plant Cell Environ.* **2009**, *32*, 497–508.
42. Mika, A.; Buck, F.; Lüthje, S. Membrane-bound class III peroxidases: Identification, biochemical properties and sequence analysis of isoenzymes purified from maize (*Zea mays* L.) roots. *J. Proteomics* **2008**, *71*, 412–424.
43. Koussevitzky, S.; Suzuki, N.; Huntington, S.; Armijo, L.; Sha, W.; Cortes, D.; Shulaev, V.; Mittler, R. Ascorbate peroxidase 1 plays a key role in the response of *Arabidopsis thaliana* to stress combination. *J. Biol. Chem.* **2008**, *283*, 34197–34203.
44. Schnable, P.S.; Ware, D.; Fulton, R.S.; Stein, J.C.; Wei, F.; Pasternak, S.; Liang, C.; Zhang, J.; Fulton, L.; Graves, T.A.; *et al.* The B73 maize genome: Complexity, diversity, and dynamics. *Science* **2009**, *326*, 1112–1115.
45. Soderlund, C.; Descour, A.; Kudrna, D.; Bomhoff, M.; Boyd, L.; Currie, J.; Angelova, A.; Collura, K.; Wissotski, M.; Ashley, E.; *et al.* Sequencing, mapping, and analysis of 27,455 maize full-length cDNAs. *PLoS Genet.* **2009**, *5*, e1000740.
46. Alexandrov, N.N.; Brover, V.V.; Freidin, S.; Troukhan, M.E.; Tatarinova, T.V.; Zhang, H.; Swaller, T.J.; Lu, Y.P.; Bouk, J.; Flavell, R.B.; *et al.* Insights into corn genes derived from large-scale cDNA sequencing. *Plant Mol. Biol.* **2009**, *69*, 179–194.
47. Phenylpropanoid biosynthesis—*Zea mays* (maize). Available online: http://www.kegg.jp/pathway/zma00940 (accessed on 28 March 2014).
48. Lignin biosynthesis, cinnamate => lignin. Available online: http://www.kegg.jp/module/zma_M00039 (accessed on 28 March 2014).
49. Lignin biosynthesis, cinnamate => lignin. Available online: http://www.kegg.jp/pathway/zma00360 (accessed on 28 March 2014).
50. Pourabdal, L.; Heidary, R.; Farboodnia, T. The effects of flooding stress on induction of oxidative stress and antioxidant enzymes activity in *Zea mays* L. seedlings. *Res. J. Biol. Sci.* **2008**, *3*, 391–394.
51. Kukavica, B.M.; Veljovic-Jovanovic, S.D.; Menckhoff, L.; Lüthje, S. Cell wall-bound cationic and anionic class III isoperoxidases of pea root: Biochemical characterization and function in root growth. *J. Exp. Bot.* **2012**, *63*, 4631–4645.
52. Kukavica, B.; Mojovic, M.; Vucinic, Z.; Maksimovic, V.; Takahama, U.; Veljovic Janovic, S. Generation of hydroxyl radical in isolated pea root cell wall, and the role of cell wall-bound peroxidase, Mn-SOD and phenoloics in their production. *Plant Cell Physiol.* **2009**, *50*, 304–317.

4

Proteomic Analysis of Responsive Proteins Induced in Japanese Birch Plantlet Treated with Salicylic Acid

Hiromu Suzuki, Yuya Takashima [†], Futoshi Ishiguri, Nobuo Yoshizawa [‡] and Shinso Yokota [*]

Department of Forest Science, Faculty of Agriculture, Utsunomiya University, 350 Mine-machi, Utsunomiya, Tochigi 321-8505, Japan; E-Mails: skaku@ba2.so-net.ne.jp (H.S.); ytakashima@ffpri.affrc.go.jp (Y.T.); ishiguri@cc.utsunomiya-u.ac.jp (F.I.)

† Current address: Forest Tree Breeding Center, Forestry and Forest Products Research Institute, 3809-1 Ishi, Juo, Hitachi, Ibaraki 319-1301, Japan.

‡ The late professor of Utsunomiya University.

* Author to whom correspondence should be addressed; E-Mail: yokotas@cc.utsunomiya-u.ac.jp

---

Abstract: The present study was performed to unravel the mechanisms of systemic acquired resistance (SAR) establishment and resistance signaling pathways against the canker-rot fungus (*Inonotus obliquus* strain IO-U1) infection in Japanese birch plantlet No.8. Modulation of protein-profile induced by salicylic acid (SA)-administration was analyzed, and SA-responsive proteins were identified. In total, 5 specifically expressed, 3 significantly increased, and 3 significantly decreased protein spots were identified using liquid chromatography/tandem mass spectrometry (LC/MS/MS) and the sequence tag method. These proteins were malate dehydrogenase, succinate dehydrogenase, phosphoglycerate kinase, diaminopimalate decarboxylase, arginase, chorismate mutase, cyclophilin, aminopeptidase, and unknown function proteins. These proteins are considered to be involved in SAR-establishment mechanisms in the Japanese birch plantlet No 8.

Keywords: Japanese birch; *Inonotus obliquus*; salicylic acid

---

1. Introduction

Japanese birch (*Betula platyphylla* var. *japonica*) belongs to the Betulaceae family and is distributed throughout the subalpine zone in Honsyu and Hokkaido, Japan. This tree is a pioneer species, and its growth is so fast that the tree is considered to be useful for biomass production [1]. Its sap is used for cosmetics and drinks. In addition, its extractives exhibit potential antioxidant and anti-cancer properties [2]. Its bark contains betulin and betulic acid the derivatives of which possess a wide spectrum of biological and pharmacological activities [3].

Inonotus obliquus (Persoon: Fries) Pilat is a white-rot fungus, causes stem heart rot of birch, and produces a black solid sclerotium referred to as sterile conk or canker-like body [4,5]. The endo-polysaccharide extracted from the mycelia of the fungus exhibits indirect anti-cancer effects [6].

Infected plant tissues induce various resistance responses, formation of physical barriers (papilla, thick cell wall, lignification, *etc.*), accumulation of antipathogenic compounds (phenolic compounds, phytoalexins, pathogenesis-related (PR) proteins, reactive oxygen species (ROS), *etc.*), and hypersensitive cell death named hypersensitive reaction (HR) [7]. The papilla is the reinforced cell wall apposition localized at the site of fungal penetration, where phenolic compounds and callose are accumulated [7]. Lignin gives the cell wall physical strength and has an antipathogenic activity [8]. Phytoalexins are low molecular weight compounds accumulated in plant tissues by pathogen infection that have an antipathogenic activity [9]. More than 200 compounds are identified as phytoalexins in various plant species [7]. PR proteins are newly expressed upon infection, although not necessarily under all pathological conditions [10]. PR proteins are divided into several groups based on their sequence and potential functions [10]. Their biochemical properties include glucanase, chitinase, peroxidase, and protease activities. Some of them have unknown functions and indirect antipathogenic activities [11]. HR is considered as one of the phenomena for programmed cell death (PCD) induced by pathogen infection. It inhibits obligate parasite growth and diffusion. Growth of other pathogens is also inhibited by antipathogenic compounds accumulated during HR induction [7]. ROS play important roles in defense responses. They act as antipathogenic and signaling compounds in plants [12]. The above-mentioned physical barriers and antipathogenic compounds inhibit pathogen growth and, consequently, induce pathogen resistance.

It is known that pathogen resistance is not only induced in infected tissues but also in systemic tissues. This resistance, referred to as systemic acquired resistance (SAR), is considered to be important in pathogen resistance reactions. Production of PR proteins is usually observed before an infectious challenge. In contrast, other reactions have been detected after an infectious challenge. For example, accumulation of phenolic compounds and higher activities of peroxidase and chitinase induced by infection of *Colletotrichum lageizarium* were observed with pretreatment of SAR-inducing molecules in *Cucumis sativus* L [13]. Thus, the systemic signals can prepare the systemic tissues for a faster defense response [14]. Salicylic acid (SA) is one of the SAR-inducing molecules that are required for SAR establishment [15]. It has been reported that various other proteins and genes are also expressed by pathogen infection and SA administration [16–18]. These proteins are considered to be important for SAR establishment.

Signal transduction mechanisms for establishment of SAR have been investigated mainly by using *Arabidopsis thaliana* and *Nicotiana tabacum* [15,19–23]. According to these studies, it has been proposed that SA induces expression of SAR-associated genes by the following mechanisms. After pathogen recognition by receptors of the plants, rapid and transient ROS production, called oxidative burst, occurs in the tissues. In a next step, the ROS induce an increase in the amount of SA in both the infected and systemic tissues, and the SA changes the cellular redox state and induces thioredoxin expression. The thioredoxin reduces the disulfide bonds in the cellular nonexpresser of PR genes 1 (NPR1) oligomers, one of the SA signal transduction factors. Reduced NPR1 oligomers release NPR1 monomers, that they bind to TGA transcription factors in the nuclei. The TGA transcription factor expresses SA-induced genes. In addition, NPR1-independent pathways also exist [23]. The translocating signal molecules between infected and systemic tissues are essential for SAR, though they have not been determined yet. It has been reported that methyl salicylate (MeSA) is considered to be a possible SAR long-distance signaling compound in *N. tabacum* [24]. However, there is a report demonstrating that SA is not the translocating signal [25] and that MeSA production is not essential for SAR in *A. thaliana* [26]. In addition, it is suggested that peptide and lipid derivatives are possible long-distance signaling molecules [27]. Moreover, a recent study has demonstrated that azelaic acid (AZA) is a possible translocating signal molecule [28].

In our previous studies of pathogen resistance in Japanese birch plantlet No.8, heat shock 60 kDa and 70 kDa proteins were found to be specifically expressed during the infection of *I. obliquus* strain IO-U1 [29]. In addition, in anatomical and histochemical observations, lignification, phenolics deposition, and necrophylactic periderm formation have occurred as infection-induced responses in Japanese birch plantlet No.8 [30] and Tohoku [31] infected with *I. obliquus* strain IO-U1.

Signal transduction mechanisms for the establishment of pathogen resistance including SAR have been investigated using herbaceous plants such as *A. thaliana* and *N. tabacum*, while there are only a few studies on woody plants.

The purpose of the present study was to unravel the mechanisms of SAR establishment and resistance signal transduction pathways against the infection of *I. obliquus* strain IO-U1 in Japanese birch plantlet No.8. The protein profile changes induced by SA-administration were analyzed, and SA-responsive proteins were identified to clarify SAR establishment mechanisms.

2. Experimental

2.1. Plant Material

Japanese birch (*Betula platyphylla* var. *japonica*), obtained from the Forestry and Forest Products Research Institute, Ibaraki, Japan was used as plant material for this study. The plantlets were grown *in vitro* on Murashige and Skoog medium [32] containing 20 g/L sucrose, 2.5 μM indole-3-acetic acid, 0.1 μM 1-naphthalenacetic acid at 25 °C under illumination at 50 $\mu mol \cdot m^{-2} \cdot s^{-1}$ for a 16-h photo period during 3 months. Axillary buds were subcultured for propagation of the plantlets on the same medium every 3 months.

2.2. Salicylic Acid (SA) Treatment

High purity grade SA (Kanto Chemical Co., Tokyo, Japan) was used for the treatment. SA aqueous solution (0.5 mM) was prepared and its pH was adjusted to around 7.0 with 0.1 M and 0.01 M NaOH.

The solution was sterilized with a membrane filter (Millex-GV, 0.22 μM, Millipore, Billerica, MA, USA) before administering it to the plantlets.

In this study, intact (C1), wounded ($C2_{SA}$), and SA infiltrated (T_{SA}) plantlets were prepared. The surface of the third node from the apex of a plantlet was cut into a V-shape with a surgical knife, and then 1 μL ultra-pure water or 1 μL SA aqueous solution was administered to $C2_{SA}$ or T_{SA} plantlets, respectively. After the treatments, the plantlets were further grown for 2 days under the conditions described above.

2.3. Preparation of Protein Samples

Protein extraction was repeated three times for each treatment. After 48 h of the treatments, each plantlet was deep-frozen with liquid nitrogen and powdered with a mortar and pestle. In a next step, the extraction buffer was added to the powdered samples in a volume of 1 mL/g fresh weight of the plantlet, and the powdered plantlet was further mashed. The extraction buffer was prepared by mixing the following EXT-1, 2, and 3 in a volume ratio of 3:2:1. EXT-1 was prepared by dissolving the Trizma Preset pH crystal 7.5 (Sigma, St. Louis, MO, USA, 7.54 g), dihydrate disodium salt of ethylenediamine-*N,N,N',N'*-tetraacetic acid (Doujin, Tokyo, Japan, 0.56 g), and glycerol (100 mL) in ultra-pure water, followed by adjusting to 250 mL. EXT-2 was prepared by dissolving Triton-X-100 (Acros Organics, Geel, Belgium, 6.4 g) in ultra-pure water and heating it at 60 ℃, followed by adjusting it to 100 mL. EXT-3 was prepared by dissolving dithiothreitol (DTT, Wako Pure Chemical Co., Osaka, Japan, 463 mg) in ultra-pure water, followed by adjusting to 50 mL.

The homogenates were sonicated with a supersonic homogenizer (Sonic Fire 250, Branson, Danbury, CT, USA). The homogenate was centrifuged at 10,000× *g* for 30 min at 4 ° C,and the supernatant was collected in another centrifuge tube (SuperClear Centrifuge tube, Labcon, Petaluma, CA, USA). The supernatant was centrifuged again under the same condition, and then the obtained supernatant was collected in a Teflon-lined centrifuge tube (Nalgene, Rochester, NY, USA). Proteins in the supernatants were precipitated with 10% (w/v) trichloroacetic acid (TCA) aqueous solution, and then the solution was incubated at −20 ° Cfor 1 h. After thawing the sample, it was centrifuged at 10,000× *g* for 30 min at 4 ° Ç and the supernatant was removed. Cold acetone was added to the obtained pellets, and the pellet was mashed and stirred with a spatula. The suspension was centrifuged at 10,000× *g* for 30 min at 4 ° C,and the supernatant was removed. This process was repeated three times. The obtained pellets were dried under nitrogen gas. Solubilization buffer (100 μL) was added to the dried pellets, and they were incubated for 30 min. The solubilization buffer was prepared by dissolving urea (GE healthcare, Little Chalfont, Buckinghamshire, UK; 4.204 g), thiourea (GE Healthcare, 1.422 g), 3-[(3-cholamidopropyl)-dimethylammonio]-1-propane sulfonate (CHAPS, GE Healthcare, 400 mg), DTT (30.8 mg), and immobilized pH gradient buffer pH 4–7 (GE Healthcare, 200 μL) in ultra-pure water, followed by adjusting to 10 mL. The suspension was centrifuged at 10,000× *g* for 30 min at 4 ° C,and the supernatant was collected in a 1.5 mL micro tube (Eppendorf, hamburg, Germany). This process was repeated two times. The protein content of the obtained sample was determined using the Bradford [33] method using ovalbumin (Sigma) as a standard. The obtained samples were stored at −20 ° C before use.

2.4. Two-Dimensional Electrophoresis (2-DE)

Two-dimensional electrophoresis (2-DE) was repeated three times for each treatment. Isoelectric focusing (IEF) was carried out as follows. The protein sample (1 mg) was put in a 1.5 mL micro tube, and the total volume was adjusted to 250 µL with rehydration buffer. The rehydration buffer was prepared by dissolving urea (12 g), CHAPS (0.5 g), immobilized pH gradient buffer pH 4–7 (125 µL), and bromophenol blue (BPB, Kanto Chemical Co. Tokyo, Japan) stock solution (50 µL) in ultra-pure water, followed by adjusting to 25 mL. BPB stock solution was prepared by dissolving BPB (100 mg) and tris-base (MP Biomedicals, Irvine, CA USA, 60 mg) in ultra-pure water, followed by adjusting to 10 mL. An Immobiline DryStrip (pH 4–7, 13 cm, GE Healthcare) was rehydrated with the above mixture at 22 °C for at least 10 h on the Ettan IPGphor II (GE Healthcare). In the preliminary experiments, the DryStrip in the pH range of 3–10 was used, and most of the protein spots were recognized in the acidic to neutral range. Thus, the DryStrip in the pH range of 4–7 was used in this study. To prevent the DryStrip from drying, it was covered with 800 µL of Immobiline DryStrip Cover Fluid (GE Healthcare). After rehydration, IEF electrode stripes (5 mm in length, GE Healthcare) were placed between DryStrip and electrodes of a strip holder. IEF was carried out at 20 °C. Maximum current was set at 50 µA per stripe. Voltage program was as follows: step-n-hold, 500 V, 1 h; gradient, 1000 V, 1 h; gradient, 8000 V, 5 h; step-n-hold, 8000 V, 8 h.

After IEF, the DryStrip was treated with 10 mL of SDS-equilibrium buffer containing 1% (w/v) DTT in a 60 mL test tube with a screw cap (IWAKI, Tokyo, Japan) for 30 min. Then the DryStripe was treated with 10 mL of SDS-equilibrium buffer containing 2.5% (w/v) iodoacetoamide in another 60 mL test tube with a screw cap for 30 min in the dark. The DryStripe was shaken in 10 mL of SDS-running buffer in another 60 mL test tube with a screw cap to remove remaining equilibrium buffer. The DryStripe was put on filter paper infiltrated with ultra-pure water to remove remaining SDS-running buffer.

SDS-polyacrylamide gel electrophoresis (SDS-PAGE) was carried out using SE600 Ruby (GE healthcare) as follows. Polyacrylamide gel was prepared at 12.5% acrylamide concentration and 1.5 mm in thickness. Running buffer was cooled with thermostatic circulator (Multi temp III, GE healthcare) set at 10 °C. The equilibrated DryStripe was fixed at the top of gel with 1 mL of agarose gel. A marker bead was prepared by mixing 60 µL of agarose gel with 30 µL of molecular weight marker (Precision Plus Protein Standards, Bio-Rad, Hercules, CA, USA). The marker bead was placed on the side end of DryStripe. Electrophoresis was carried out at 600 V of maximum voltage and 15 mA of constant current per gel for 15 min. Then constant current was set at 30 mA per gel, and electrophoresis was carried out until BPB reached the bottom of the gel.

2.5. Staining 2-DE Gel

The 2-DE gel was stained with Coomassie Brilliant Blue (CBB) as follows. The gel was shaken in 250 mL of CBB (PhastGel Blue R-350, GE Healthcare) staining solution for 3 h. After that the 2-DE gel was shaken in 250 mL of destaining solution two times for 10 and 3 h, respectively. The destaining solution was prepared by mixing 150 mL ethanol and 50 mL acetic acid with 300 mL distilled water. In the case of image analysis of 2-DE gel, the destained gel was shaken in 250 mL of preservation solution for 30 min twice. The preservation solution was prepared by dissolving 150 mL ethanol and 23 mL glycerol in distilled water, followed by adjusting to 500 mL with distilled water. The 2-DE gel was preserved with Gel Drying kit (Promega, Madison, WI, USA).

2.6. Image Analysis of 2-DE Gels

The 2-DE gel images were captured with a scanner (GT-9700, EPSON, Gen, Suwa, Japan) at a 300 dpi resolution and analyzed with ImageMaster 2D Platinum ver. 5.0 (GE Healthcare). Protein spots without reproduction were excluded from the analysis. Protein spots only expressed in each treatment were considered as specifically expressed ones. Those protein spots were analyzed, which were considered as specifically expressed in T_{SA} gel and were expressed in T_{SA} gel in significantly different amounts compared to the corresponding spots in both C1 and $C2_{SA}$ gels. These protein spots were regarded as SA-responsive protein spots. Protein expression was evaluated in percent intensity for statistical significance ($p < 0.05$) using Student's *t*-test provided with the software.

2.7. In-Gel Digestion

In-gel digestion of the protein spots was carried out as follows: The spots of SA-responsive proteins were cut out from stained gels of T_{SA} and put in 1.5 mL micro tubes (Eppendorf, Hamburg, Germany). The cut gel was treated with 200 μL of destaining solution under agitation for 10 min to overnight. The gel was soaked in 100 μL of acetonitrile under agitation for 5 min. The acetonitrile was removed, and the gel was dried with a centrifugal evaporator (CE-1, Hitachi, Tokyo, Japan) for 15 min. The gel was infiltrated in 100 μL of reducing solution (10 μL of 1 M DTT solution and 25 μL of 1 M ammonium hydrogen carbonate in 965 μL ultra-pure water) under agitation at 56 ℃ for 1 h. The gel was washed with100μL of washing buffer (75 μL of 1 M ammonium hydrogen carbonate in 2925 μL ultra-pure water) under agitation for 10 min. The gel was treated with 100 μL of alkylating solution (10 mg iodeacetamide (Wako Pure Chemical Co., Osaka, Japan) in 1 mL washing buffer) under agitation for 45 min in the dark.

The gel waswashedwith100μLofwashingbufferunderagitationfor10min.Thegelwaswashedwith200 μL of washing solution (5 mL methanol and 5 mL of 10% acetic acid) under agitation four times at least for 1 h. The gel was soaked in 100 μL of equilibration buffer (100 μL of 1 M ammonium hydrogen carbonate in 900 μL ultra-pure water) under agitation for 5 min. The gel was treated with 200 μL of dehydration solution (2 mL acetonitrile and 100 μL of 1 M ammonium hydrogen carbonate in 1.9 mL ultra-pure water) under agitation twice for 10 min. The dehydration solution was removed, and the gel was dried using a centrifugal evaporator for 15 min. The gel was incubated in 20 μL of trypsin (gold mass spectrometry grade, Promega) solution on ice for 30 min. The excessive trypsin solution was removed, and the gel was incubated at 37 ° Covernight. The gel was sonicated in 50 μL of extraction buffer (mixture of 500 μL acetonitrile and 500 μL 0.1% trifluroacetic acid (Wako Pure Chemical Co.)) using an ultrasonicator (3510, Branson) for 10 min. The extract was collected in a 1.5 mL micro tube. This process was repeated again with 25 μL of extraction buffer. The collected peptide sample was dried with a centrifugal evaporator until it was completely dried.Thepeptidesamplewas desalted with ZipTip μ-C18 (Millipore, Billerica, MA, USA).

2.8. LC/MS/MS of the Peptide Sample

A nanospray LTQ XL Orbitrap MS (Thermo Fisher Scientific, Waltham, MA, USA) was operated in data-dependent acquisition mode using the installed XCalibur software. Using an Ultimate 3000 nanoLC (Dionex, Sunnyvale, CA, USA), peptide samples in 0.1% formic acid were loaded onto a 300 μm ID × 5 mm C18 PepMap trap column. The peptide samples were eluted from the trap column and their separation and spraying were done using 0.1% formic acid in acetonitrile at a flow rate of 200 nL/min on a nano-capillary column (NTTC-360, Nikkyo Technos, Tokyo, Japan) with a spray voltage of 1.8 kV. Full scan mass spectra were obtained in the orbitrap over 150–2000 *m/z* with a resolution of 15,000. The three most intense ions above the 1000 threshold were selected for collision-induced fragmentation in the linear ion trap at a normalized collision energy of 35% after accumulation to a target value of 1000. Dynamic exclusion was employed within 30 s to prevent repetitive selection of peptides. Obtained MS/MS spectra were converted to individual DTA files using BioWorks software ver. 3.3.1 (Thermo Fisher Scientific). The following parameters were set to create a list of peaks: parent ions in the mass range with no limitation, one grouping of MS/MS scans, and the threshold at 100.

2.9. Database Search

The database search was carried out by sequence tag method with the MASCOT search engine [34] (Matrix Science, Boston, MA, USA) using obtained data files. The proteins with the highest scores were considered. The following parameters were set for the search: Database, NCBInr; taxonomy, green plants; digest enzyme, trypsin; maximum missed cleavage, one; fixed modification, carbamidomethylation of cysteine; variable modification, oxidation of methionine; peptide mass tolerance, 10 ppm; fragment mass tolerance, 0.2 Da; peptide charge, +1, +2, +3; statistical significance, $p < 0.05$.

3. Results and Discussion

3.1. 2-DE and Image Analysis

The 2-DE gel images are shown in Figures 1–3. Of these spots, the numbers of total reproducing spots in C1, $C2_{SA}$, and T_{SA} gels were 718, 719, and 763, respectively (Figure 4). The numbers of specifically expressed spots were 47, 34, and 23, respectively. IDs were designated to each SA-responsive protein spot (Figure 5).

Most of the T_{SA}-specifically-expressed spots were detected in the ranges of pI 5.5–7.0 and 30–50 kDa on the gel (Figure 3). C1-specifically-expressed spots were more frequently detected on the acidic side (Figure 1). $C2_{SA}$-specifically-expressed spots were detected in the middle ranges of the gel (Figure 2).

Figure 6 shows the mean spot percent intensities of significantly increased or decreased proteins. The numbers of significantly increased spots in the T_{SA} gel compared to both C1 and $C2_{SA}$ gels, to the C1 gel, and to the $C2_{SA}$ gel were two, one, and two, respectively. The numbers of significantly decreased spots in the T_{SA} gel compared to both the C1 and $C2_{SA}$ gels, to the C1 gel, and to the $C2_{SA}$ gel were one, one, and two, respectively. The U1 spot of T_{SA} increased about two-fold compared to both C1 and $C2_{SA}$. The U3 and U5 of T_{SA} increased more than two-fold compared to C1 and $C2_{SA}$, respectively. The D1 decreased to less than one half compared to both C1 and $C2_{SA}$. The D2, D3, and D4 decreased to less than one half compared to C1 or $C2_{SA}$.

Figure 1. Coomassie Brilliant Blue (CBB) stained two-dimensional electrophoresis gel of C1 plantlet. ○: Specifically expressed protein spots in C1 gel.

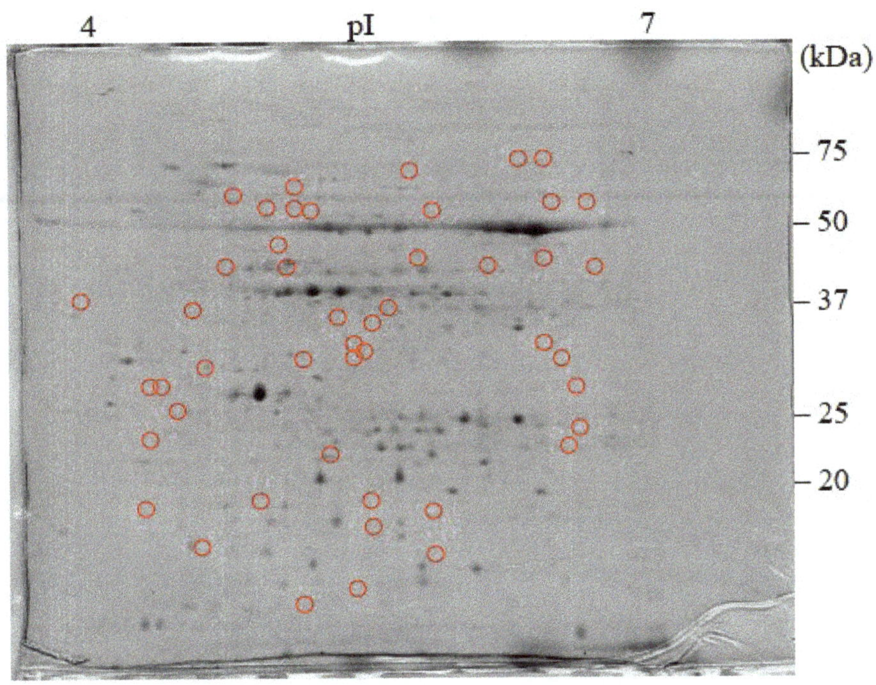

Figure 2. CBB stained two-dimensional electrophoresis gel of $C2_{SA}$ plantlet. ○: Specifically expressed protein spots in $C2_{SA}$ gel.

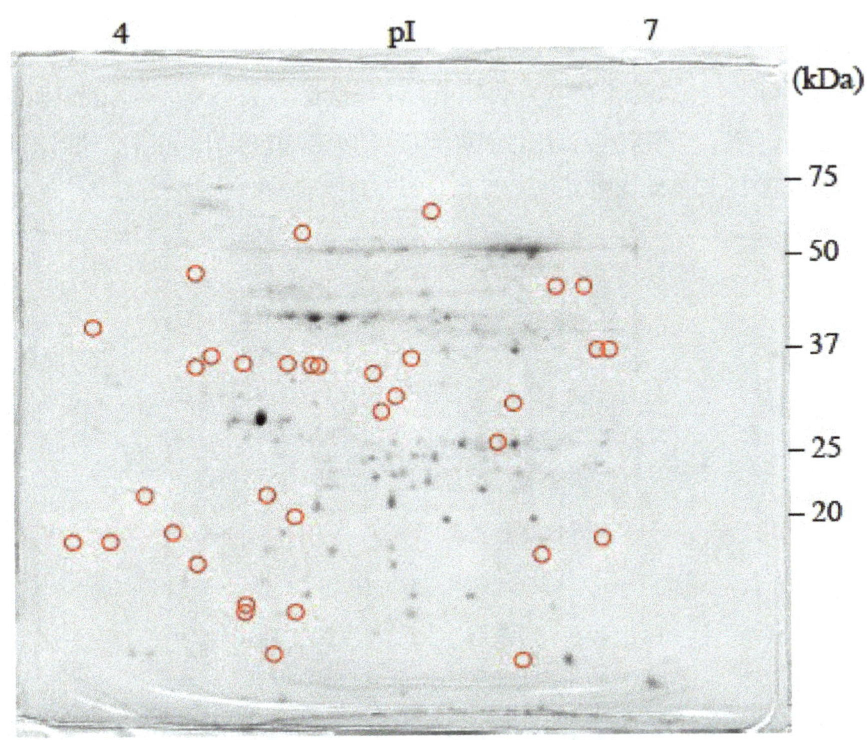

Figure 3. CBB stained two-dimensional electrophoresis gel of T_{SA} plantlet. ○: Specifically expressed protein spot in T_{SA} gel. ○: Significantly increased protein spots in T_{SA} gel. ○: Significantly decreased protein spots in T_{SA} gel.

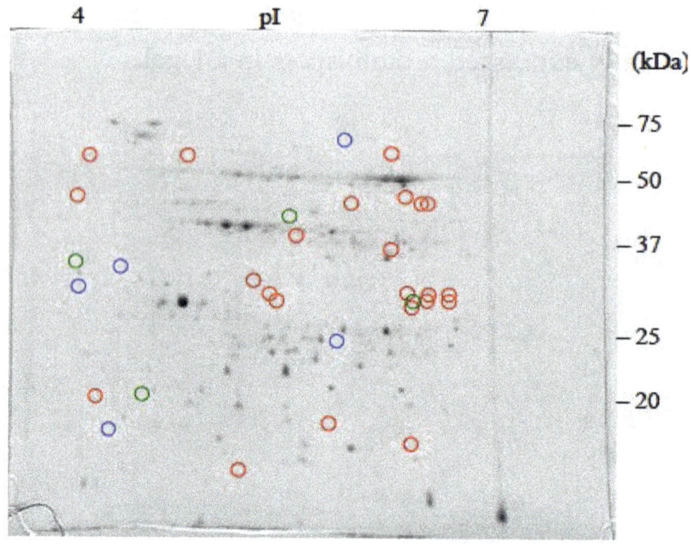

Figure 4. The number of reproducing spots detected in each treatment.

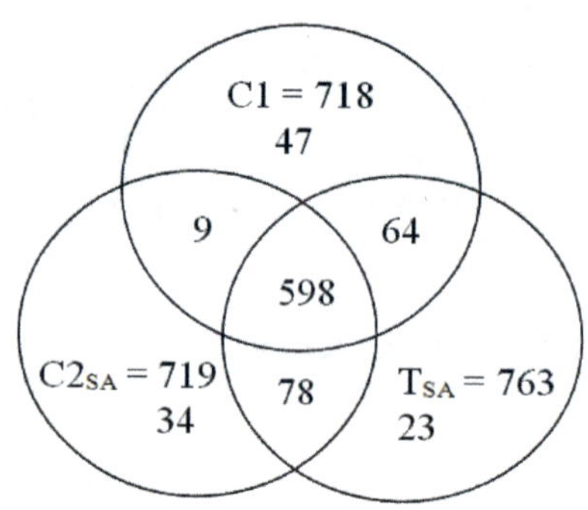

Figure 5. IDs of salicylic acid (SA)-responsive protein spots. +: Specifically expressed protein spots in T_{SA} gel; +: Significantly increased protein spots in T_{SA} gel; +: Significantly decreased protein spots in T_{SA} gel.

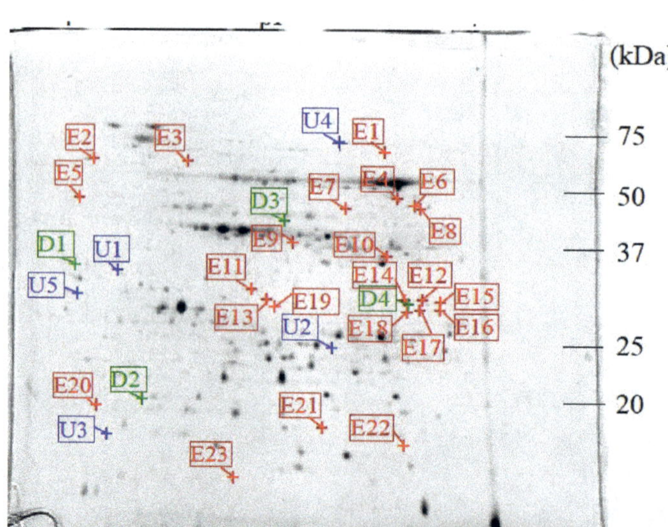

Figure 6. Mean spot percent intensities of significantly changed proteins. Bar represents mean ± standard error of percent intensity of each protein spot in three repetitive gels from different plantlets.

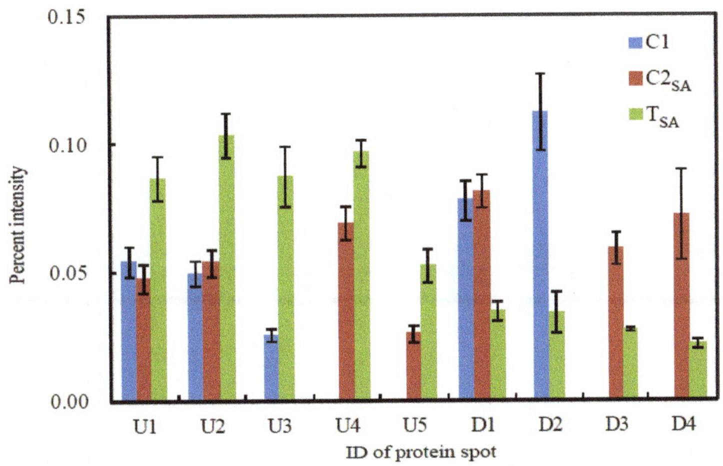

3.2. Protein Identification

In total, 5 specifically expressed, 3 significantly increased, and 3 significantly decreased spots were identified by liquid chromatography/tandem mass spectrometry (LC/MS/MS) (Figure 7) and the sequence tag method (Table 1). These proteins were categorized into four groups: energy production, metabolism, protein synthesis, and unknown function.

Figure 7. The number of SA responsive protein spots. The IDs in red color correspond to the protein spots analyzed by liquid chromatography/tandem mass spectrometry (LC/MS/MS). Percent intensity: The mean percent intensity of each SA-responsive protein spot in T_{SA} gel. T_{SA}/C1, T_{SA}/$C2_{SA}$: The ratios of the mean percent intensity of each SA-responsive protein spot in T_{SA} gel to the mean percent intensity of each corresponding protein spot in C1 and $C2_{SA}$ gels.

ID	%intensity			Ratio		MW (kDa)	pI
	C1	$C2_{SA}$	T_{SA}	T_{SA}/C1	T_{SA}/$C2_{SA}$		
U3	0.0256	-	0.0876	3.42	-	14.88	4.59
D2	0.1121	-	0.0341	0.30	-	19.00	4.79

ID	%intensity			Ratio		MW (kDa)	pI
	C1	$C2_{SA}$	T_{SA}	T_{SA}/C1	T_{SA}/$C2_{SA}$		
U1	0.0543	0.0476	0.0868	1.60	1.82	32.90	4.66
U2	0.0494	0.0536	0.1034	2.09	1.93	25.53	6.09
D1	0.0778	0.0814	0.0349	0.45	0.43	33.33	4.26

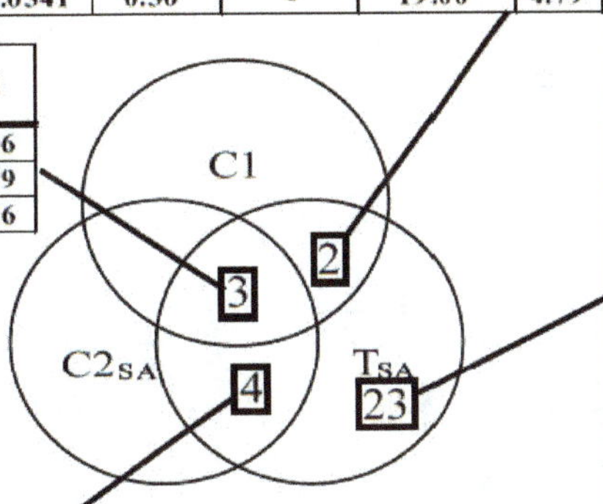

ID	%Intensity			Ratio		MW (kDa)	pI
	C1	$C2_{SA}$	T_{SA}	T_{SA}/C1	T_{SA}/$C2_{SA}$		
U4	-	0.0689	0.0965	-	1.40	71.52	6.06
U5	-	0.0258	0.0523	-	2.02	29.74	4.39
D3	-	0.0588	0.0271	-	0.46	65.92	5.38
D4	-	0.0721	0.0216	-	0.30	30.70	6.60

ID	%Intensity	MW (kDa)	pI
E1	0.0629	68.19	6.42
E2	0.0540	61.17	4.46
E3	0.0750	60.34	5.06
E4	0.0559	48.96	6.49
E5	0.0293	46.73	4.38
E6	0.0515	47.49	6.60
E7	0.0174	46.28	6.17
E8	0.0896	47.26	6.63
E9	0.1300	38.46	5.82
E10	0.0812	36.57	6.45
E11	0.0337	31.76	5.57
E12	0.0893	31.14	6.71
E13	0.0339	30.81	5.68
E14	0.0374	31.03	6.82
E15	0.0582	31.17	6.56
E16	0.0886	30.20	6.80
E17	0.1045	29.81	6.65
E18	0.0362	29.43	6.56
E19	0.0417	30.02	5.73
E20	0.0720	17.20	4.54
E21	0.0163	16.98	6.04
E22	0.0346	15.33	6.56
E23	0.0371	12.49	5.48

Table 1. Identified proteins by LC/MS/MS and sequence tag method.

ID	Protein	No. in NCBI	Score	% cover	Mw (kDa) Observed/Theoretical	pI Observed/Theoretical	Ratio $T_{SA}/C1$	Ratio $T_{SA}/C2_{SA}$	Plant
					Energy production				
E10	Malate dehydrogenase	gi|255585544	139	14	36.57/28.84	6.45/5.44	-	-	*Ricinus communis*
U4	SDH1-1; ATP binding/succinate dehydrogenase	gi|15240075	43	2	71.52/70.24	6.06/5.86	-	1.40	*Arabidopsis thaliana*
D3	Phosphoglycerate kinase, chloroplastic; Flags; Precursor	gi|2499497	85	11	65.92/50.31	5.38/8.48	-	0.46	*Nicotiana tabacum*
					Metabolism				
E6	Diaminopimelate decarboxylase putative	gi|255543757	79	3	47.34/54.55	6.62/6.58	-	-	*Ricinus communis*
E9	Arginase	gi|148828535	440	21	42.29/37.15	5.76/6.14	-	-	*Malus hupehensis*
U1	Chorismate mutase precursor	gi|429153	39	2	32.90/38.05	4.66/5.90	1.60	1.82	*Arabidopsis thaliana*
					Protein synthesis				
E22	Peptidylprolyl isomerase (cyclophilin)	gi|21886603	44	8	15.33/18.51	6.56/8.68	-	-	*Betula pendula*
U5	Aminopeptidase	gi|255080640	35	1	29.74/63.62	4.39/7.97	-	2.02	*Micromonas* sp. RCC299
					Unknown				
E12	Predicted protein	gi|168008150	23	1	31.14/71.46	6.71/8.15	-	-	*Physcomitrella patens* subsp. *Patens*
D1	Hypothetical protein	gi|2342730	16	1	33.33/54.76	4.26/6.28	0.45	0.43	*Arabidopsis thaliana*
D2	Predicted: hypothetical protein	gi|225432620	101	14	19.00/22.52	4.79/9.73	0.30	-	*Vitis vinifera*

Note: $T_{SA}/C1$, $T_{SA}/C2_{SA}$: The ratios of the mean percent intensity of each SA-responsive protein spot in T_{SA} gel to the mean percent intensity of each corresponding protein spot in C1 and $C2_{SA}$ gels.

3.2.1. The Proteins Related to Energy Production

The E10 protein was identified as malate dehydrogenase (MDH). MDH catalyzes the oxidation of malate to oxaloacetate, producing NADH [35]. This protein was induced in *C. sativus* L. cv Negrito by *Trichoderma asperellum* inoculation with increase in SA [36]. *Phellinus sulphurascens* inoculation also induced this protein in *Pseudotsuga menziesii* [37]. Mitochondrial MDH lowers leaf respiration and alters photorespiration and plant growth by partitioning carbon and energy in leaves of Arabidopsis [38]. In addition, it was reported that over-expression of cytosolic NAD-dependent MDH gene resulted in promotion of growth and tolerance to cold and high salinity in apple callus and tomato [39]. Thus, MDH expression may increase the resistance by altering the distribution of carbon and energy in Japanese birch plantlet No.8 treated with SA.

The U4 protein was identified as SDH1-1; ATP binding/succinate dehydrogenase. Succinate dehydrogenase catalyzes the oxidation of succinate to fumarate by reducing FAD to $FADH_2$ in the citric acid cycle [40]. This protein expression was induced in a *Hordeum vulgare* L.-susceptible line by *Erysiphe graminis* inoculation [41], and some isozymes were expressed in *A. thaliana* seeds with SA administration [16]. $FADH_2$ has an important role in oxidation-reduction reactions as well as NADPH metabolism. Increase of this protein may also increase $FADH_2$ production and supply fumaric acid to the TCA cycle leading to the production of resistant compounds in Japanese birch plantlet No.8.

D3 was identified as phosphoglycerate kinase. Phosphoglycerate kinase participates in the glycolytic pathway. This enzyme catalyzes the formation of 3-phosphoglycerate from 1,3-bisphosphoglycerate [42]. Some isozyme expressions were decreased in *A. thaliana* seeds and *G. barbadense* seedlings with SA administration and *Verticillium dahliae* infection, respectively [16,18]. In the case of *G. barbadense*, phosphoglycerate kinase expression decreased, whereas the expressions of other enzymes related to the glycolytic pathways increased. It is, therefore, proposed that the expression change of these proteins redirects the metabolic flux from the glycolysis to the pentose phosphate pathway, producing more intermediates, such as 1-deoxy-D-xylulose 5-phosphate for the generation of isoprenoid [19]. Isoprenoids are known to be important antipathogenic compounds. Decrease of this enzyme, therefore, may affect the biosynthetic pathways of antipathogenic compounds in Japanese birch plantlet No.8.

3.2.2. The Proteins Related to Metabolism

The E6 protein was identified as diaminopimelate decarboxylase. This enzyme decarboxylates *meso*-2,6-diaminopimelic acid to lysine in the diaminopimelic acid pathway [43]. The function of this enzyme on pathogen resistance is not known. Increase of this protein may increase lysine production, and lysine may be used for synthesis of other proteins involved in producing antipathogenic and/or signaling compounds in Japanese birch plantlet No.8.

The E9 protein was identified as arginase. This enzyme hydrolyzes L-arginine to L-ornithine and urea [44]. This activity and the expression of this enzyme are increased by wounds, treatment with jasmonic acid or coronatine [45], and *P. syringae* infection [46]. It is suggested that this protein has an important role for polyamine production, because L-ornithine is a precursor for polyamine. Polyamines are aliphatic hydrocarbons possessing more than two amino groups, and they have various physiological activities [47]. Polyamines also play an important role in plant-pathogen interactions.

Polyamines are usually conjugated with cinnamic acids, and the resulting conjugates are known as hydroxycinnamic acid amides (HCAAs). It has been reported that good correlations exist between the accumulation of HCCAs and pathogen resistance [48]. Modulation of host polyamine levels has led to significant changes in host susceptibility to different kinds of pathogens in *N. tabacum* [49]. In addition, polyamine is a source of H_2O_2 through its degradation, and it induces hypersensitive response-like cell death in *N. tabacum* [50]. It has been reported that SA treatment increases the amount of putrescine, one of polyamines in *Zea mays* L., hybrid Norma [51]. This protein expression induced by SA may lead to an increase polyamine production through an increase of L-ornithine in Japanese birch plantlet No.8.

The U1 protein was identified as chorismate mutase. This enzyme transforms chorismic acid to prephenic acid in the shikimic acid pathway [52]. This enzyme expression was increased by elicitor treatment and infections with *Fusarium oxysporum* and *Alternaria raphani* in *A. thaliana* [53]. Phenylalanine derived from the shikimic acid pathway is a precursor of important compounds for defense responses, such as phenolic compounds containing monolignols and isoflavonoids [54]. SA is also derived from phenylalanine, and it has been suggested that the SA-mediated signaling pathway is regulated by a positive feedback loop [55–57]. In addition, SA treatment has increased the amount of phenolic compounds and induced SAR in *Vigna mungo* [20]. Increase of this protein expression is considered to increase the amount of phenolic compounds including SA for antipathogenic responses and signaling pathways in Japanese birch plantlet No.8.

3.2.3. The Proteins Related to Protein Synthesis

The E22 protein was identified as cyclophilin. This protein catalyzes rotation of proline-peptide bonds and is considered to participate in the protein folding process as a chaperone [58,59]. This protein expression was induced by various biotic and abiotic stresses in *Phaseolus vulgaris* L. cv. Saxa. As SA administration also induces its expression, it has been proposed that cyclophilin may function as a chaperone-like molecule in diseased plants in order to decrease the risks of degradation or to avoid aggregation of proteins, the reactions that take place under stress [60]. In this study, cyclophilin was induced, suggesting that this protein folds pathogen-induced proteins and protects other proteins from degradation and aggregation in Japanese birch plantlet No.8.

The U5 protein was identified as aminopeptidase. This protein family removes an amino acid from the N-terminal of a peptide, and includes leucine aminopeptidase, prolineiminopeptidase, serine aminopeptidase, and so on [61]. The aminopeptidase family protein and leucin aminopeptidase were increased in their expression with SA treatment in *A. thaliana* suspension culture [62]. Expression of aminopeptidase was also increased with *P. syringae* DC3000 inoculation in *A. thaliana* [46]. SA involves in the interactions between *A. thaliana* and *P. syringae* DC3000 [63]. N-terminal residues of the protein are correlated with a protein half-life through degradation by S26 proteasome in animals, yeast, and prokaryotes. In addition, it has been suggested that this mechanism is also active and important in plants. The ability of aminopeptidases to remove N-terminal residues and to reveal penultimate residues could influence a protein's interactions between the N-terminal residues and the ubiquitin-proteasomal system, and could, therefore, influence protein stability [61]. In addition, it has been reported that cysteine proteases play an instrumental role in programmed cell death (PCD) triggered by oxidative stress in *Glycine max* [64]. As oxidative stress is correlated with the SA signaling pathway [21], it is suggested that SA is involved in the degradation of proteins induced by PCD. The increased aminopepetidase found in this study may influence protein stability and signaling pathways by removing the N-terminal residues of proteins in Japanese birch plantlet No.8.

3.2.4. Unknown Function Protein

The E12 was identified as predicted protein. In addition, the D1 and D2 were identified as hypothetical proteins. Although their functions are unknown, changes in expression of these proteins may be involved in SAR establishment in Japanese birch plantlet No.8.

4. Conclusions

In this study, protein profile changes induced by SA-treatment were analyzed, and SA-responsive proteins were identified to unravel the mechanisms of SAR establishment in Japanese birch plantlet No.8. In total 5 specifically expressed, 3 significantly increased, and 3 significantly decreased proteins were identified by LC/MS/MS and sequence tag method. These proteins were categorized into energy production, metabolism, protein synthesis, and unknown function, and were considered to be involved in SAR establishment in Japanese birch plantlet No.8.

Acknowledgments

The authors are grateful to Afshin Salavati, Tehran University, College of Agriculture and Natural Resources, Karaj, Iran, and Setsuko Komatsu, National Agriculture and Food Research Organization, National Institute of Crop Science, Tsukuba, Japan, for their kind help in LC/MS/MS measurement and database search to identify salicylic acid responsive proteins. This study was supported by JSPS KAKENHI Grant Numbers 19580162 and 22580156.

Author Contributions

Conception of experimental design was performed by S.Y. H.S. and Y.T. performed the experiments. F.I., N.Y. and S.Y. supervised the project. S.Y. wrote the paper.

References

1. Hosoi, Y. *Betula platyphylla* var. *japonica*. In *Current Biotechnology Complete Book: Propagation and Breeding of Woody Plant*; Ohyama, K., Saito, A., Eds.; Nougyo Tosyo: Tokyo, Japan, 1989; Volume 6, pp. 157–160.
2. Ju, E.M.; Lee, S.E.; Hwang, H.J.; Kim, J.H. Antioxidant and anticancer activity of extract from *Betula platyphylla* var. *japonica*. *Life Sci.* **2004**, *74*, 1013–1026.

3. Sami, A.; Taru, M.; Salme, K.; Jari, Y.K. Pharmacological properties of the ubiquitous natural product betulin. *Eur. J. Pharm. Sci.* **2006**, *29*, 1–13.
4. Zabel, R.A. Basidiocarp development in *Inonotus obliquus* and its inhibition by stem treatment. *Forest Sci.* **1976**, *22*, 431–437.
5. True, R.P.; King, J.F. Cankers and decays of birch associated with two *Poria* species. *J. Forest* **1995**, *53*, 412–415.
6. Kim, Y.O.; Park, H.W.; Kim, J.H.; Lee, J.Y.; Moon, S.H.; Shin, C.S. Anti-cancer effect and structural characterization of endo-polysaccharide from cultivated mycelia of *Inonotus obliquus*. *Life Sci.* **2006**, *79*, 72–80.
7. Schumann, G.L.; D'Arcy, C.J. Induced (active) defenses. In *Essential Plant Pathology*; APS Press: St. Paul, MN, USA, 2006; pp. 203–206.
8. Boerjan, W.; Ralph, J.; Baucher, M. Lignin biosynthesis. *Annu. Rev. Plant Biol.* **2003**, *54*, 23–61.
9. Lattanzio, V.; Lattanzio, V.M.T.; Cardinali, A. Role of phenolics in the resistance mechanisms of plants against fungal pathogens and insects. In *Neural Proteoglycan*; Maeda, N., Ed.; Research Signpost: Kerala, India, 2006; pp. 23–67.
10. Van Loon, L.C.; Rep, M.; Pieterse, C.M.J. Significance of inducible defense-related proteins in infected plants. *Annu. Rev. Phytopathol.* **2006**, *44*, 135–162.
11. Ferreira, R.B.; Monteiro, S.; Freitas, R.; Santos, C.N.; Chen, Z.; Batista, L.M.; Duarte, J.; Borges, A.; Teixeira, A.R. The role of plant defence proteins in fungal pathogenesis. *Mol. Plant Pathol.* **2007**, *8*, 677–700.
12. Mehdy, M.C. Active oxygen species in plant defense against pathogens. *Plant Physiol.* **1994**, *105*, 467–472.
13. Siegrist, J.; Jeblick, W.; Kauss, H. Defense responses in infected and elicited cucumber (*Cucumis sativus* L.) hypocotyl segments exhibiting acquired resistance. *Plant Physiol.* **1994**, *105*, 1365–1374.
14. Métraux, J.P.; Nawrath, C.; Genoud, T. Systemic acquired resistance. *Euphytica* **2002**, *124*, 237–243.
15. Gaffney, T.; Friedrich, L.; Vernooij, B.; Negrotto, D.; Nye, G.; Uknes, S.; Ward, E.; Kessmann, H.; Ryals, J. Requirement of salicylic acid for the induction of systemic acquired resistance. *Science* **1993**, *261*, 754–756.
16. Rajjou, L.; Belghazi, M.; Huguet, R.; Robin, C.; Moreau, A.; Job, C.; Job, D. Proteomic investigation of the effect of salicylic acid on *Arabidopsis* seed germination and establishment of early defense mechanisms. *Plant Physiol.* **2006**, *141*, 910–923.
17. Kundu, S.; Chakraborty, D.; Pal, A. Proteomic analysis of salicylic acid induced resistance to mungbean yellow mosaic India virus in *Vigna mungo*. *J. Proteomics* **2011**, *74*, 337–349.
18. Wang, F.X.; Ma, Y.P.; Yang, C.L.; Zhao, P.M.; Yao, Y.; Jian, G.L.; Luo, Y.M.; Xia, G.X. Proteomic analysis of the sea-island cotton roots infected by wilt pathogen *Verticillium dahliae*. *Proteomics* **2011**, *11*, 4296–4309.
19. Cao, H.; Bowling, S.A.; Gordon, A.S.; Dong, X. Characterization of an *Arabidopsis* mutant that is nonresponsive to inducers of systemic acquired resistance. *Plant Cell* **1994**, *6*, 1583–1592.
20. Alverez, M.E.; Pennell, R.I.; Meijer, P.J.; Ishikawa, A.; Dixon, R.A.; Lamb, C. Reactive oxygen intermediates mediate a systemic signal network in the establishment of plant immunity. *Cell* **1998**, *92*, 773–784.

21. Martinez, C.; Baccou, J.C.; Bresson, E.; Baissac, Y.; Daniel, J.F.; Jalloul, A.; Montillet, J.L.; Geiger, J.P.; Assigbetsé, K.; Nicole, M. Salicylic acid mediated by the oxidative burst is a key molecule in local and systemic responses of cotton challenged by an avirulent race of *Xanthomonas campestris* pv *malvacearum*. *Plant Physiol.* **2000**, *122*, 757–766.
22. Tada, Y.; Spoel, S.H.; Pajerowska-Mukhtar, K.; Mou, Z.; Song, J.; Wang, C.; Zuo, J.; Dong, X. Plant immunity requires conformational charges of NPR1 via S-nitrosylation and thioredoxins. *Science* **2008**, *321*, 952–956.
23. Blanco, F.; Salinas, P.; Cecchini, N.M.; Jordana, X.; van Hummelen, P.; Alvarez, M.E.; Holuigue, L. Early genomic responses to salicylic acid in *Arabidopsis*. *Plant Mol. Biol.* **2009**, *70*, 79–102.
24. Park, S.W.; Kaimoyo, E.; Kumar, D.; Mosher, S.; Klessig, D.F. Methyl salicylate is a critical mobile signal for plant systemic acquired resistance. *Science* **2007**, *318*, 113–116.
25. Vernooij, B.; Friedrich, L.; Morse, A.; Reist, R.; Kolditz-Jawhar, R.; Ward, E.; Uknes, S.; Kessmann, H.; Ryals, J. Salicylic acid is not the translocated signal responsible for inducing systemic acquired resistance but is required in signal transduction. *Plant Cell* **1994**, *6*, 959–965.
26. Attaran, E.; Zeier, T.E.; Griebel, T.; Zeier, J. Methyl salicylate production and jasmonate signaling are not essential for systemic acquired resistance in *Arabidopsis*. *Plant Cell* **2009**, *21*, 954–971.
27. Maldonado, A.M.; Doerner, P.; Dixon, R.A.; Lamb, C.J.; Cameron, R.K. A putative lipid transfer protein involved in systemic resistance signaling in *Arabidopsis*. *Nature* **2002**, *419*, 399–403.
28. Jung, H.W.; Tschaplinski, T.J.; Wang, L.; Glazebrook, J.; Greenberg, J.T. Priming in systemic plant immunity. *Science* **2009**, *324*, 89–91.
29. Takashima, Y.; Ishiguri, F.; Iizuka, K.; Yoshizawa, N.; Yokota, S. Proteome analysis of infection-specific proteins from Japanese birch (*Betula platyphylla* var. *japonica*) plantlet No.8 infected with Inonotus obliquus strain IO-U1. *Plant Biotechnol.* **2013**, *30*, 83–87.
30. Takashima, Y.; Suzuki, M.; Ishiguri, F.; Iizuka, K.; Yoshizawa, N.; Yokota, S. Cationic peroxidase related to basal resistance of *Betula platyphylla* var. *japonica* plantlet No.8 against canker-rot fungus *Inonotus obliquus* strain IO-U1. *Plant Biotechnol.* **2013**, *30*, 199–205.
31. Rahman, M.M.; Ishiguri, F.; Takashima, Y.; Azad, M.A.K.; Iizuka, K.; Yoshizawa, N.; Yokota, S. Anatomical and histochemical characteristics of Japanese birch (Tohoku) plantlets infected with the *Inonotus obliquus* IO-U1 strain. *Plant Biotechnol.* **2008**, *25*, 183–189.
32. Murashige, T.; Skoog, F. A revised medium for rapid growth and bio assays with tobacco tissue cultures. *Physiol. Plant.* **1962**, *15*, 473–497.
33. Bradford, M.M. A rapid and sensitive method for the quantitation of microgram quantities of protein utilizing the principle of protein-dye binding. *Anal. Biochem.* **1976**, *72*, 248–254.
34. MS/MS Ions Search. Available online: http:www.matrixscience.com (accessed on 6 July 2011).
35. Siedow, J.N.; Day, D.A. Respiration and photorespiration. In *Biochemistry & Molecular Biology of Plants*; Buchanan, B.B., Gruissem, W.G., Jones, R.L., Eds.; American Society of Plant Physiologists: Rockville, MD, USA, 2000; p. 685.
36. Segarra, G.; Casanova, E.; Bellido, D.; Odena, M.A.; Oliveira, E.; Trillas, I. Proteome, salicylic acid, and jasmonic acid changes in cucumber plants inoculated with *Trichoderma asperellum* strain T34. *Proteomics* **2007**, *7*, 3943–3952.

37. Islam, M.A.; Sturrock, R.N.; Ekramoddoullah, A.K.M. A proteomics approach to identify proteins differentially expressed in Douglas-fir seedlings infected by *Phellinus sulphurascens*. *J. Proteomics* **2008**, *71*, 425–438.
38. Tomaz, T.; Bagard, M.; Pracharoenwattana, I.; Lindén, P.; Lee, C.P.; Carroll, A.J.; Ströher, E.; Smith, S.M.; Gardeström, P.; Millar, A.H. Mitochondrial malate dehydrogenase lowers leaf respiration and alters photorespiration and plant growth in *Arabidopsis*. *Plant Physiol.* **2010**, *154*, 1143–1157.
39. Yao, Y.-X.; Dong, Q.-L.; Zhai, H.; You, C.-X.; Hao, Y.-J. The functions of an apple *cytosolic malate dehydrogenase* gene in growth and tolerance to cold and salt stresses. *Plant Physiol. Biochem.* **2011**, *49*, 257–264.
40. Siedow, J.N.; Day, D.A. Respiration and photorespiration. In *Biochemistry & Molecular Biology of Plants*; Buchanan, B.B., Gruissem, W.G., Jones, R.L., Eds.; American Society of Plant Physiologists: Rockville, MD, USA, 2000; p. 684.
41. Sako, N.; Stahmann, M.A. Multiple molecular forms of enzymes in barley leaves infected with *Erysiphe graminis* f.sp. *hordei*. *Physiol. Plant Pathol.* **1972**, *2*, 217–226.
42. Dennis, D.T.; Blakeley, S.D. Carbohydrate metabolism. In *Biochemistry & Molecular Biology of Plants*; Buchanan, B.B., Gruissem, W.G., Jones, R.L., Eds.; American Society of Plant Physiologists: Rockville, MD, USA, 2000; p. 668.
43. Azevedo, R.A.; Lea, P.J. Lysine metabolism in higher plants. *Amino Acids* **2001**, *20*, 261–279.
44. Jenkinson, C.P.; Grody, W.W.; Cederbaum, S.D. Comparative properties of arginases. *Comp. Biochem. Physiol.* **1996**, *114B*, 107–132.
45. Chen, H.; McCaig, B.C.; Melotto, M.; He, S.Y.; Howe, G.A. Regulation of plant arginase by wounding, jasmonate, and the phtotoxin coronatine. *J. Biol. Chem.* **2004**, *279*, 45998–46007.
46. Jones, A.M.E.; Thomas, V.; Bennett, M.H.; Mansfield, J.; Grant, M. Modifications to the *Arabidopsis* defense proteome occur prior to significant transcriptional change in response to inoculation with *Pseudomonas syringae*. *Plant Physiol.* **2006**, *142*, 1603–1620.
47. Kusano, T.; Yamaguchi, K.; Berberich, T.; Takahashi, Y. Advances in polyamine research in 2007. *J. Plant Res.* **2007**, *120*, 345–350.
48. Walters, D.R. Polyamines and plant disease. *Phytochemistry* **2003**, *64*, 97–107.
49. Marina, M.; Maiale, S.J.; Rossi, F.R.; Romero, M.F.; Rivas, E.I.; Gárriz, A.; Ruiz, O.A.; Pieckenstain, F.L. Apoplastic polyamine oxidation plays different roles in local responses of tabacco to infection by necrotrophic fungus *Sclerotinia sclerotiorum* and the biotrophic bacterium *Pseudomonas viridiflava*. *Plant Physiol.* **2008**, *147*, 2164–2178.
50. Yoda, H.; Fujimura, K.; Takahashi, H.; Munemura, I.; Uchimiya, H.; Sano, H. Polyamines as a common source of hydrogen peroxide in host- and nonhost hypersensitive response during pathogen infection. *Plant Mol. Biol.* **2009**, *70*, 103–112.
51. Németh, M.; Janda, T.; Horváth, E.; Páldi, E.; Szalai, G. Exogenous salicylic acid increases polyamine content but may decrease drought tolerance in maize. *Plant Sci.* **2002**, *162*, 569–574.
52. Tzin, V.; Galili, G. New insights into the shikimate and aromatic amino acids biosynthesis pathways in plants. *Mol. Plant* **2010**, *3*, 956–972.

53. Eberhard, J.; Ehrler, T.T.; Epple, P.; Felix, G.; Raesecke, H.R.; Amrhein, N.; Schmid, J. Cytosolic and plastidic chorismate mutase isozymes from *Arabidopsis thaliana*: Molecular characterization and enzymatic properties. *Plant J.* **1996**, *10*, 815–821.
54. Dixon, R.A. Natural products and plant disease resistance. *Nature* **2001**, *411*, 843–847.
55. Pallas, J.A.; Paiva, N.L.; Lamb, C.; Dixon, R.A. Tobacco plants epigenetically suppressed in phenylalanine ammonia-lyase expression do not develop systemic acquired resistance in response to infection by tobacco mosaic virus. *Plant J.* **1996**, *10*, 281–293.
56. Shirasu, K.; Nakajima, H.; Rajasekhar, V.K.; Dixon, R.A.; Lamb, C. Salicylic acid potentiates an agonist-dependent gain control that amplifies pathogen signals in the activation of defense mechanisms. *Plant Cell* **1997**, *9*, 261–270.
57. Smith-Becker, J.; Marois, E.; Huguet, E.J.; Midland, S.L.; Sims, J.J.; Keen, N.T. Accumulation of salicylic acid and 4-hydroxybenzoic acid in phloem fluids of cucumber during systemic acquired resistance is preceded by a transient increase in phenylalanine ammonia-lyase activity in petioles and stems. *Plant Physiol.* **1998**, *116*, 231–238.
58. Bächinger, H.P. The influence of peptidyl-prolyl *cis-trans* isomerase on the *in vitro* folding of type III collagen. *J. Biol. Chem.* **1987**, *262*, 17144–17148.
59. Miernyk, J.A. Protein folding in the plant cell. *Plant Physiol.* **1999**, *121*, 695–703.
60. Marvet, J.; Margis-Pinheiro, M.; Frendo, P.; Burkard, G. Bean cyclophilin gene expression during plant development and stress conditions. *Plant Mol. Biol.* **1994**, *26*, 1181–1189.
61. Walling, L.L. Recycling or regulation? The role of amino-terminal modifying enzymes. *Curr. Opin. Plant Biol.* **2006**, *9*, 227–233.
62. Cheng, F.Y.; Blackburn, K.; Lin, Y.M.; Goshe, M.B.; Williamson, J.D. Absolute protein quantification by LC/MSn for global analysis of salicylic acid-induced plant protein secretion responses. *J. Proteome Res.* **2009**, *8*, 82–93.
63. Delaney, T.P.; Uknes, S.; Vernooij, B.; Friendrich, L.; Weymann, K.; Negrotto, D.; Gaffney, T.; Gut-Rella, M.; Kessmann, H.; Ward, E.; *et al.* A central role of salicylic acid in plant disease resistance. *Science* **1994**, *266*, 1247–1250.
64. Solomon, M.; Belenghi, B.; Delledonne, M.; Menachem, E.; Levine, A. The involvement of cysteine proteases and protease inhibitor genes in the regulation of programmed cell death in plants. *Plant Cell* **1999**, *11*, 431–443.

5

Protein Profiling Reveals Novel Proteins in Pollen and Pistil of W22 (ga1; Ga1) in Maize

Jin Yu [1,†], Swapan Kumar Roy [1,†], Abu Hena Mostafa Kamal [1,†,‡], Kun Cho [2], Soo-Jeong Kwon [1], Seong-Woo Cho [3], Yoon-Sup So [1], James B. Holland [4] and Sun Hee Woo [1,*]

1. Department of Crop Science, Chungbuk National University, Cheong-ju 361-763, Korea; E-Mails: edjlsa8603@naver.com (J.Y.); swapankhulna@gmail.com (S.K.R.); abuhena7@gmail.com (A.H.M.K.); kwonsj1220@naver.com (S.-J.K.); yoonsupso@cbnu.ac.kr (Y.-S.S.)
2. Division of Mass Spectrometry Research, Korea Basic Science Institute, Chungbuk 863-883, Korea; E-Mail: chokun@kbsi.re.kr
3. Lab of Molecular Breeding, Arid land Research Center, Tottori University, Tottori 680-8550, Japan; E-Mail: whtjddn0435@chungbuk.ac.kr
4. USDA-ARS Plant Science Research Unit, Department of Crop Science, Box 7620, North Carolina State University, Raleigh, NC 27695, USA; E-Mail: Jim.Holland@ars.usda.gov

† These authors contributed equally to this work.

‡ Current address: National Institute of Crop Science, NARO, Tsukuba 305-8517, Japan.

* Author to whom correspondence should be addressed; E-Mail: shwoo@chungbuk.ac.kr

---

Abstract: Gametophytic factors mediate pollen-pistil interactions in maize (*Zea mays* L.) and play active roles in limiting gene flow among maize populations and between maize and teosinte. This study was carried out to identify proteins and investigate the mechanism of gametophytic factors using protein analysis. W22 (ga1); which did not carry a gametophytic factor and W22 (Ga1), a near iso-genic line, were used for the proteome investigation. SDS-PAGE was executed to investigate proteins in the pollen and pistil of W22 (ga1) and W22 (Ga1). A total of 44 differentially expressed proteins were identified in the pollen and pistil on SDS-PAGE using LTQ-FTICR MS. Among the 44 proteins, a total of 24 proteins were identified in the pollen of W22 (ga1) and W22 (Ga1) whereas 20 differentially expressed proteins were identified from the pistil of W22 (ga1) and W22 (Ga1). However, in pollen, 2 proteins were identified only in the W22 (ga1) and 12 proteins only in the W22 (Ga1) whereas 10 proteins were confirmed from the both of W22 (ga1) and W22 (Ga1). In contrary, 10 proteins were appeared only in the pistil of W22 (ga1) and 7 proteins from W22 (Ga1) while 3 proteins confirmed in the both of W22 (ga1) and W22 (Ga1). Moreover, the identified proteins were generally involved in hydrolase activity, nucleic acid binding and nucleotide binding. These results help to reveal the mechanism of gametophytic factors and provide a valuable clue for the pollen and pistil research in maize.

Keywords: maize; pollen; pistil; gametophytic factors; proteomics

1. Introduction

In angiosperms, pollen-pistil interactions are important for the subsequent successful reproduction and formation of seed [1]. In flowering plants, interaction between pollen and pistil ascertains reproductive compatibility [2]. Gametophytic factors are important, especially those known as pollen killer genes, gametocidal genes, gamete eliminators and gamete aborters. They have been introduced in several economically important plant species such as maize [3,4], tobacco [5], wheat [6], tomato [7], lima beans [8] and barley [9]. The first gametophytic factor (gametophytic factor 1) related to segregation distortion was reported in maize. However, the pollination with Ga1 pollen only or with ga1 pollen only, led to normal genotype ratios. Due to the fastening of pollen-tube growth in pollen with Ga1 than with ga1, a mixture of Ga1 and ga1 pollen resulted in an excess of the genotypes with the linked Su allele [3].

Maize (*Zea mays* L.) is a model species for investigating pollen-pistil interactions, and is one of the most essential cereal crops in the world [10]. However, several maize genotypes carry genes referred to as gametophytic factors that mediate pollen-pistil interactions and subsequently impair the success of fertilization [11]. Pollen-pistil interactions are essential for the seed and fruit formation, revealing that their mechanisms are of great importance, especially for understanding the completion of the plant life cycle and for accelerating agricultural production. Recent transcriptomic and proteomic studies have improved our knowledge regarding pollen/pistil gene and protein expression and eventually, the desirable genes are possibly involved in the pollen-pistil interactions [12].

The high-throughput proteomics approach is thought to be a powerful tool for the analysis of proteins related to gametophytic factors. The proteomes of pollen have been described previously [13–18], whereas the proteome analysis of pollen and pistil is relatively not well studied. The pollen and pistil protein of maize (*Zea mays* L.) were analyzed using SDS-PAGE combined with MS identification. However, two-dimensional gel electrophoresis (2-DE) combined with MS analysis have provided the most potential and reliable method for proteomic investigations. Previously, 2-DE techniques combined with MALDI-TOF (matrix assisted laser desorption ionization/time of flight) MS or ESI Q-TOF (electrospray ionization quadrupole-TOF) MS/MS have been executed to investigate the proteomes during pollen development within the anther as well as proteomes of mature and germinated pollen in various plant species [13–18]. These proteomic studies have significantly promoted our knowledge of the regulation of pollen and pollen tube development at the molecular level.

In addition, in F_1 hybrid production, the use of gametophytic factors has played a crucial role because it prevents pollen from contamination. However, the mechanism of gametophytic factors is still unknown. For understanding the mechanism of gametophytic factors, proteomics has been employed to analyze proteins from pollen and pistil of maize. This study was carried out to identify proteins related to gametophytic factors using protein analysis.

2. Experimental

2.1. Plant Materials and Genetic Background

W22 (ga1, ga1) is a common inbred line developed by university of Wisconsin. It has a normal dent genotype; therefore, it does not carry a gametophytic factor. W22 (Ga1-s) line was created by Dr. Kermicle from university of Wisconsin by crossing with a popcorn (white cloud variety) which carries Ga1-s to W22, followed by five successive backcrossing to W22, while selecting for Ga1-s.

Seeds of W22 (ga1) and W22 (Ga1) were sown in a greenhouse in the seedling tray. After 3~4 days, it was transplanted to 6 pots and the paper bag was covered to prohibit fertilization between the pollen and pistil of same maize line during anthesis period. Within a week, one gram of maize pollen and pistil were incubated in 50 mL of 0.1 M NH_4HCO_3 buffer (pH 8.0) for 30 min. Then, the soluble fraction was isolated by centrifugation at 17,000× *g* for 30 min and dialyzed against double-distilled water overnight. The extract was lyophilized and stored at 4 °C for further use. All experiments are replicated 3 times.

2.2. Protein Extraction and Electrophoresis

A portion (0.5 g) of pollen and pistil was ground in liquid nitrogen. Using a modified method, the proteins were extracted from the pollen and pistil according to previously described methods [19]. The seeds were then suspended in Solution I [(10% trichloroacetic acid (TCA) in acetone containing and 0.07% 2-mercaptoethanol (2-ME)] and then sonicate for 5–10 min. Solution II [0.07% 2-mercaptoethanol (2-ME) in acetone containing] was added in the pellets and the vortex, and then centrifuged at 20,000× *g* at 4 °C for 5 min. This step was repeated and the pellets were dried by vacuum centrifugation for 10 min. The dried powder was diluted with lysis buffer (7 M urea, 2 M thiourea, 5% CHAPS, and 2 mM tributylphosphine), incubate at 37 °C for 2 h and then centrifuged at 20,000× *g* at 4 °C for 20 min. The supernatants were collected to 1.5 mL tube. The protein concentrations were determined by RC/DC assay and then it was stored at −80 °C for further utilization.

Proteins were extracted from the pollen and pistil according to TCA/acetone precipitation method prior to SDS-PAGE. Proteins were separated on 16 × 16 cm SDS-PAGE gels (gradient 14%–16% acrylamide) as described previously [20]. The electrophoresis conditions were set and run at 50 m. A for 2 h until the sample buffer dye reached the lower part of the gel. The experiment is biologically triplicate. The gels were stained with coomassie brilliant blue R-250 and scanned using a scanner (HP Scanjet G4010, Palo Alto, CA, USA).

2.3. In-gel Digestion

CBB-stained gel slices were washed several times with 30% methanol until the colors were completely removed. Then the gel slices were destained with 10 mM NH_4HCO_3 in 50% ACN (Acetonitrile), squeezed for 10 min with 100% ACN (Acetonitrile) and dried by vacuum centrifugation. After destaining steps, the gel slices were reduced with 10 mM DTT in 100 mM NH_4HCO_3 at 56 °C for 1 h and then alkylated with 55 mM Iodoacetamide (IAA) in 100 mM NH_4HCO_3 in the dark for 40 min. Then the gel slices were digested with 50 μL trypsin (10 ng/μL) (Promega Corporation, Madison, WI, USA) and incubated at 37 °C for 16 h. After digestion steps, the peptides were extracted with 50 mM ammonium bi-carbonate and repeated these steps several times with a solution containing 0.1% formic acid in 50% ACN (acetonitrile) until 200~250 μL. The solution containing eluted peptides was concentrated up to drying by vacuum centrifugation and the resultant extracts were confirmed by LTQ-FTICR mass spectrometry. The dried samples were stored at 4 °C prior to mass spectrometry analysis.

2.4. MS/MS Analysis and Bioinformatics

All MS experiments for peptide identification were performed on a Nano-LC/MS system consisting of a Surveyor HPLC system and a 7-tesla Finningan LTQ-FTICR mass spectrometer (Therm Electron, Bremen, Germany) equipped with a nano-ESI source. Ten microliters of each sample were loaded by a Surveyor auto sampler (Surveyor) onto a C18 trap column for desalting and concentration at a flow rate of 20 μL/min. The mass spectrometer was operated in the data-dependent mode to automatically switch between MS and MS acquisition. General mass spectrometric conditions included spray voltage, 2.2 kV; no sheath and auxiliary gas flow; ion transfer tube temperature, 220 °C; collision gas pressure, 1.3 millitorrs; normalized collision energy using wide band activation mode; and 35% of MS. Ion selection threshold was 500 counts for MS/MS. An activation $q = 0.25$ and an activation time of 30 ms were applied in MS/MS acquisitions. Acquired MS spectra were searched using an in-house licensed MASCOT search engine (Mascot version 2.2.04; Matrix Science, London, UK). To identify the peptides, MASCOT (version 2.3.01, Matrix Science, London, UK), operated on a local server, was used to search the maize (*Zea mays*) database. MASCOT was used to the monoisotopic mass selected, a peptide mass tolerance of 10 ppm, and a fragment ion mass tolerance of 0.8 Da. Trypsin was selected as enzyme, with one potential missed cleavage. ESI-FTICR was selected as instrument type, and carbamidomethyl cysteine and oxidized methionine were chosen as variable modifications. All proteins identified by high-scoring peptides were considered true matches, and at least two peptide matches. The high-scoring peptides corresponded with the peptides that were above the threshold in our MASCOT search (expected $p < 0.05$).

3. Results and Discussion

3.1. Protein Expression on SDS-PAGE

SDS-PAGE was performed in order to profile proteins from the pollen and pistil. Pollen and pistil were collected from wild type W22 (ga1) and near-isogenic lines W22 (Ga1), respectively. Pollen lanes were well separated whereas pistil lanes were not clearly visual (Figure 1, Supplementary Figure 1). However, high performance LTQ-FTICR MS was used to excise the gel into 10 pieces.

Figure 1. SDS-PAGE pattern in the pollen and pistil of W22 (ga1; Ga1) in maize. Samples were analyzed triplicate as described in the method section and gels were stained using Coomassie Brilliant Blue (CBB) staining. Standard molecular weight (kDa) is on the left.

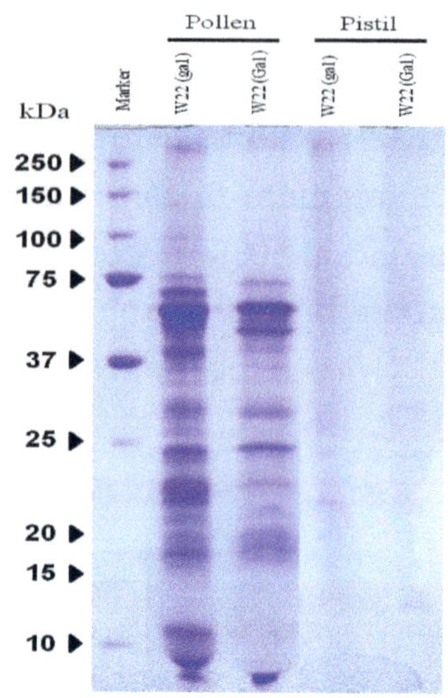

3.2. Specific Protein Analysis of Identifying Proteins from Pollen

Twenty four differential expressed proteins were identified from pollen on SDS-PAGE using LTQ-FTICR MS. Two proteins namely chaperonin CPN60-2 and albumin b-32 were only identified from the pollen of W22 (ga1), whereas 12 proteins were only appeared in the pollen of W22 (Ga1) such as adagio protein 3, ATP synthase subunit alpha, ATP synthase subunit beta, histone H2B.4, 1-Cys peroxiredoxin PER1, glucose-6-phosphate isomerase, ADP, ATP carrier protein 2, cysteine synthase, ferredoxin-dependent glutamate synthase, expansin-B9, expansin-B1, peptidyl-prolyl *cis*-trnas isomerase (Table 1). However, 10 proteins were commonly shared from both of the W22 (ga) and W22 (Ga1) like elongation factor 1 alpha, exopolygalacturonase (3 subunits), expasin B-10, profilin-3, endochitinase A, endochinase B, expansin B-11 and ribosome-inactivating protein 3.

3.3. Specific Protein Analysis of Identified Proteins from Pistil

Out of 20 proteins, 10 proteins (catalase isozyme 1, catalase isozyme 3, acetolactate synthase 2, ADP-ATP carrier protein 1, asparagine synthetase, histone H2B.1, histone H2A, histone-lysine *N*-methyltransferase EZ1, 14-3-3-like protein GF14 12, polycomb group protein FIE2) were identified in the pistil of W22 (ga1), where as 7 proteins (14-3-3 like protein GF14-6, 3-hydroxy-3-methylglutaryl-coenzyme A reductase, Glyceraldehyde-3-phosphate dehydrogenase A, Glyceraldehyde-3-phosphate dehydrogenase, casein kinase II subunit alpha, late embryogenesis abundant protein EMB564, Cell number regulator 2) were identified in the pistil of W22 (Ga1) (Table 1). However, 3 proteins like 2,3-bisphosphoglycerate-independent phosphoglycerate mutase, alpha-amylase/trypsin inhibitor, histone-lysine *N*-methyltransferase EZ1 were identified in both of the pistil of W22 (ga1) and W22 (Ga1) (Table 1).

Table 1. Features of the identified proteins in the pollen and pistil of W22 (ga1; Ga1) using Linear Quadruple Trap-Fourier-Transform Ion Cyclotron Resonance mass spectrometer (LTQ-FTICR-MS).

AN [1]	Protein Description	W22 (ga1)		W22 (Ga1)		PS [2]	MW [3]	PM [4]	pI [5]	PC [6]	MS-MS Ion Score
		Pollen	Pistil	Pollen	Pistil						
Hydrolase activity											
P26216	Exopolygalacturonase	√	×	√	×	79	43,416	6.95	4	11.7	64.93
P35339	Exopolygalacturonase	√	×	√	×	63	43,269	8.44	5	17.8	67.85
P35338	Exopolygalacturonase	√	×	√	×	79	43,387	6.59	5	15.1	66.03
Q41803	Elongation factor 1-alpha	√	×	√	×	96	49,202	9.19	7	23	46.8
P29022	Endochitinase A	√	×	√	×	103	29,106	8.3	3	22.5	69.56
P29023	Endochitinase B (Fragment)	√	×	√	×	61	28,147	8.94	2	14.9	59.82
P10593	Albumin b-32	√	×	×	×	36	32,408	5.38	3	20.8	46.86
P25891	Ribosome-inactivating protein 3	√	×	√	×	30	33,236	5.83	3	15.3	36.78
Nucleotide binding											
Q43298	Chaperonin CPN60-2, mitochondrial	√	×	×	×	31	60,897	5.67	2	6.2	9.26
P49094	Asparagine synthetase	×	√	×	×	21	66,535	5.83	2	6.1	21.43
P05494	ATP synthase subunit alpha, mitochondrial	×	×	√	×	102	55,146	8	5.85	19.3	34.9
P49106	14-3-3-like protein GF14-6	×	×	×	√	157	29,644	3	4.76	17.6	48.74
O24594	3-hydroxy-3-methylglutaryl-coenzyme A reductase	×	×	×	√	28	60,892	2	6.77	6	3.07
Nucleic acid binding											
P30755	Histone H2B.1	×	√	×	×	32	16,410	10	5	39.1	16.68
P40280	Histone H2A	×	√	√	×	65	16,417	10.59	2	28.3	70.53
Q8S4P5	Histone-lysine N-methyltransferase EZ2	×	√	×	√	24	99,916	8.47	7	11	9.82
Q8S4P6	Histone-lysine N-methyltransferase EZ1	×	√	×	×	14	103,703	8.85	3	4.7	2.64
P49120	Histone H2B.4	×	×	√	×	65	15,173	4	10.02	31.4	4.97
Catalytic activity											
P80608	Cysteine synthase	×	×	√	×	31	34,185	2	5.91	10.5	5.12
P23225	Ferredoxin-dependent glutamate synthase, chloroplastic	×	×	√	×	18	175,063	5	6.21	5.8	8.66
Q41769	Acetolactate synthase 2	×	√	×	×	31	68,982	6.48	3	9.1	29.41
P30792	2,3-bisphosphoglycerate-independent phosphoglycerate mutase	×	√	×	√	59	60,582	5.29	3	9.7	67.15
Antioxidant activity											
P18122	Catalase isozyme 1	×	√	×	×	27	56,841	7.4	2	9.8	26.82
P18123	Catalase isozyme 3	×	√	×	×	27	56,760	6.47	2	4.8	5.02
A2SZW8	1-Cys peroxiredoxin PER1	×	×	√	×	28	24,890	2	6.31	10.5	38.51

Table 1. *Cont.*

AN [1]	Protein Description	W22 (ga1)		W22 (Ga1)		PS [2]	MW [3]	PM [4]	p*I* [5]	PC [6]	MS-MS Ion Score
		Pollen	Pistil	Pollen	Pistil						
Hydrolase activity											
P04709	ADP, ATP carrier protein 1	×	√	×	×	41	42,365	9.85	4	11.1	22.01
857	ADP, ATP carrier protein 2	×	×	√	×	56	42,306	4	9.85	12.4	4.23
Protein binding											
P35083	Profilin-3	√	×	×	×	60	14,228	4.91	2	29	60.41
Q01526	14-3-3-like protein GF14-12	×	√	×	×	149	29,618	4.75	3	18	8.99
Oxidoreductase activity											
P09315	Glyceraldehyde-3-phosphate dehydrogenase A, chloroplastic	×	×	×	√	31	42,840	2	7	8.7	30.94
P08735	Glyceraldehyde-3-phosphate dehydrogenase, cytosolic 1	×	×	×	√	44	36,500	3	6.46	13.9	5.2
Transporter activity											
P19023	ATP synthase subunit beta, mitochondrial	×	×	√	×	128	59,067	8	6.01	25.9	25.94
P28523	Casein kinase II subunit alpha	×	×	×	√	16	39,205	3	8.41	14.8	16.2
Isomerase activity											
P21569	Peptidyl-prolyl cis-trans isomerase	×	×	√	×	127	18,337	2	8.91	18	36.38
P49105	Glucose-6-phosphate isomerase	×	√	√	×	36	62,198	2	6.96	4.1	1.62
Enzyme regulator activity											
P13867	Alpha-amylase/trypsin inhibitor	×	√	×	√	64	22,060	8.16	2	15	62.82
Signal transducer activity											
Q9C9W9	Adagio protein 3	×	×	√	×	23	69,019	3	6.06	2.3	9.6
Unknown											
P0C1Y5	Expansin-B11	√	×	√	×	174	28,943	8.44	4	13.8	39.64
Q8VZY6	Polycomb group protein FIE2	×	√	×	×	27	42,475	5.89	2	15.3	9.65
Q07154	Expansin-B9	×	×	√	×	200	29,062	5	9.01	21.6	39.5
P58738	Expansin-B1	×	×	√	×	31	29,066	4	8.99	18.6	39.5
P46517	Late embryogenesis abundant protein EMB564	×	×	×	√	55	9678	2	6.6	27.5	55.46
B6TYV8	Cell number regulator 2	×	×	×	√	18	19,222	2	7.37	17.7	18.17

[1] AN: Accession Number, [2] PS: Protein Score, [3] MW: Molecular Weight, [4] PM: Protein Matches, [5] p*I*: Iso-electric Point, [6] PC: Protein Coverage.

3.4. Cross-Correlation and Functional Distribution of Identified Proteins from Pollen and Pistil

The cross-correlation was clarified of total identified proteins between pollen and pistil. Two proteins were identified from the pollen of W22 (ga1) whereas 12 proteins from the pollen of W22 (Ga1). However, 10 proteins shared from both of the pollen of W22 (ga1) and W22 (Ga1) (Figure 2). Furthermore, 10 proteins were identified from the pistil of W22 (ga1) whereas 7 proteins confirmed from the pistil of W22 (Ga1). However, 3 proteins shared from both of the pistil of W22 (ga1) and W22 (Ga1) (Figure 2).

Figure 2. Cross-correlation of the identified protein between pollen and pistil of W22 (ga1; Ga1) in maize.

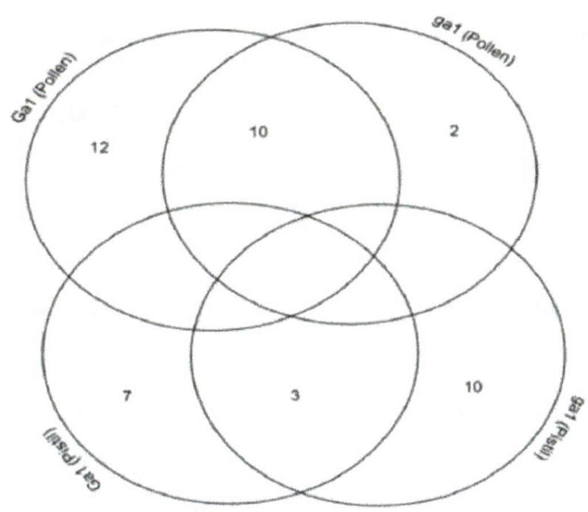

A total of 44 differentially expressed proteins were classified into 13 possible functional categories by using Protein Information Resources (PIR) shown in Figure 3. Out of 44 unique proteins, most of them involved in hydrolase activity (18%), nucleotide binding (11%), nucleic acid binding (11%), catalytic activity (9%), antioxidant activity (7%), isomerase activity (5%), oxidoreductase activity (5%), transporter activity (5%), ion binding (5%), protein binding (4%), enzyme regulator activity (2%), signal transducer activity (2%) and unknown (16%) (Figure 3).

Figure 3. Functional classification of the total identified proteins in the pollen and pistil of W22 (ga1; Ga1) in maize.

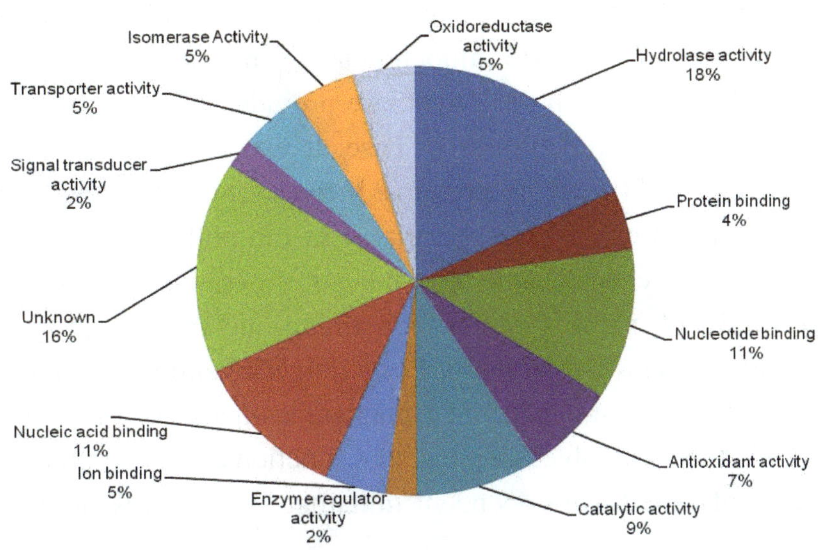

3.5. The Implication of Differentially Expressed Proteins from Pollen and Pistil of Maize

A total of 44 proteins were identified from the pollen and pistil of W22 (ga1) and W22 (Ga1) of which 24 proteins were confirmed from the pollen of W22 (ga1) and W22 (Ga1) and 20 proteins from the pistil of W22 (ga1) and W22 (Ga1). However, two proteins were identified from the pollen of W22 (ga1) whereas albumin b-32 (32.4 kDa, p*I* 5.38) regarded as the protein of maize endosperm that is a monomeric albumin with an apparent molecular weight of about 32 kDa with a p*I* of 5.38. Di Fonzo *et al.*, 1988 [21] found that the two variants expose similar amino acid composition but minor differences are appeared by their tryptic peptide maps. They also noticed that the protein is localized in the soluble part of the cytoplasm and does not bind to any particular structure.

In addition, 12 proteins were identified only from the pollen of W22 (Ga1). However, ATP synthase subunit alpha (55.1 kDa, p*I* 5.85) and ATP synthase subunit beta (59.0 kDa, p*I* 6.01) were confirmed in our investigation. ATP synthesis is membrane-bound enzyme complexes/ion transporters that accelerate ATP synthesis and/or hydrolysis with the transport of protons across a membrane. It can harness the energy from a proton gradient, using the flux of ions across the membrane via the ATPase proton channel to drive the synthesis of ATP. The alpha/A and beta/B subunits can each be divided into three regions, or domains, centered on the ATP-binding protein, and based on structure and function. The central domain contains the nucleotide-binding residues that make direct contact with the ADP/ATP molecule [22]. 1-Cys peroxiredoxin PER1 (24.8 kDa, p*I* 6.31) was considered as the antioxidant protein which seems to contribute to the inhibition of germination during stress. It was prevailed that overexpression of rice 1-cys-peroxiredoxin in transgenic tobacco accelerated oxidative stress tolerance, but dormancy was not affected [23]. Glucose-6-phosphate isomerase (62.1 kDa, p*I* 6.96) was identified in the pollen of W22 (Ga1) that catalyzes the conversion of glucose-6-phosphate into fructose 6-phosphate in the second step of glycolysis. This protein has various functions inside and outside the cell. This protein is also involved in the glycolysis and gluconeogenesis within the cytoplasm, while outside the cell it acts as a neurotrophic factor for spinal and sensory neurons. In *Ananas comosus*, it was prevailed that the mitochondria may produce this protein to allow cytoplasmic conversion of glucose-6 phosphate into fructose-6 phosphate in the second step of glycolysis [24]. ADP, ATP carrier protein 2 (42.3 kDa, p*I* 9.85) was found in the pollen of W22 (Ga1) that catalyzes the exchange of ADP and ATP over the mitochondrial inner membrane. An ADP/ATP carrier protein was found in *K. pinnata* mitochondria. The protein may be employed in the mitochondrial energy synthesis in which ATP synthase provides ATP via oxidative phosphorylation, and may work in reverse as a proton-pumping ATPase. It was revealed at *K. pinnata* that ADP and ATP could sustain via ADP/ATP carrier proteins between mitochondrial membranes and other organelles [24]. Cysteine synthase (CS) was identified in the pollen of W22 (Ga1) with molecular weight 34.1 kDa and p*I* 5.91. CS catalyzes the biosynthesis of cysteine in plants [25]; cysteine acts as a precursor for the synthesis of various sulfur containing metabolites [26], whereas glutathione represents the most important one which employed as a universal antioxidant and detoxifier for coping with various stresses [27]. Ferredoxin-dependent glutamate synthase (175 kDa, p*I* 6.21) was identified in the pollen of W22 (Ga1) that is involved in metabolic function especially in nitrogen assimilation. In Arabidopsis, the increase of ferredoxin dependent glutamate synthase is probably a consequence of limited electron transport and may affect feedback regulation to compete for electrons required for nitrogen assimilation [28].

In pistil, 10 proteins were identified in the W22 (ga1). However, catalase isozyme 1 (24.8 kDa, p*I* 6.31) was detected in the W22 (ga1) that occurs in almost all aerobically respiring organisms and serves to protect cells from the toxic effects of hydrogen peroxide. In the early stage of drought stress, catalase activities were found to increase or be stable, and then decrease with further increase in magnitude of water stress [29]. Furthermore, catalase isozyme 1 was only increased at 10 DPA under drought stress in Kauz which indicated that catalase might be activated to diminish toxic compounds during the early stage while the plant acclimatize the drought stress [30]. However, the catalase isozyme 3 (56.7 kDa, p*I* 6.47) was identified in the pistil of W22 (ga1) and it occurs also in almost all aerobically respiring organisms and serves to protect cells from the toxic effects of hydrogen peroxide. Its levels are highest in the light period and are lowest in the dark period. Therefore, it may be important for scavenging hydrogen peroxide at night, rather than during the day. Acetolactate synthase 2 (68.9 kDa, p*I* 6.48) was confirmed in the pistil of W22 (ga1). The acetolactate synthase (ALS) enzyme is a protein observed in plants and micro-organisms. ALS catalyzes the first step in the synthesis of the branched-chain amino acids (valine, leucine, and isoleucine) [31]. This protein is well known enzyme that is involved in catalytic activity, especially a part of the biosynthesis of various amino acids. However, in plants, it is located in the chloroplasts to assist the metabolic processes. It has been found in several experiments that mutated strands of *Escherichia coli* K-12 without the enzyme were not able to grow in the presence of only acetate as the only carbon sources [32]. Asparagine synthetase (66.5 kDa, p*I* 5.83) is an enzyme that generates asparagine from aspartate and arises only in the pistil of W22 (ga1). This reaction is similar to that accelerated by glutamine synthetase. It is also possible that asparagine synthetase poses its effects by fulfilling an as yet unknown function in the cell that is independent of its catalytic activity [33]. 14-3-3 like protein GF14-6 (29.6 kDa, p*I* 4.76) was confirmed in the pistil of W22 (ga1) that is associated with a DNA binding complex to bind to the G box, a well-characterized *cis*-acting DNA regulatory element found in plant genes. The functional properties of 14-3-3s are to bind and activate tyrosine and tryptophane hydroxylase in bovine brain in the presence of Ca^{2+}/calmodulin-dependent protein kinase type II [34].

However, seven proteins were detected only in the pistil of W22 (Ga1). Glyceraldehyde 3-phosphate dehydrogenase (36.4 kDa, p*I* 7.01) were identified from the pistil of W22 (ga1) that catalyzes the conversion of glyceraldehyde 3-phosphate to D-glycerate 1, 3-bisphosphate. In soybean, glyceraldehyde 3-phosphate dehydrogenase (GAPDH) was identified as down-regulated at both the mRNA and protein levels in response to NaCl treatment, suggesting that it plays a role in salt stress and can be used as a target gene in soybean seedlings [35]. The main role of this gene is the tolerance and its relationship to improving salt tolerance in plants [36]. It is revealed that the ATP production will be reduced by the down-regulation of glyceraldehydes-3-phosphate dehydrogenase and eventually there will be a decrease in plant growth under salt stress. Late embryogenesis abundant protein EMB564 (9.6 kDa, p*I* 6.6) constitutes a set of proteins that participate in plant stress responses. During exposure to abiotic challenges, late embryogenesis abundant (LEA) proteins accumulate naturally in desiccation-tolerant structures, such as seed or pollen grains, and in plant vegetative tissues. However, Emb564 acts for displaying a complex combination of different PTMs, including phosphorylation, acetylation, methylation and deamination in the native protein, which may be relevant for its seed-specific role [37].

Furthermore, 10 proteins were detected from the pollen which shared both W22 (ga1) and W22 (Ga1) and three proteins were shared from both W22 (ga1) and W22 (Ga1) of pistil. Profilin-3 (14.2 kDa, p*I* 4.91) was confirmed in the pollen of both W22 (ga1) and W22 (Ga1). Profilins generate a large and diverse protein family. Multiple isoforms of profilins are available in many species, being encoded by separate genes, or in some cases translated from mRNA splice variants. In the case of animals and higher plants, isoforms may be exposed in a tissue-specific manner. Moreover, profilins are identified at different subcellular locations [38]; in particular, enrichment of the dynamic plasma membranes was ascertained for various cells types. Also, profilins were investigated in association with internal membranes that implicated in vesicular transport [39]. It revealed that the overall functional properties of different profilins are similar and eventually, one isoform can be interchanged with another one from quite a distant source [40]. Endochitinase A (29.3 kDa, p*I* 8.3) was identified from both of the pollen of W22 (ga1) and W22 (Ga1) that defends against chitin containing fungal pathogens. In *Arabidopsis thaliana*, a basic endochitinase (At3g12500) was confirmed that was involved in the ethylene/jasmonic acid-mediated signaling pathway during systemic acquired resistance [41].

4. Conclusions

The protein analysis of pollen and pistil in maize was accomplished to profile proteins related to gametophytic factors. Using SDS-PAGE, a total of 24 proteins from pollen and 20 proteins from pistil were identified following LTQ-FTICR MS. However, 2 proteins were only found in the pollen of W22 (ga1) whereas 10 proteins were revealed in the pollen of W22 (Ga1). In the case of the pistil, 10 proteins appeared in W22 (ga1) and 7 proteins were distinctly observed in W22 (Ga1). The proteins were mostly involved in the hydrolase activity, nucleic acid binding and nucleotide binding. More extensive studies are needed to fully understand the mechanism of gametophytic factors underlying the pollen and pistil of the lines.

Acknowledgments

This work was supported by the research grant of the Chungbuk National University in 2011.

Author Contributions

All the authors have contributed to the writing and editing of the paper. Each of them has specifically contributed to the following parts: experiment design and funding (SHW), proteomics (YJ), mass analysis (KC), coordination of writing and editing (SHW, YJ, SKR, AHMK, SJK, SWC, YSS, JBH).

References

1. Cheung, A.Y. Pollen-pistil interactions in compatible pollination. *Proc. Natl. Acad. Sci. USA* **1995**, *92*, 3077–3080.
2. Kermicle, J.L.; Evans, M.M.S. The *Zea mays* sexual compatibility gene ga2: Naturally occurring alleles, their distribution, and role in reproductive isolation. *J. Hered.* **2010**, *101*, 737–749.
3. Mangelsdorf, P.C.; Jones, D.F. The expression of mendelian factors in the gametophyte of maize. *Genetics* **1926**, *11*, 423–455.
4. Burnham, C.R. Cytogenetic studies of a case of pollen abortion in maize. *Genetics* **1941**, *26*, 460–468.
5. Cameron, D.R.; Moav, R.M. Inheritance in *Nicotiana tabacum* XXVII. Pollen killer, an alien genetic locus inducing abortion of microspores not carrying it. *Genetics* **1957**, *42*, 326–335.
6. Loegering, W.Q.; Sears, E.R. Distorted inheritance of stem-rust resistance of timstein wheat caused by a pollen-killing gene. *Can. J. Genet. Cytol.* **1963**, *5*, 65–72.
7. Rick, C.M. Abortion of male and female gametes in the tomato determined by allelic interaction. *Genetics* **1966**, *53*, 85–96.
8. Allard, R.W. An additional gametophyte factor in the lima bean. *Der Zücht.* **1963**, *33*, 212–216.
9. Ramage, R.T. Heterosis and hybrid seed production in barley. In *Heterosis*; Frankel, R., Ed.; Springer Berlin Heidelberg: Berlin/Heidelberg, Germany, 1983; Volume 6, pp. 71–93.
10. Dresselhaus, T.; Lausser, A.; Márton, M.L. Using maize as a model to study pollen tube growth and guidance, cross-incompatibility and sperm delivery in grasses. *Ann. Bot.* **2011**, *108*, 727–737.
11. Nelson, O. The gametophyte factors of maize. In *The Maize Handbook*; Freeling, M., Walbot, V., Eds.; Springer New York: New York, NY, USA, 1994; pp. 496–503.
12. Li, X.M.; Sang, Y.L.; Zhao, X.Y.; Zhang, X.S. High-throughput sequencing of small RNAs from pollen and silk and characterization of miRNAs as candidate factors involved in pollen-silk interactions in maize. *PLoS One* **2013**, *8*, e72852.
13. Chen, Y.; Chen, T.; Shen, S.; Zheng, M.; Guo, Y.; Lin, J.; Baluška, F.; Šamaj, J. Differential display proteomic analysis of *Picea meyeri* pollen germination and pollen-tube growth after inhibition of actin polymerization by latrunculin B. *Plant J.* **2006**, *47*, 174–195.
14. Fernando, D.D. Characterization of pollen tube development in *Pinus strobus* (eastern white pine) through proteomic analysis of differentially expressed proteins. *Proteomics* **2005**, *5*, 4917–4926.
15. Holmes-Davis, R.; Tanaka, C.K.; Vensel, W.H.; Hurkman, W.J.; McCormick, S. Proteome mapping of mature pollen of *Arabidopsis thaliana*. *Proteomics* **2005**, *5*, 4864–4884.
16. Imin, N.; Kerim, T.; Weinman, J.J.; Rolfe, B.G. Characterisation of rice anther proteins expressed at the young microspore stage. *Proteomics* **2001**, *1*, 1149–1161.
17. Imin, N.; Kerim, T.; Weinman, J.J.; Rolfe, B.G. Low temperature treatment at the young microspore stage induces protein changes in rice anthers. *Mol. Cell. Proteomics* **2006**, *5*, 274–292.
18. Noir, S.; Bräutigam, A.; Colby, T.; Schmidt, J.; Panstruga, R. A reference map of the *Arabidopsis thaliana* mature pollen proteins. *Biochem. Biophys. Res. Commun.* **2005**, *337*, 1257–1266.
19. Kamal, A.H.M.; Jang, I.D.; Kim, D.E.; Suzuki, T.; Chung, K.Y.; Choi, J.S.; Lee, M.S.; Park, C.H.; Park, S.U.; Lee, S.H.; *et al.* Proteomics analysis of embryo and endosperm from mature common buckwheat seeds. *J. Plant Biol.* **2011**, *54*, 81–91.

20. Schägger, H.; von Jagow, G. Tricine-sodium dodecyl sulfate-polyacrylamide gel electrophoresis for the separation of proteins in the range from 1 to 100 kDa. *Anal. Biochem.* **1987**, *166*, 368–379.
21. Fonzo, N.; Hartings, H.; Brembilla, M.; Motto, M.; Soave, C.; Navarro, E.; Palau, J.; Rhode, W.; Salamini, F. The b-32 protein from maize endosperm, an albumin regulated by the o2 locus: Nucleic acid (cDNA) and amino acid sequences. *Mol. Gen. Genet. MCG* **1988**, *212*, 481–487.
22. Antes, I.; Chandler, D.; Wang, H.; Oster, G. The unbinding of ATP from F1-ATPase. *Biophys. J.* **2003**, *85*, 695–706.
23. Lee, K.O.; Jang, H.H.; Jung, B.G.; Chi, Y.H.; Lee, J.Y.; Choi, Y.O.; Lee, J.R.; Lim, C.O.; Cho, M.J.; Lee, S.Y. Rice 1cys-peroxiredoxin over-expressed in transgenic tobacco does not maintain dormancy but enhances antioxidant activity. *FEBS Lett.* **2000**, *486*, 103–106.
24. Hoang, T.K.H.; Akihiro, N. Mitochondrial proteomic analysis of cam plants, *Ananas comosus* and *Kalanchoe pinnata*. *Ann. Biol. Res.* **2012**, *3*, 88–97.
25. Takahashi, H.; Saito, K. Subcellular localization of spinach cysteine synthase isoforms and regulation of their gene expression by nitrogen and sulfur. *Plant Physiol.* **1996**, *112*, 273–280.
26. Mendoza-Cózatl, D.; Loza-Tavera, H.; Hernández-Navarro, A.; Moreno-Sánchez, R. Sulfur assimilation and glutathione metabolism under cadmium stress in yeast, protists and plants. *FEMS Microbiol. Rev.* **2005**, *29*, 653–671.
27. Noctor, G.; Foyer, C.H. Ascorbate and glutathione: Keeping active oxygen under control. *Annu. Rev. Plant Physiol. Plant Mol. Biol.* **1998**, *49*, 249–279.
28. Zybailov, B.; Friso, G.; Kim, J.; Rudella, A.; Rodríguez, V.R.; Asakura, Y.; Sun, Q.; van Wijk, K.J. Large scale comparative proteomics of a chloroplast Clp protease mutant reveals folding stress, altered protein homeostasis, and feedback regulation of metabolism. *Mol. Cell. Proteomics* **2009**, *8*, 1789–1810.
29. Mittler, R.; Zilinskas, B.A. Regulation of pea cytosolic ascorbate peroxidase and other antioxidant enzymes during the progression of drought stress and following recovery from drought. *Plant J.* **1994**, *5*, 397–405.
30. Jiang, S.-S.; Liang, X.-N.; Li, X.; Wang, S.-L.; Lv, D.-W.; Ma, C.-Y.; Li, X.-H.; Ma, W.-J.; Yan, Y.-M. Wheat drought-responsive grain proteome analysis by linear and nonlinear 2-DE and MALDI-TOF mass spectrometry. *Int. J. Mol. Sci.* **2012**, *13*, 16065–16083.
31. Chipman, D.; Barak, Z.E.; Schloss, J.V. Biosynthesis of 2-aceto-2-hydroxy acids: Acetolactate synthases and acetohydroxyacid synthases. *Biochim. Biophys. Acta (BBA) Protein Struct. Mol. Enzymol.* **1998**, *1385*, 401–419.
32. Dailey, F.E.; Cronan, J.E. Acetohydroxy acid synthase I, a required enzyme for isoleucine and valine biosynthesis in *Escherichia coli* K-12 during growth on acetate as the sole carbon source. *J. Bacteriol.* **1986**, *165*, 453–460.
33. Abbatiello, S.E.; Pan, Y.-X.; Zhou, M.; Wayne, A.S.; Veenstra, T.D.; Hunger, S.P.; Kilberg, M.S.; Eyler, J.R.; Richards, N.G.J.; Conrads, T.P. Mass spectrometric quantification of asparagine synthetase in circulating leukemia cells from acute lymphoblastic leukemia patients. *J. Proteomics* **2008**, *71*, 61–70.
34. Ichimura, T.; Isobe, T.; Okuyama, T.; Takahashi, N.; Araki, K.; Kuwano, R.; Takahashi, Y. Molecular cloning of cdna coding for brain-specific 14-3-3 protein, a protein kinase-dependent activator of tyrosine and tryptophan hydroxylases. *Proc. Natl. Acad. Sci. USA* **1988**, *85*, 7084–7088.

35. Sobhanian, H.; Razavizadeh, R.; Nanjo, Y.; Ehsanpour, A.A.; Jazii, F.R.; Motamed, N.; Komatsu, S. Proteome analysis of soybean leaves, hypocotyls and roots under salt stress. *Proteome Sci.* **2010**, *8*, 1477–5956.
36. Holmberg, N.; Bülow, L. Improving stress tolerance in plants by gene transfer. *Trends Plant Sci.* **1998**, *3*, 61–66.
37. Amara, I.; Odena, A.; Oliveira, E.; Moreno, A.; Masmoudi, K.; Pagès, M.; Goday, A. Insights into maize lea proteins: From proteomics to functional approaches. *Plant Cell Physiol.* **2012**, *53*, 312–329.
38. Jockusch, B.M.; Murk, K.; Rothkegel, M. The profile of profilins. *Rev. Physiol. Biochem. Pharmacol.* **2007**, *159*, 131–149.
39. Dong, J.; Radau, B.; Otto, A.; Müller, E.-C.; Lindschau, C.; Westermann, P. Profilin I attached to the golgi is required for the formation of constitutive transport vesicles at the trans-golgi network. *Biochim. Biophys. Acta (BBA) Mol. Cell. Res.* **2000**, *1497*, 253–260.
40. Rothkegel, M.; Mayboroda, O.; Rohde, M.; Wucherpfennig, C.; Valenta, R.; Jockusch, B.M. Plant and animal profilins are functionally equivalent and stabilize microfilaments in living animal cells. *J. Cell Sci.* **1996**, *109*, 83–90.
41. Jaquinod, M.; Villiers, F.; Kieffer-Jaquinod, S.; Hugouvieux, V.; Bruley, C.; Garin, J.; Bourguignon, J. A proteomics dissection of *Arabidopsis thaliana* vacuoles isolated from cell culture. *Mol. Cell. Proteomics* **2007**, *6*, 394–412.

6

Potentiality of Soybean Proteomics in Untying the Mechanism of Flood and Drought Stress Tolerance

Zahed Hossain [1,*] and Setsuko Komatsu [2,*]

[1] Plant Stress Biology Lab, Department of Botany, West Bengal State University, Kolkata-700126, India

[2] National Institute of Crop Science, National Agriculture and Food Research Organization, Tsukuba 305-8518, Japan

* Authors to whom correspondence should be addressed; E-Mails: zahed_kly@yahoo.com (Z.H.); skomatsu@affrc.go.jp (S.K.)

Abstract: Dissecting molecular pathways at protein level is essential for comprehensive understanding of plant stress response mechanism. Like other legume crops, soybean, the world's most widely grown seed legume and an inexpensive source of protein and vegetable oil, is also extremely sensitive to abiotic stressors including flood and drought. Irrespective of the kind and severity of the water stress, soybean exhibits a tight control over the carbon metabolism to meet the cells required energy demand for alleviating stress effects. The present review summarizes the major proteomic findings related to changes in soybean proteomes in response to flood and drought stresses to get a clear insight into the complex mechanisms of stress tolerance. Furthermore, advantages and disadvantages of different protein extraction protocols and challenges and future prospects of soybean proteome study are discussed in detail to comprehend the underlying mechanism of water stress acclimation.

Keywords: soybean; proteomics; drought; flood; review

1. Introduction

Plants, being sessile organisms, are prone to various environmental stresses. Flooding and drought are the two different forms of water stress that constitute major limiting factors for plant growth, development and quality crop production. Soybean, the world's most widely grown seed legume, provides an inexpensive source of protein and vegetable oil for human consumption. This important legume crop is adapted to be grown in a wide range of climatic conditions; nevertheless, at seedling stage its growth is significantly affected by several abiotic stressors, including flooding [1–11] and drought [12,13].

Dissecting stress tolerance mechanism at molecular level has always been a priority in any crop development program. Stress-induced changes in gene expression modulate metabolic processes through alteration of cellular protein abundance and function. Therefore, understanding how the function of proteins changes under stressed conditions is crucial for clarifying the molecular mechanisms underlying stress tolerance and crop injury. Identification and understanding the biological function of any novel gene conferring such tolerance is a more ambitious goal than merely determining its sequence. Due to lack of correlation between mRNAs' expression levels and the abundance of their corresponding proteins, proteomic techniques provide one of the best options for the functional analysis of translated regions of the genome. Furthermore, several proteins undergo post-translational modifications such as removal of signal peptides, phosphorylation and glycosylation, that are extremely important for protein function. Hence, a proteomics approach, complemented with genome-sequence data and modern bioinformatics, offers a powerful tool to identify and characterize novel proteins and to follow temporal changes in protein relative abundances under adverse environmental conditions.

Conventional gel-based proteomic approaches, and gel free-mass spectrometry (MS)-based methods involving label-based and label-free protein quantification have been extensively used for characterization of stress-responsive proteins in soybean [5,6,9–11,14–17]. The present review provides an overview of the major findings related to changes in soybean proteomes in response to flooding and drought stresses to get a clear insight into the complex mechanisms involved in plants stress response. Furthermore, strengths and weaknesses of different proteomic methodologies of extracting complete proteome and challenges and future prospects of soybean proteome study are discussed in detail to comprehend the underlying mechanism of water stress tolerance.

2. Protein Extraction

The choice of method for protein extraction largely depends on the type of plant organelle and organs, and/or the nature of desired proteins to be extracted (Table 1). Presence of various interfering substances, such as phenolic compounds, proteolytic and oxidative enzymes, terpenes, organic acids, and carbohydrates create complications during the process of protein extraction, resulting in inferior results such as proteolytic breakdown, streaking, smearing and charge heterogeneity [18]. Elimination of these disturbing compounds during protein extraction is thus necessary to get the optimum result.

Table 1. Summary of soybean proteome analyses in response to flood and drought.

Stress	Cultivar/ Stress exposure	Organ/ Organelle	Protein extraction buffer	Protein solubilization/ lysis buffer	Proteomic methodologies	Spot resolved Proteins	Differentially abundant protein classification: Function	Differentially abundant protein classification: Localization	Ref.
Flooding	Enrei (5 days)	Leaf Hypocotyl Root	10% TCA, 0.07% 2-ME in acetone	8 M urea, 2 M thiourea, 5% CHAPS, 2 mM tributyl-phosphine, 0.4% Ampholytes pH 3–10	IEF, SDS-PAGE, nanoLC-MS/MS	577 (L): 24↑26↓ 555 (H): 35↑31↓ 515 (R): 20↑27↓	Met, Ene, ProtDesSt, DisDef, ProtSyn	Mito, Nucl, Cyto, Extr, ER, Cysk, PM	[9]
	Enrei (2 days)	Hypocotyl Root mitochondria	-	8 M urea, 2% NP-40, 5% 2-ME, 5% PVP 40, 0.4% Ampholytes pH 3–10	IPG, SDS-PAGE, BN-PAGE, nanoLC-MS/MS	Matrix 327 29↑7↓ Membrane 72 5↑11↓	Ene, DisDef	Mito, Chlo	[5]
	Enrei (2 days)	Hypocotyl Root cell wall	-	8 M urea, 2% NP-40, 0.8% Ampholine pH 3.5–10, 5% 2-ME and 5% PVP 40	IEF, SDS-PAGE, MALDI-TOF MS, nanoLC-MS/MS, protein sequencing	204 4↑12↓	Met, ProtDesSt, DisDef	Sec	[7]
	Enrei (1–4 days)	Hypocotyl Root	Phosphate saline buffer pH 7.6, 400 mM NaCl, 3 mM NaN_3 followed by 10% TCA	8 M urea, 2% NP-40, 0.8% Ampholine (pH 3.5–10), 5% 2-ME and 5% PVP 40	IEF/IPG, SDS-PAGE, MALDI-TOF MS, protein sequencing	803 21↑7↓	ProtDesSt, DisDef, Ene, Pmet, CellSt, Trans	-	[8]
	Asoagari (3, 7 days)	Root	Cold acetone containing 10% TCA, 0.07% 2-ME	8 M urea, 1% CHAPS, 0.5% IPG buffer pH 4–7, 20 mM DTT, BPB	IPG, SDS-PAGE, MALDI-TOF MS, ESI-MS/MS	~900 14↑5↓5 Newly induced	Met, Ene, DisDef, ProtSyn	-	[19]
	Enrei (12–48 h)	Hypocotyl Root	-	9.5M urea, 2% NP-40, 2% Ampholines pH 3–10, 5% 2-ME	IEF/IPG tube gel, 2-DE, MALDI-TOF MS, nanoLC-MS/MS, protein sequencing	799 14↑20↓	Ene, DisDef, Pmet, CellSt, Secmet, Sgnl	-	[20]

Table 1. *Cont.*

Stress	Cultivar/ Stress exposure	Organ/ Organelle	Protein extraction buffer	Protein solubilization/ lysis buffer	Proteomic methodologies	Spot resolved Proteins	Differentially abundant protein classification		Ref.
							Function	Localization	
	Enrei (1 days)	Hypocotyl Root plasma membrane	-	8 M urea, 2% NP-40, 0.8% Ampholine pH 3.5–10, 5% 2-ME and 5% PVP 40	IEF tube gel, 2-DE, MALDI-TOF MS, nanoLC-MS/MS, protein sequencing	150 12↑2↓	ProtDesSt, ProtSyn, DisDef, CellDiv, Trans, Pmet, Ene, Secmet, Sgnl	-	[21]
Flooding Low oxygen	Enrei (3, 6 days Low oxygen)	Root	10% TCA, 0.07% 2-ME in acetone	8 M urea, 2 M thiourea, 5% CHAPS, 2 mM tributyl-phosphine, 0.4% Ampholytes pH 3–10	IEF, SDS-PAGE , MALDI-TOF MS, nanoLC-MS/MS	1,233 F: 4↑12↓ LO: 2↓	Met, Ene, ProtDesSt, Sgnl, ProtSyn, DisDef	Cyto, Chlo, Nucl	[10]
Drought	Enrei (Stop watering 10% PEG 4 days)	Leaf Hypocotyl Root	10% TCA, 0.07% 2-ME in acetone	8 M urea, 2 M thiourea, 5% CHAPS, and 2 mM tributyl-phosphine, 0.4% Ampholytes pH 3–10	IPG, SDS-PAGE, nanoLC-MS/MS	549 (L): PEG: 20↑17↓ Drought: 20↑21↓ 451 (H): PEG: 20↑13↓ Drought: 18↑19↓ 632 (R): PEG: 20↑10↓ Drought: 33↑16↓	Met, Ene, ProtSyn, DisDef	Chlo, Cyto, Nucl, Mito	[12]
	Taegwang (withholding water - 5 days, rewatering - 4 days)	Root	Mg/NP-40 buffer [0.5 M Tris-HCl (pH 8.3), 2% NP-40, 20 mM $MgCl_2$, 1 mM PMSF , 2% 2-ME, 1% PVP], water-saturated phenol, followed by ammonium acetate in methanol	8 M urea, 1% CHAPS, 0.5% IPG buffer (pH 4–7), 20 mM DTT, BPB	IPG, SDS-PAGE, MALDI-TOF MS	1,350 6↑20↓2 New	Met, Ene, Sgnl, DisDef, CellSt,	-	[13]

Table 1. *Cont.*

Stress	Cultivar/ Stress exposure	Organ/ Organelle	Protein extraction buffer	Protein solubilization/ lysis buffer	Proteomic methodologies	Spot resolved Proteins	Differentially abundant protein classification		Ref.
							Function	Localization	
Osmotic stress	Enrei (10% PEG 1–4 days)	Hypocotyl Root plasma membrane	Plasma membrane proteins precipitated by TCA followed by cold acetone washing	7 M urea, 0.2 M thiourea, 0.2mM tributylphosphine, 5% PVP-40, 0.4% CHAPS, 0.2% Ampholytes (pH 3.0–10.0)	IEF tube gel, SDS-PAGE, LC MS/MS, nanoLC-MS/MS	202 11↑75↓	Sgnl, Met, ProtSyn, DisDef, Trans	-	[22]
Osmotic stress	Enrei (0, 5, 10, 20% PEG 1–4 days)	Root	Phosphate saline buffer (pH 7.6): 65 mM K_2HPO_4, 2.6 mM KH_2PO_4, 400 mM NaCl and 3 mM NaN_3 followed by 10% TCA	8 M urea, 2% NP-40, 0.8% Ampholine (pH 3.5–10), 5% 2-ME and 5% PVP 40	IEF tube gel, SDS-PAGE, MALDI-TOF MS, protein sequencing	415 19↑18↓	DisDef, Ene, ProtDesSt, Met, CellSt, Secmet.	-	[23]

Up and down arrows indicate stress-induced increase and decrease in protein abundance respectively. Abbreviations: BPB, bromophenol blue; CBB, Coomassie brilliant blue; DTT, dithiothreitol; IPG, immobilized pH gradient; IEF, isoelectric focusing; LC, liquid chromatography; MS, mass spectrometry; PMSF, phenyl methyl sulfonyl fluoride; PVP, Polyvinylpyrrolidone; TCA, trichloroacetic acid; 2-ME, 2-mercaptoethanol. L, H and R represent leaf, hypocotyl and root respectively. Functional classification: Met, metabolism; Ene, energy; ProtDesSt, protein destination/storage; ProtSyn, protein synthesis; DisDef, disease/defense; CellDiv, cell division; Trans, transporter; Pmet, primary metabolism; Secmet, secondary metabolism; CellSt, cell structure; Sgnl, signal transduction; GrDev, growth and development; TrStr, translocation and storage. Subcellular localization: Chlo, chloroplast; Mito, mitochondria; PM, plasma membrane; Nucl, nuclear; Cyto, cytoplasm; Extr, extracellular matrix; ER, endoplasmic reticulum; Cysk, cytoskeleton; Vacu, vacuolar; Sec, secretory pathway.

Soybean seeds contain a large amount of secondary metabolites like kaempferol and quercetin which not only hampers high-quality protein extraction, but also impedes protein spot separation in high resolution two-dimensional polyacrylamide gel electrophoresis (2-DE) gels, resulting in a significant reduction in the number of distinctly resolved protein spots [24,25]. Furthermore, presence of abundant storage proteins such as β-conglycinin and glycinin often hinders the isolation and characterization of less abundant seed proteins. Sample fractionation technique has proved to be an efficient strategy for successful removal of such highly abundant storage proteins. With the simple addition of 10 mM calcium chloride to the salt soluble soybean seed protein extract in low ionic strength buffer, the α, α', and β subunits of β-conglycinin and the acidic and basic subunits of glycinin were found to be reduced significantly from the total protein extract [26]. For extracting soybean seed proteins both at mature [27] and seed filling stages [28], phenol based protein extraction method was reported to be more effective. As compared to the trichloroacetic acid (TCA)/acetone or Tris-HCl buffer, protein extracted in buffer comprises of 50% phenol, 0.45 M sucrose, 5 mM EDTA, 0.2% 2-mercaptoethanol, 50 mM Tris–HCl (pH 8.8) produced a large number of reproducible protein spots. Natarajan *et al.* [29] also compared four different protein extraction/solubilization methods-urea, thiourea/urea, phenol, and a modified TCA/acetone to determine their effectiveness in separating soybean seed proteins by 2-DE. The thiourea/urea and TCA methods were found to be more suitable in resolving less abundant and high molecular weight proteins. In addition, these two methods exhibited higher protein resolution and spot intensity as compared to the rest of the methods. Recently, Barbosa *et al.* [30] successfully analysed mature seed proteome by extracting proteins in 50 mM Tris-HCl (pH 8.8), 1.5 mM KCl, 10 mM dithiothreitol (DTT), 1.0 mM phenylmethylsulfonyl fluoride (PMSF), and 0.1% SDS followed by precipitation in 0.1 M ammonium acetate in methanol.

To compare soybean leaf and flower proteomes at different developmental stages, Ahsan and Komatsu [31] evaluated three different protein extraction protocols-TCA precipitation [32], phenol extraction method [33] with modifications and direct tissue homogenizing in suitable protein solubilization buffers. To optimize protein pellet solubilization buffer, A-buffer containing 8 M urea, 2% Nonidet P-40, 2% ampholine (pH 3.5–10), 5% 2-mercaptoethanol, and 5% polyvinylpyrrolidone (PVP)-40; B-buffer [32] containing 7 M urea, 0.2 M thiourea, 0.2 mM tributylphosphine (TBP), 0.4% CHAPS, 5% PVP-40, and 2% ampholine (pH 3–10); and C-buffer containing 8.5 M urea, 2.5 M thiourea, 5% CHAPS, 1% DTT, 1% Triton X-100, and 0.5% ampholines (pH 3–10 and 5–8) were tested. A combination of the phenol-based method with C-solubilization buffer generated high quality proteome maps in terms of well-separated resolved spots, spot intensity, and the number of proteins in the 2-DE gels with no horizontal streaking and high background noise levels.

For root proteomic analysis, TCA/acetone precipitation is the most widely used protein extraction method. Root proteins extracted in 10% TCA and 0.07% 2-mercaptoethanol in acetone followed by subsequent solubilization in the lysis buffer containing 8 M urea, 2 M thiourea, 5% CHAPS, and 2 mM TBP results in a high quality gel with a good number of resolved protein spots [16,34]. Addition of DTT and PVP in the soybean protein extraction buffer was found to be effective in enhancing the number of resolved spots in gels [35,36]. Ahsan and Komatsu [31] reported that treatment of root with Mg/Nonidet P-40 buffer followed by extraction with alkaline phenol and methanol/ammonium acetate produced high-quality proteome maps consisting of numerous well-separated spots with high intensity, on 2-DE gels.

On the whole, instead of having physicochemical limitations of each and every protocol, the TCA/acetone precipitation and phenol-based protocols are the most reliable and efficient protein extraction methods for various soybean organs to obtain high quality gels [37,38] (Table 1).

3. Changes in Soybean Proteome in Response to Flooding

Soil oxygen deprivation, the most inevitable consequence of flooding, forces submerged plants to shift from aerobic to anaerobic respiration [39,40]. This metabolic swing helps plants to regenerate NAD^+ through ethanol fermentation by selectively synthesizing flooding-inducible proteins involved in sucrose breakdown, glycolysis, and fermentation [41]. The suppressed energy metabolisms accelerate energy depletion resulting in growth retardation, and render flooded plants vulnerable to other biotic and abiotic stresses.

Different physiological and molecular aspects of plant response toward flooding stress are well documented. In this section, contribution of proteomic studies to flooding stress mediated modulation of protein networks have been summarized for better understanding of flood sensing and tolerance mechanism both at organ and whole plant level (Figure 1).

Organ-specific proteome response of soybean seedlings under flooding stress has been well analyzed [1,2,5–10,16,42] (Table 1). Root represents the first organ of a plant in sensing waterlogged condition. Thus, root has always been a target of proteomic investigation to elucidate the plants' flood response mechanism. Root proteome study of submerged young soybean seedlings revealed that glycolysis related proteins including UDP-glucose pyrophosphorylase and fructose-bisphosphate aldolase, disease/defense-related proteins such as ROS (reactive oxygen species) scavengers, chaperones, hemoglobin, and/or acid phosphatase were mostly affected [16,20,42]. A separate study by Alam *et al.* [19] has shown higher expression of glycolysis and fermentation pathways related proteins in roots of three-week-old seedlings.

Analysis of enzyme activities and carbohydrate contents in flooded seedlings further confirmed that glucose degradation and sucrose accumulation accelerated during flooding due to activation of glycolysis and decrease of sucrose degrading enzymes [16]. In addition, the methylgly . oxal pathway, the detoxification route linked to glycolysis, was found to be increased under flooding.

Flood-specific accumulation of alcohol dehydrogenase (ADH2) in roots of soybean indicates activation of alcohol fermentation pathway to cope with the hypoxic condition [6]. A recent proteomic study on flooding-tolerant mutant line showing better root growth under flooded condition revealed higher abundances of fermentation-related proteins including different types of ADHs and pyruvate decarboxylase isozymes on exposure to submergence [1]. Additionally, no changes in the cell wall loosening-related proteins were observed under flooding stress, thereby preserving the viability of the root tip and permitting rapid growth at post-stress period.

Figure 1. Water stress mediated changes in metabolic pathways. Blue and red arrows indicate changes in protein abundance (upward arrows indicate increase and downward arrows indicate decrease) in response to flooding and drought, respectively.

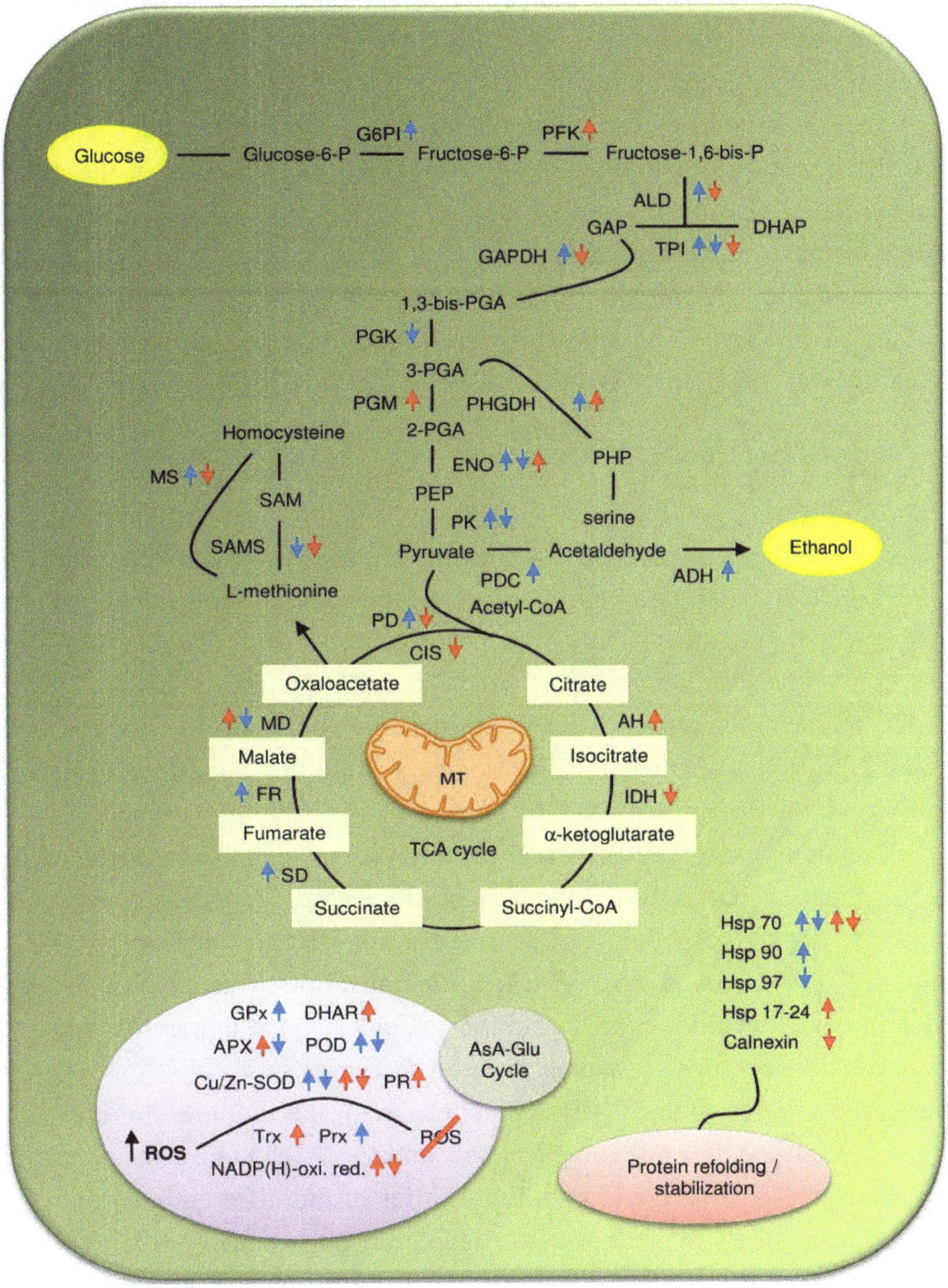

Abbreviations: ADH; alcohol dehydrogenase; AH, aconitate hydratase; ALD, aldolase; APX, ascorbate peroxidase; DHAR, dehydroascorbate reductase; ENO, enolase; FR, fumarase; GAPDH, glyceraldehyde 3-phosphate dehydrogenase; G6PI, Glucose-6-phosphate isomerase; GPx, glutathione peroxidase; GR, glutathione reductase; Hsp, heat shock proteins; IDH, isocitrate dehydrogenase; MD, malate dehydrogenase; PD, pyruvate dehydrogenase; PDC, pyruvate decarboxylase; PFK, Phosphofructokinase; PGK, phosphoglycerate kinase; PGM, phosphoglycerate mutase; PHGDH, 3-Phosphoglycerate dehydrogenase; PHP, 3-phosphohydroxypyruvate; PK, pyruvate kinase; PR, pathogenesis-related; POD, peroxidase; Prx, Peroxidoxin; ROS, reactive oxygen species; SAMS, S-adenosylmethionine synthetase; SAM, S-adenosylmethionine; SD, succinate dehydrogenase; TPI, triose-phosphate isomerase; Trx, thioredoxin.

Within the root system, the tip portion of the primary roots plays an essential role in seedling establishment. Flooding induced cell death in the root tip region and a subsequent suppression in root elongation have been reported in flooded soybean seedlings [17]. Predominant proteins involved in stress response, glycolysis, redox homeostasis, and protein processing found to be located in differentiated root zones including root apex with different abundances [36]. Gel free MS based quantitative proteomics and phosphoproteomics approaches have been well exploited to enumerate the altered protein relative abundance profiles of soybean root tips under flooding stress [17]. Classification of differentially accumulated proteins revealed that majority of the proteins involved in glycolysis, fermentation, cell wall metabolism and nucleotide metabolism were increased; while, the relative abundance of most of the proteins involved in amino acid metabolism and cell organization were decreased. In addition, few proteins including sucrose-binding protein, phosphatidylinositol-4-phosphate 5-kinases, actins, and alpha-tubulins, were found to be accumulated specifically in the root tip region. Accumulation of sucrose-binding proteins in flooded soybean root tips suggests an enhanced sucrose accumulation. This observation is in agreement with the finding, reported previously in soybean roots and hypocotyls [16]. Furthermore, Yanagawa and Komatsu [43] reported that flooding, and not the hypoxic condition, was responsible for the root tip degradation resulting from ubiquitin/proteasome-mediated proteolysis, as these injuries were independent of the oxygen concentration. It is believed that the Ub/proteasome-mediated proteolysis of enzymes involved in glycolysis and fermentation pathways may be negatively controlled under the hypoxic condition caused by flooding [3]. Previous gel and gel-free MS based proteomic study by Nanjo *et al.* [16] has also revealed differential regulation of 20S proteasome subunits in flooded soybean. Altered expression of each 20S proteasome subunit in response to flooding stress may thus affect the amount as well as the activity of the 26S proteasome, thereby altering flooding tolerance.

Among the differentially expressed ROS scavenger proteins, cytosolic ascorbic peroxidase (cAPX) and superoxide dismutase (SOD) were found to be decreased in response to flooding. Proteomic screening of six different soybean cultivars revealed a significant decrease in cAPX 2 proteins on exposure to flooding [44]. Abundance of cAPX 2 transcripts was also found to be decreased significantly after flooding, as did the APX activity. Results suggest that cytosolic APX 2 plays a key role in flood-induced stress response of young soybean seedlings.

The post-stress recovery period is equally critical phase for the ultimate survival of a stressed plant. Salavati *et al.* [34] examined the proteome change under post-flooding recovery stage in soybean roots. Clustering analysis based on the expression profiles of the differentially abundant protein spots revealed that flooding resulted in a decrease of ion transport-related proteins and an increase of proteins involved in cytoskeletal reorganization, cell expansion, and programmed cell death. The observed changes in protein relative abundance suggest that the regulation of root growth through cell wall modification and the synthesis of S-adenosylmethionine-related metabolites may be involved in post-flooding recovery processes in soybean seedlings.

Flood-induced reduction in plant biomass is directly related to stomatal limitations on net photosynthesis that result in reduced carbon assimilation [45]. Restriction in photosynthetic activity is also influenced by changes in the photosynthetic components, such as ribulose-1,5-bisphosphate carboxylase/oxygenase (RuBisCO) and other photosynthesis-related proteins [46]. Leaf proteome analysis of soybean seedlings revealed that most of decreased proteins were involved in energy production and primary/secondary metabolism [47]. This observation is in agreement with the results of a recent gel-based organ-specific proteomic study by Khatoon *et al.* [9]. As compared to the roots and hypocotyls, more metabolism, energy and disease/defense related proteins were found to be decreased in leaves. The reduced levels of isoflavone reductase and other disease/defense-related proteins (SOD, CAT) in the roots and leaves of flooded seedlings compared to non-stressed seedlings indicate that the defense response is highly suppressed in soybean seedlings under flooding stress. Furthermore, a decreased relative abundance of chlorophyll *a-b* binding proteins were recorded. Overall, reduced photosynthetic activity along with low expression of ROS scavenging proteins lead to suppression of seedling growth under flooding.

As compared to whole organ proteome study, an in-depth investigation of subcellular organelles proteomes generates much detailed information about the intrinsic mechanism of stress response as it correlates the possible relationship between the protein abundance and plant stress tolerance. The intracellular organelles and compartments and their interactions during the stressed condition represent the primary defense response. Among the organelles, mitochondria have been a target for subcellular proteomic study, as most of the abiotic stresses primarily impair mitochondrial electron transport chain resulting in excess ROS generation. Proteomic technique coupled with metabolomics has been successfully used to study the flooding stress effects on mitochondrial function of flooded soybeans [5]. Flooding stress caused a considerable impairment of the electron transport chain in the roots and hypocotyls of soybean seedlings. Abundance of inner membrane carrier proteins and proteins related to complexes III, IV, and V of the electron transport chain were found to be decreased, while proteins and metabolites related to TCA and γ-amino butyrate (GABA) shunt were increased under flooding stress resulting in high NADH production. In addition, succinate-semialdehyde dehydrogenase and GABA were significantly increased by flooding stress, as was 2-oxoglutarate dehydrogenase, suggesting that the GABA shunt is involved in a replenishment of intermediates needed for energy production that have been depleted by flooding stress.

Plant cell wall plays an essential role in stress sensing and signal transduction between the apoplast and symplast. Investigation on the function of the cell wall of flooded soybean seedlings revealed decrease in lipoxygenases, germin-like protein precursors, stem glycoprotein precursors and Cu–Zn SOD [7]. Proteome analysis suggested that flooding caused suppression of lignifications in roots through a decrease of ROS scavenging enzymes and jasmonate biosynthesis. Similarly, alterations in the plasma membrane proteins of soybean exposed to flooding stress were analyzed using gel-based and gel-free proteomics techniques [21]. Plasma membrane acts as a primary interface between the cellular cytoplasm and the extracellular environment and thus plays a vital role in cell communication. Among the stress induced novel proteins, SOD was found to be remarkably increased, suggesting that the antioxidative system may play a crucial role in protecting cells from oxidative damage following exposure to flooding stress. In addition, flood induced an enhanced accumulation of heat shock cognate 70 kDa protein which might protect proteins from denaturation and degradation during flooding stress.

In a recent gel-free proteomic study by Komatsu *et al.* [2], exogenous application of phytohormone abscisic acid (ABA) at early seedling stage has been found to be effective in enhancing flood tolerance in soybean. The abundance of 34 nuclear proteins such as histone deacetylase and U2 small nuclear ribonucleoprotein increased by ABA supplementation under flooding; while, 35 nuclear proteins such as importin alpha, chromatin remodeling factor, zinc finger protein, transducin, and cell division 5 protein were decreased. In addition, mRNA expression levels of cell division cycle 5 protein, C_2H_2 zinc finger protein SERRATE, CCCH type zinc finger family protein, and transducin were found to be down-regulated under the ABA treatment. Authors suggested that ABA might be involved in the enhancement of flooding tolerance through the control of energy conservation via glycolytic system and the regulation by zinc finger proteins, cell division cycle 5 protein and transducin. Similar nuclear proteomic analysis by Oh *et al.* [48] reported acceleration of protein poly-ADP-ribosylation and suppression of RNA metabolism in root tips of young soybean seedlings under flooding stress. A separate proteomic study on endoplasmic reticulum (ER)-enriched fraction of flooded soybean root tips revealed decreased abundances of proteins involved in stress, hormone metabolism, cell wall and DNA repairing [4]. Additionally, expression of luminal-binding protein 5 was specifically induced under flood stress, while arabinogalactan protein 2 and methyltransferase PMT2 were found to be down-regulated. Overall, results indicate that flooding predominantly affects protein synthesis and glycosylation in the ER of soybean root tips.

Taken together, these results suggest that the tight metabolic regulation over the energy consumption and quick activation of plant defense system are essential to conquer the flooding stress.

4. Drought Induced Modulation of Soybean Proteome Composition

Drought constitutes another form of water stress that results from scarcity of water around the root zone. Like other legumes, soybean is also sensitive to drought condition. Decline in photosynthetic carbon gain as a result of stomatal closure or due to a decrease in RuBisCO activity is one of the major reasons behind the loss of crop productivity during drought phase [49]. The activity of the photosynthetic electron transport chain is finely tuned to the availability of CO_2, and photosystem II activities often decline in parallel under drought conditions [50]. In soybean, photosynthesis decreases by about 70% during severe water stress, although the respiration rate is not that much affected [51]. Drought tolerance has always been considered as one of the top priorities for soybean improvement [52]. The genetic complexity of drought tolerance, the lack of efficient selection technique, environmental variability, and the strong interactions between genotype and water availability are some of the key limiting factors for designing drought tolerant soybean cultivars [13].

Different aspects of plant response toward dehydration stress have been well documented. However, information on drought sensing and tolerance mechanism at the proteome level is very limited. In this section, published proteomic works on dehydration stress mediated changes in soybean proteomes are summarized for better understanding of the drought stress responsive mechanism (Figure 1). The functional categorization revealed that most of the drought-responsive proteins were chiefly involved in redox regulation, oxidative stress response, signal transduction, protein folding, secondary metabolism, and photosynthesis.

Root is found to be the most drought-responsive organ showing maximum changes in protein abundance in response to stress. Polyethylene glycol (PEG), a high molecular weight osmotic substance, is frequently used to simulate drought stress in soybean lowering the water potential in a similar way to soil drying [22,23]. Changes in relative abundance of metabolism-related proteins were shown to be increased in leaves of both PEG-treated and drought-stressed seedlings, while proteins related to energy production- and protein synthesis were decreased [12]. In a separate study, abundance of proteins associated with a wide variety of cellular functions, including carbohydrate and nitrogen metabolism, cell wall modification, signal transduction, cell defense and programmed cell death were found to be highly affected in soybean roots subjected to severe but recoverable drought stress at seedling stage [13].

Toorchi *et al.* [23] studied the PEG-induced osmotic stress related proteins in soybean roots using a 2-DE gel based proteomic approach. Osmotic stress is just a simulation of drought condition, where high concentration of osmolyte e.g. PEG stimulates the roots to look for unexplored water. This results in continuing root growth and a delay in root lignification. Protein identification revealed a decrease in caffeoyl-CoA 3-*O*-methyltransferase, Hsp-70, S-adenosylmethionine synthetase with high abundance of disease/defense associated proteins [23]. In plant cell wall, lignin is the major structural component of secondary thickening that imparts mechanical strength to stems and roots, and hydrophobicity to water-conducting vascular elements. Caffeoyl-CoA-O-methyltransferase is involved in the lignification process and its decreased abundance in soybean roots under osmotic stress thus might result in the reduction of lignin content as an adaptive response to osmotic stress. In separate proteomic investigation, comparative analysis of plasma membrane proteins of two-day-old soybeans under PEG-mediated osmotic stress revealed an increase in transporter proteins, indicating a high rate of ion efflux by the plasma membrane bound H^+-ATPase [22].

In addition, calnexin protein was found to be highly increased under stress. Nevertheless, decreased expression of calnexin was reported in 14-day-old soybean roots under 10% PEG treatment [53]. Calnexin is an ER-localized molecular chaperone protein, involved in folding and quality control of proteins. This protein interacts with many nascent membrane and soluble proteins of the secretory pathway and participates in the folding and quality control of newly synthesized glycoproteins [54]. Authors suggested that calnexin interacts with a 70 kDa heat shock cognate protein and probably functions as a molecular chaperone under PEG-induced osmotic stress.

Overall, drought or PEG-mediated osmotic stress at seedling stage affects a wide range of cellular functions, including carbohydrate and nitrogen metabolism, cell wall modification, signal transduction, cell defense and programmed cell death in soybean. Proteomic findings of drought stressed soybean indicate that proteins associated with osmotic adjustment, defense signaling and programmed cell death play key roles in drought adaptation.

5. Novel Methodological Approaches to Study Plant Proteomes

The ultimate success of any proteomic approach depends upon various factors including isolation of full component of proteins, separation, visualization and their accurate identification. In spite of recent advancement, more emphasis needs to be given on the protein extraction protocols, in particular for very low and high abundance proteins. In soybean root, identification of low abundance of signaling proteins, transcription factors and their protein complexes is often a challenge for 2-DE based proteomic techniques. Nanjo *et al.* [17] adapted a gel free analysis of complete root-tip proteome, in which protein samples were reduced and alkylated in a denaturing solution followed by trypsin digestion. Trypsin-digested samples were then injected on nanoLC coupled to MS/MS. This method allows detection of MS peaks with up to 5000 times differences in abundance. In order to determine the composition of plant protein complexes, Smaczniak *et al.* [55] used another, rather more sensitive fluorophore-tagged protein immunoprecipitation and label-free MS-based quantification techniques to facilitate identification of low abundance signaling and regulatory protein complexes from native plant tissues. Furthermore, an advanced technique like laser-capture micro-dissection [56] for tissue proteomics could be used further for accurate identification of tissue- and cell-specific proteins involved in plant response to abiotic stresses. Gil-Quintana *et al.* [57] reported proteomic analysis of root nodules of drought stressed soybean using shotgun proteomics technique. In order to have complete proteome of root nodules, including all low abundance proteins, protein digests were analysed via shotgun nano-LC-ultra using a monolithic reversed-phase column directly coupled to an Orbitrap XL mass spectrometer [58].

Similarly, in leaf, presence of extremely abundant photosynthetic CO_2 fixation enzyme RuBisCO not only limits the dynamic resolution and yield of low abundance proteins of interest but also masks other proteins or affects the electrophoretic migration of neighboring protein species [59]. Different fractionation techniques based upon different physiological or biochemical principles have been proposed to deplete or reduce a substantial portion of RuBisCO from total leaf protein extract [60,61]. Ahsan *et al.* [62] used a PEG-fractionation method to eliminate RuBisCO during protein extraction from tomato leaves. In this method, proteins were first extracted using Mg/Nonidet P-40 buffer consisting of 0.5 M Tris-HCl, 2% Nonidet P-40, 20 mM $MgCl_2$, 2% 2-mercaptoethanol, 1 mM PMSF, and 1% PVP, and were then fractionated with 15% PEG. Furthermore, anti-RuBisCO LSU antibody affinity column with protein A-Sepharose as a resin has been successfully used for effective elimination of RuBisCO [63]. In comparison of these complex and lengthy methods, Krishnan and Natarajan [64] developed a fast and simple fractionation technique using 10 mM Ca^{2+} and 10 mM phytate to precipitate 85% of the RuBisCO from soybean leaf soluble protein extract. Recently, Khan *et al.* [65] also reported a modified protein extraction method for effective removal of RuBisCO. In this method, leaves were homogenized in buffer mixture containing 50% extraction buffer (100 mM Tris-HCl (pH 8.0), 100 mM EDTA, 50 mM borax, 50 mM vitamin C, 1% PVP-40, 1% triton X-100, 2% 2-mercaptoethanol and 30% sucrose) and 50% solubilization buffer (8.5 M urea, 2.5 M thiourea, 5% CHAPS, 1% DTT), 1% triton X-100 and 0.5% ampholin (pH 3–10 and 5–8)) followed by incubation on ice for 1 h and precipitation by adding 20% TCA. The 2-DE pattern revealed displaced RuBisCO LSU to a new position with low molecular weight and pI value.

As compared to whole organ, in-depth sub-cellular organelle proteome study generates much detailed information on the intrinsic mechanism of plants' abiotic stress responses. One of the most challenging aspects of subcellular proteomics is the proper isolation of the concerned organelle from the total tissue extract. The conventional methods of subcellular fractionation typically involve differential and density-gradient centrifugation, using a series of centrifugation steps to separate different populations of cellular compartments or organelles from cell homogenates based on their mass and/or density. Nevertheless, the resolving power of differential centrifugation is comparatively poor and may result in fractions containing different organelles having similar sedimentation velocities [66]. In contrast, density-gradient centrifugation has been extensively used in organellar proteomics studies. This method separates organelles based on continuous or discontinuous gradients using various media, such as sucrose, Ficoll, Percoll, Nycodenz and Metrizamide of different osmolarities, viscosities or densities.

Free-flow electrophoresis (FFE) is another alternative strategy for fractionation of organelles based on their net global isoelectric charges or electrophoretic mobilities. Immunoaffinity purification is a more advanced technique to isolate organelles with specificity and in adequate yields [67]. Both affinity purification and immunoprecipitation methods are based on principle of binding immobilized ligands (such as antibodies) with that of targets (organelle of interest). Fluorescent-assisted organelle sorting (FAOS) is the most emerging sophisticated organelle isolation technique that works on the principle of flow cytometry. This organelle specific marker protein based approach has been found to be effective in mitochondria [68] and vesicles [69] sorting for proteomic analysis. Similarly, subtractive proteomics approach is capable of precise assigning and identifying proteins to their specific subcellular locations. This method effectively eliminates target organelles contamination from co-purifying organelles. It compares and subtracts the identified protein constituents of the contaminated fraction containing the organelle of interest against that of a crude preparation [70].

Protein phosphorylation is the best-studied posttranslational modification that plays a pivotal role in signal transduction cascade. Identification of kinases, their substrates, and the specific site of phosphorylation is thus a key to molecular understanding of stress signaling. The MS-based phosphoproteomic technology has become an invaluable tool for the identification of phosphoproteins and mapping of phosphorylation sites. Nevertheless, identification of *in vivo* phosphorylation sites of individual proteins of interest, necessity for their functional characterization is a big challenge for any phosphoproteomic study. Phosphorylated proteins represent only a small fraction of the whole proteome, thus demanding an effective enrichment method prior to quantification and identification [71]. Immobilized metal affinity chromatography (IMAC) and immunoprecipitation using antibodies against phosphorylated amino acids are the two well known pre- fractionation techniques largely employed before MS analysis. Much progress has been made in quantitative and dynamic analysis of mapped phosphorylation sites in recent time. This method comprised of the isolation of phosphopeptides by IMAC followed by MS/MS or MS(n) analysis has enabled detection of hundreds of *in vivo* phosphorylation sites [72]. Phosphopeptides have been successfully isolated from complex mixtures with strong cationic exchange (SCX) chromatography [73] or strong anionic exchange (SAX) chromatography followed by IMAC [74]. Phosphoproteomics analysis of the Arabidopsis plasma membrane led to the identification and characterization of more than 300 phosphorylation sites [75]. The majority of phosphorylation sites of the membrane transporters have been found to be conserved among putative orthologs and to a lesser extent among some members of the same protein family. On the other hand, affinity purification of phosphoproteins with phospho-specific antibodies such as anti-phosphoserine/threonine prior to MS has limited applications in plants. Recent development in the specific labelling techniques greatly helps in the quantification of phosphorylation profiles and their stress-induced changes with the passage of treatment time. The iTRAQ and SILAC labelling have been found to be most successful in combination with IMAC and MS [72]. These techniques label peptides at the final stage before MS *in vitro* or label proteins during cell growth *in vivo*, respectively, and enable the measurement of changes of individual phosphorylation sites during a time-course stress experiment. Hsu *et al.* [76] compared both label-free LC-MS and stable isotope labelling LC-MS methods for quantitative analysis of phosphorylation sites in membrane fractions of salt stressed Arabidopsis. The functional phosphoproteomic analysis led to a successful identification of novel salt stress-responsive protein phosphorylation sites from membrane isolates of salt-stressed plants by membrane shaving followed by Zirconium ion-charged magnetic beads, and tandem MS analyses.

Moreover, introduction of Pro-Q Diamond dye based fluorescence-linked assay has opened new avenues in a large-scale quantitative analysis of phosphoproteins. Pro-Q Diamond has been successfully used to specifically label and detect phosphoserine-, phosphothreonine-, and phosphotyrosine-containing proteins directly in SDS-polyacrylamide gels and 2-DE gels. Nanjo *et al.* [17] successfully exploited Pro-Q Diamond phosphoprotein dye technology in determining flooding induced changes in phosphorylation status of proteins involved in energy generation, protein synthesis and cell structure maintenance in root tips of soybean seedlings.

Compared to phosphoproteomics, plant redox proteomics study of oxidatively modified proteins is more challenging, due to technical limitations such as maintaining the *in vivo* redox states of proteins and the lability of certain PTMs during sample preparation and mass spectrometric analysis [77]. To balance redox metabolism, cells possess a redox signaling network that can sense environmentally induced redox imbalances and initiates compensatory responses either to readjust redox homeostasis or to repair oxidative damage [78]. Within plant cell, chloroplasts, mitochondria and peroxisomes are the primary sites of ROS/ RNS (reactive nitrogen species) generation and the NADPH oxidase located at the plasmalemma, and the cell wall/apoplast peroxidases, amine oxidases, and oxalate oxidases are important components of the ROS-generating system. The highly dynamic and robust ROS gene network that encodes both ROS-producing and ROS-scavenging proteins plays an essential role in monitoring and controlling cellular ROS levels in addition to ROS mediated signalling [79]. Oxidative or nitrosative stress leads to redox modifications of proteins, and may be reversible such as oxidation of cysteines to disulphides or sulphenic acids or irreversible modifications, e.g., carbonylation, oxidation of cysteines to sulphonic acids, oxidation of tryptophan [80].

Among the available high-throughput techniques, redox proteomics has been found to be the best-suited approach for identifying and quantifying redox-based changes within the plant proteome under oxidative stress conditions. A typical redox proteome labeling method uses either direct labelling of free reduced thiols or blocking labeling of disulfides/reversibly oxidized thiol groups with an alkylating agent such as *N*-ethylmaleimide (NEM) to block free cysteines, followed by DTT mediated reduction to reduce oxidized cysteines to Cys-SH and a subsequent labelling with fluorescent dye such as 5-iodoacetamidofluorescein (IAF) or monobromobimane (mBBr) [79]. Proteins are then separated by 2-DE and identified by LC-MS/MS technique. One advantage of using a dye like mBBr or IAF is a direct visualization of separated redox active proteins on the UV transilluminator. Moreover, shotgun proteomics approach has been exploited for identification of thiol-containing proteins selected as sub-proteomes trapped on activated thiol sepharose (ATS) beads [81]. Gel-free method exploiting derivatization of carbonylated proteins with 2,4-dinitrophenylhydrazine (DNPH) followed by tryptic digestion and enrichment by reversed phase chromatography coupled with MS/MS (RPC-MS/MS) or ion exchange and reversed phase chromatography coupled with MS/MS (IEC/RPC-MS/MS) has been successfully used for identification of carbonylated proteins and their oxidation sites [82].

Another major challenge for quantitative soybean proteomics is separation and identification of protein isoforms/species. During the course of evolution, soybean genome has undergone two rounds of whole genome duplication and many tandem duplication events [83]. Due to higher gene duplication and recombination process, so many protein isoforms exist in soybean as compared to rice and Arabidopsis. The 2-DE based proteomic techniques have a wide application in identifying these isoforms. Protein species occupy different positions on the 2-DE gel matrix based on their individual isoelectric point (p*I*) and relative molecular weight (MW), but share the same identification. Over the gel-based proteomic approach, bottom-up LC-MS/MS technique offers more advantages in identifying protein species. This method comprises of unambiguous identification of a single protein species relies on the identification of at least one peptide sequence that is uniquely found in that protein species [84]. Proper selection of the database would further facilitate the identification of such protein species with accuracy.

6. Conclusions

Instead of several limitations and challenges, soybean proteomics has proved itself as a valuable tool for identifying stress responsive target proteins with a clear picture of translational and post translational modification. More research works at the proteome level need to be undertaken for better understanding the minute changes in a cell's protein signature to cope with the flooding and drought stress. Comparative organelle proteomes studies would be a great contribution towards understanding the cross-talk between stress signaling pathways. The convergence of diverse MS techniques coupled with bioinformatics technology with improved sample preparation and fractionation strategies is further needed to get a more precise and comprehensive picture of plant stress response mechanisms.

Acknowledgements

This work was supported by the Grants from National Agriculture and Food Research Organization, Japan.

Author Contributions

Z.H. and S.K. had equal contributions in writing this review.

References

1. Komatsu, S.; Nanjo, Y.; Nishimura, M. Proteomic analysis of the flooding tolerance mechanism in mutant soybean. *J. Proteomics* **2013**, *79*, 231–250.
2. Komatsu, S.; Han, C.; Nanjo, Y.; Altaf-Un-Nahar, M.; Wang, K.; He, D.; Yang, P. Label-free quantitative proteomic analysis of abscisic acid effect in early-stage soybean under flooding. *J. Proteome Res.* **2013**, *12*, 4769–4784.
3. Komatsu, S.; Hiraga, S.; Yanagawa, Y. Proteomics techniques for the development of flood tolerant crops. *J. Proteome Res.* **2012**, *11*, 68–78.
4. Komatsu, S.; Kuji, R.; Nanjo, Y.; Hiraga, S.; Furukawa, K. Comprehensive analysis of endoplasmic reticulum-enriched fraction in root tips of soybean under flooding stress using proteomics techniques. *J. Proteomics* **2012**, *77*, 531–560.

5. Komatsu, S.; Yamamoto, A.; Nakamura, T.; Nouri, M.Z.; Nanjo, Y.; Nishizawa, K.; Furukawa, K. Comprehensive analysis of mitochondria in roots and hypocotyls of soybean under flooding stress using proteomics and metabolomics techniques. *J. Proteome Res.* **2011**, *10*, 3993–4004.
6. Komatsu, S.; Deschamps, T.; Hiraga, S.; Kato, M.; Chiba, M.; Hashiguchi, A.; Tougou, M.; Shimamura, S.; Yasue, H. Characterization of a novel flooding stress-responsive alcohol dehydrogenase expressed in soybean roots. *Plant Mol. Biol.* **2011**, *77*, 309–322.
7. Komatsu, S.; Kobayashi, Y.; Nishizawa, K.; Nanjo, Y.; Furukawa, K. Comparative proteomics analysis of differentially expressed proteins in soybean cell wall during flooding stress. *Amino Acids* **2010**, *39*, 1435–1449.
8. Komatsu, S.; Sugimoto, T.; Hoshino, T.; Nanjo, Y.; Furukawa, K. Identification of flooding stress responsible cascades in root and hypocotyl of soybean using proteome analysis. *Amino Acids* **2010**, *38*, 729–738.
9. Khatoon, A.; Rehman, S.; Hiraga, S.; Makino, T.; Komatsu, S. Organ-specific proteomics analysis for response mechanism in soybean seedlings under flooding stress. *J. Proteomics* **2012**, *75*, 5706–5723.
10. Khatoon, A.; Rehman, S.; Oh, M.W.; Woo, S.H.; Komatsu, S. Analysis of response mechanism in soybean under low oxygen and flooding stresses using gel-base proteomics technique. *Mol. Biol. Rep.* **2012**, *39*, 10581–10594.
11. Nanjo, Y.; Nakamura, T.; Komatsu, S. Identification of indicator proteins associated with flooding injury in soybean seedlings using label-free quantitative proteomics. *J. Proteome Res.* **2013**, *12*, 4785–4798.
12. Mohammadi, P.P.; Moieni, A.; Hiraga, S.; Komatsu, S. Organ-specific proteomic analysis of drought-stressed soybean seedlings. *J. Proteomics* **2012**, *75*, 1906–1923.
13. Alam, I.; Sharmin, S.A.; Kim, K.H.; Yang, J.K.; Choi, M.S.; Lee, B.H. Proteome analysis of soybean roots subjected to short-term drought stress. *Plant Soil* **2010**, *333*, 491–505.
14. Hossain, Z.; Hajika, M.; Komatsu, S. Comparative proteome analysis of high and low cadmium accumulating soybeans under cadmium stress. *Amino Acids* **2012**, *43*, 2393–2416.
15. Swigonska, S.; Weidner, S. Proteomic analysis of response to long-term continuous stress in roots of germinating soybean seeds. *J. Plant Physiol.* **2013**, *170*, 470–479.
16. Nanjo, Y.; Skultety, L.; Ashraf, Y.; Komatsu, S. Comparative proteomic analysis of early-stage soybean seedlings responses to flooding by using gel and gel-free techniques. *J. Proteome Res.* **2010**, *9*, 3989–4002.
17. Nanjo, Y.; Skultety, L.; Uváčková, L.; Klubicová, K.; Hajduch, M.; Komatsu, S. Mass spectrometry-based analysis of proteomic changes in the root tips of flooded soybean seedlings. *J. Proteome Res.* **2012**, *11*, 372–385.
18. Komatsu, S.; Ahsan, N. Soybean proteomics and its application to functional analysis. *J. Proteomics* **2009**, *72*, 325–336.
19. Alam, I.; Lee, D.G.; Kim, K.H.; Park, C.H.; Sharmin, S.A.; Lee, H.; Oh, K.W.; Yun, B.W.; Lee, B.H. Proteome analysis of soybean roots under waterlogging stress at an early vegetative stage. *J. Biosci.* **2010**, *35*, 49–62.
20. Komatsu, S.; Yamamoto, R.; Nanjo, Y.; Mikami, Y.; Yunokawa, H.; Sakata, K. A comprehensive analysis of the soybean genes and proteins expressed under flooding stress using transcriptome and proteome techniques. *J. Proteome Res.* **2009**, *8*, 4766–4778.

21. Komatsu, S.; Wada, T.; Abaléa, Y.; Nouri, M. Z.; Nanjo, Y.; Nakayama, N.; Shimamura, S.; Yamamoto, R.; Nakamura, T.; Furukawa, K. Analysis of plasma membrane proteome in soybean and application to flooding stress response. *J. Proteome Res.* **2009**, *8*, 4487–4499.
22. Nouri, M.Z.; Komatsu, S. Comparative analysis of soybean plasma membrane proteins under osmotic stress using gel-based and LC MS/MS-based proteomics approaches. *Proteomics* **2010**, *10*, 1930–1945.
23. Toorchi, M.; Yukawa, K.; Nouri, M.Z.; Komatsu, S. Proteomics approach for identifying osmotic-stress-related proteins in soybean roots. *Peptides* **2009**, *30*, 2108–2117.
24. Buttery, B.R.; Buzzell, R.I. Soybean flavonol glycosides: Identification and biochemical genetics. *Can. J. Bot.* **1973**, *53*, 309–313.
25. Saravanan, R.S.; Rose, J.K. A critical evaluation of sample extraction techniques for enhanced proteomic analysis of recalcitrant plant tissues. *Proteomics* **2004**, *4*, 2522–2532.
26. Krishnan, H.B.; Oehrle, N.W.; Natarajan, S.S. A rapid and simple procedure for the depletion of abundant storage proteins from legume seeds to advance proteome analysis: A case study using *Glycine max*. *Proteomics* **2009**, *9*, 3174–3188.
27. Mooney, B.P.; Thelen, J.J. High-throughput peptide mass fingerprinting of soybean seed proteins: Automated workflow and utility of UniGene expressed sequence tag databases for protein identification. *Phytochemistry* **2004**, *65*, 1733–1744.
28. Hajduch, M.; Ganapathy, A.; Stein, J.W.; Thelen, J.J. A systematic proteomic study of seed filling in soybean. Establishment of high-resolution two-dimensional reference maps, expression profiles, and an interactive proteome database. *Plant Physiol.* **2005**, *137*, 1397–1419.
29. Natarajan, S.; Xu, C.; Caperna, T.J.; Garrett, W.M. Comparison of protein solubilization methods suitable for proteomic analysis of soybean seed proteins. *Anal. Biochem.* **2005**, *342*, 214–220.
30. Barbosa, H.S.; Arruda, S.C.; Azevedo, R.A.; Arruda, M.A. New insights on proteomics of transgenic soybean seeds: Evaluation of differential expressions of enzymes and proteins. *Anal. Bioanal. Chem.* **2012**, *402*, 299–314.
31. Ahsan, N.; Komatsu, S. Comparative analyses of the proteomes of leaves and flowers at various stages of development reveal organ-specific functional differentiation of proteins in soybean. *Proteomics* **2009**, *9*, 4889–4907.
32. Toorchi, M.; Nouri, M.Z.; Tsumura, M.; Komatsu, S. Acoustic technology for high-performance disruption and extraction of plant proteins. *J. Proteome Res.* **2008**, *7*, 3035–3041.
33. Hurkman, W.J.; Tanaka, C.K. Solubilization of plant membrane proteins for analysis by two-dimensional gel electrophoresis. *Plant Physiol.* **1986**, *81*, 802–806.
34. Salavati, A.; Khatoon, A.; Nanjo, Y.; Komatsu, S. Analysis of proteomic changes in roots of soybean seedlings during recovery after flooding. *J. Proteomics* **2012**, *75*, 878–893.
35. Zhen, Y.; Qi, J.L.; Wang, S.S.; Su, J.; Xu, G.H.; Zhang, M.S.; Miao, L.; Peng, X.X.; Tian, D.; Yang, Y.H. Comparative proteome analysis of differentially expressed proteins induced by Al toxicity in soybean. *Physiol. Plant* **2007**, *131*, 542–554.
36. Mathesius, U.; Djordjevic, M.A.; Oakes, M.; Goffard, N.; Haerizadeh, F.; Weiller, G.F.; Singh, M.B.; Bhalla, P.L. Comparative proteomic profiles of the soybean (*Glycine max*) root apex and differentiated root zone. *Proteomics* **2011**, *11*, 1707–1719.

37. Rose, J.C.; Bashir, S.; Giovannoni, J.J.; Jahn, M.M.; Saravanan, R.S. Tackling the plant proteome: Practical approaches, hurdles and experimental tools. *Plant J.* **2004**, *39*, 715–733.
38. Espagne, C.; Martinez, A.; Valot, B.; Meinnel, T.; Giglione, C. Alternative and effective proteomic analysis in *Arabidopsis*. *Proteomics* **2007**, *7*, 3788–3799.
39. Hossain, Z.; López-Climent, M.F.; Arbona, V.; Pérez-Clemente, R.M.; Gómez-Cadenas, A. Modulation of the antioxidant system in *Citrus* under waterlogging and subsequent drainage. *J. Plant Physiol.* **2009**, *166*, 1391–404.
40. Hole, D.J.; Greg Cobb, B; Hole, P.S.; Drew, M.C. Enhancement of anaerobic respiration in root tips of *Zea mays* following low oxygen (hypoxic) acclimation. *Plant Physiol.* **1992**, *99*, 213–218.
41. Bailey-Serres, J.; Voesenek, L.A.C.J. Flooding stress: Acclimations and genetic diversity. *Annu. Rev. Plant Biol.* **2008**, *59*, 313–339.
42. Hashiguchi, A.; Sakata, K.; Komatsu, S. Proteome analysis of early-stage soybean seedlings under flooding stress. *J. Proteome Res.* **2009**, *8*, 2058–2069.
43. Yanagawa, Y.; Komatsu, S. Ubiquitin/proteasome-mediated proteolysis is involved in the response to flooding stress in soybean roots, independent of oxygen limitation. *Plant Sci.* **2012**, *185–186*, 250–258
44. Shi, F.; Yamamoto, R.; Shimamura, S.; Hiraga, S.; Nakayama, N.; Nakamura, T.; Yukawa, K.; Hachinohe M.; Matsumoto, H.; Komatsu, S. Cytosolic ascorbate peroxidase 2 (cAPX 2) is involved in the soybean response to flooding. *Phytochemistry* **2008**, *69*, 1295–1303.
45. Mielke, M.S.; de Almeida, A.A.F.; Gomes, F.P.; Aguilarb, M.A.G.; Mangabeiraa, P.A.O. Leaf gas exchange, chlorophyll fluorescence and growth responses of *Genipa americana* seedlings to soil flooding. *Envinron. Exp. Bot.* **2003**, *50*, 221–231.
46. Maayan, I.; Shaya, F.; Ratner, K.; Mani, Y.; Lavee, S.; Avidan, B.; Shahak, Y.; Ostersetzer-Biran, O. Photosynthetic activity during olive (*Olea europaea*) leaf development correlates with plastid biogenesis and RuBisCO levels. *Physiol. Plant.* **2008**, *134*, 547–558.
47. Donnelly, B.E.; Madden, R.D.; Ayoubi, P.; Porter, D.R.; Dillwith, J.W. The wheat (*Triticum aestivum* L.) leaf proteome. *Proteomics* **2005**, *5*, 1624–1633.
48. Oh, M.; Nanjo, Y.; Komatsu, S. Identification of nuclear proteins in soybean under flooding stress using proteomic technique. *Protein Pept. Lett.* **2014**, *21*, 458–467.
49. Bota, J.; Flexas, J.; Medrano, H. Is photosynthesis limited by decreased Rubisco activity and RuBP content under progressive water stress? *New Phytol.* **2004**, *162*, 671–681.
50. Loreto, F.; Tricoli, D.; di Marco, G. On the relationship between electron transport rate and photosynthesis in leaves of the C4 plant *Sorghum bicolor* exposed to water stress, temperature changes and carbon metabolism inhibition. *Aust. J. Plant Physiol.* **1995**, *22*, 885–892.
51. Ribas-Carbo, M.; Taylor, N.L.; Giles, L.V.; Busquets, S.; Finnegan, P.M.; Day, D.A. Effects of water stress on respiration in soybean leaves. *Plant Physiol.* **2005**, *139*, 466–473.
52. Stacey, G.; Libault, M.; Brechenmacher, L.; Wan, J.; May, G.D. Genetics and functional genomics of legume nodulation. *Curr. Opin. Plant Biol.* **2006**, *9*, 110–121.
53. Nouri, M.Z.; Hiraga, S.; Yanagawa, Y.; Sunohara, Y.; Matsumoto, H.; Komatsu, S. Characterization of calnexin in soybean roots and hypocotyls under osmotic stress. *Phytochemistry* **2012**, *74*, 20–29.
54. Brockmeier, A.; Williams, D.B. Potent lectin-independent chaperone function of calnexin under conditions prevalent within the lumen of the endoplasmic reticulum. *Biochemistry* **2006**, *45* 12906–12916.

55. Smaczniak, C.; Li, N.; Boeren, S.; America, T.; van Dongen, W.; Goerdayal, S.S.; de Vries, S.; Angenent, G.C.; Kaufmann, K. Proteomics-based identification of low-abundance signaling and regulatory protein complexes in native plant tissues. *Nat. Protoc.* **2012**, *7*, 2144–2158.
56. Dembinsky, D.; Woll, K.; Saleem, M.; Liu, Y.; Fu, Y.; Borsuk, L.A.; Lamkemeyer, T.; Fladerer, C.; Madlung, J.; Barbazuk, B.; *et al.* Transcriptomic and proteomic analyses of pericycle cells of the maize primary root. *Plant Physiol.* **2007**, *145*, 575–588.
57. Gil-Quintana, E.; Larrainzar, E.; Seminario, A.; Díaz-Leal, J.L.; Alamillo, J.M.; Pineda, M.; Arrese-Igor, C.; Wienkoop, S.; González, E.M. Local inhibition of nitrogen fixation and nodule metabolism in drought-stressed soybean. *J. Exp. Bot.* **2013**, *64*, 2171–2182.
58. Larrainzar, E.; Wienkoop, S.; Weckwerth, W.; Ladrera, R.; Arrese-Igor, C.; González, E.M. Medicago truncatula root nodule proteome analysis reveals differential plant and bacteroid responses to drought stress. *Plant Physiol.* **2007**, *144*, 1495–1507.
59. Herman, E.M.; Helm, R.M.; Jung, R.; Kinney, A.J. Genetic modification removes an immunodominant allergen from soybean. *Plant Physiol.* **2003**, *132*, 36–43.
60. Cho, J.H.; Hwang, H.; Cho, M.H.; Kwon, Y.K.; Jeon, J.S.; Bhoo, S.H.; Hahn, T.R. The effect of DTT in protein preparations for proteomic analysis: Removal of a highly abundant plant enzyme, ribulose bisphosphate carboxylase/oxygenase. *J. Plant Biol.* **2008**, *51*, 297–301.
61. Widjaja, I.; Naumann, K.; Roth, U.; Wolf, N.; Mackey, D.; Dangl, J.L.; Scheel, D.; Lee, J. Combining subproteome enrichment and Rubisco depletion enables identification of low abundance proteins differentially regulated during plant defense. *Proteomics* **2009**, *9*, 138–147.
62. Ahsan, N.; Lee, D.G.; Lee, S.H.; Kang, K.Y.; Bahk, J.D.; Choi, M.S.; Lee, I.J.; Renaut, J.; Lee, B.H. A comparative proteomic analysis of tomato leaves in response to waterlogging stress. *Physiol. Plant* **2007**, *131*, 555–570.
63. Hashimoto, M.; Komatsu, S. Proteomic analysis of rice seedlings during cold stress. *Proteomics* **2007**, *7*, 1293–1302.
64. Krishnan, H.B.; Natarajan, S.S. A rapid method for depletion of Rubisco from soybean (*Glycine max*) leaf for proteomic analysis of lower abundance proteins. *Phytochemistry* **2009**, *70*, 1958–1964.
65. Khan, N.A.; Komatsu, S.; Sawada, H.; Nouri, M.Z.; Kohno, Y. Analysis of proteins associated with ozone stress response in soybean cultivars. *Protein Pept. Lett.* **2013**, *20*, 1144–1152.
66. Lee, Y.H.; Tan, H.T.; Chung, M.C. Subcellular fractionation methods and strategies for proteomics. *Proteomics* **2010**, *10*, 3935–3956.
67. Ackermann, B.L.; Berna, M.J. Coupling immunoaffinity techniques with MS for quantitative analysis of low abundance protein biomarkers. *Exp. Rev. Proteomics* **2007**, *4*, 175–186.
68. Cossarizza, A.; Ceccarelli, D.; Masini, A. Functional heterogeneity of an isolated mitochondrial population revealed by cytofluorometric analysis at the single organelle level. *Exp. Cell Res.* **1996**, *222*, 84–94.
69. Cao, Z.; Li, C.; Higginbotham, J.N.; Franklin, J.L.; Tabb, D.L.; Graves-Deal, R.; Hill, S.; Cheek, K.; Jerome, W.G.; Lapierre, L.A.; *et al.* Use of fluorescence-activated vesicle sorting for isolation of Naked2-associated, basolaterally targeted exocytic vesicles for proteomics analysis. *Mol. Cell. Proteomics* **2008**, *7*, 1651–1667.

70. Zhou, Z.; Licklider, L.J.; Gygi, S.P.; Reed, R. Comprehensive proteomic analysis of the human spliceosome. *Nature* **2002**, *419*, 182–185.
71. Matros, A.; Kaspar, S.; Witzel, K.; Mock, H.P. Recent progress in liquid chromatography-based separation and label-free quantitative plantproteomics. *Phytochemistry* **2011**, *72*, 963–974.
72. Bentem, S.; Roitinger, E.; Anrather, D.; Csaszar, E.; Hirt, H. Phosphoproteomics as a tool to unravel plant regulatory mechanisms. *Physiol. Plant.* **2006**, *126*, 110–119.
73. Ballif, B.A.; Villen, J.; Beausoleil, S.A.; Schwartz, D.; Gygi, S.P. Phosphoproteomic analysis of the developing mouse brain. *Mol. Cell Proteomics* **2004**, *3*, 1093–1101.
74. Nuhse, T.S.; Stensballe, A.; Jensen, O.N.; Peck, S.C. Large scale analysis of *in vivo* phosphorylated membrane proteins by immobilized metal ion affinity chromatography and mass spectrometry. *Mol. Cell Proteomics* **2003**, *2*, 1234–1243.
75. Nuhse, T.S.; Stensballe, A.; Jensen, O.N.; Peck, S.C. Phosphoproteomics of the *Arabidopsis* plasma membrane and a new phosphorylation site database. *Plant Cell* **2004**, *16*, 2394–2405.
76. Hsu, J.L.; Wang, L.Y.; Wang, S.Y.; Lin, C.H.; Ho, K.C.; Shi, F.K.; Chang, I.F. Functional phosphoproteomic profiling of phosphorylation sites in membrane fractions of salt-stressed *Arabidopsis thaliana*. *Proteome Sci.* **2009**, *7*, 1–16.
77. Navrot, N.; Finnie, C.; Svensson, B.; Hägglund, P. Plant redox proteomics. *J. Proteomics* **2011**, *74*, 1450–1462.
78. Dietz, K.J.; Pfannschmidt, T. Novel regulators in photosynthetic redox control of plant metabolism and gene expression. *Plant Physiol.* **2011**, *155*, 1477–1485.
79. Bykova, N.V.; Rampitsch, C. Modulating protein function through reversible oxidation: Redox-mediated processes in plants revealed through proteomics. *Proteomics* **2013**, *13*, 579–596.
80. Braconi, D.; Bernardini, G.; Santucci, A. Linking protein oxidation to environmental pollutants: redox proteomic approaches. *J. Proteomics* **2011**, *19*, 74, 2324–2337.
81. Hu, W.; Tedesco, S.; McDonagh, B.; Sheehan, D. Shotgun redox proteomics in sub-proteomes trapped on functionalized beads: Identification of proteins targeted by oxidative stress. *Mar. Environ. Res.* **2010**, *69*, S25–S27.
82. Madian, A.G.; Regnier, F.E. Proteomic identification of carbonylated proteins and their oxidation sites. *J. Proteome Res.* **2010**, *9*, 3766–3780.
83. Schmutz, J.; Cannon, S.B.; Schlueter, J.; Ma, J.; Mitros, T.; Nelson, W.; Hyten, D.L.; Song, Q.; Thelen, J.J.; Cheng, J.; *et al.* Genome sequence of the palaeopolyploid soybean. *Nature* **2010**, *463*, 178–183.
84. Sobkowiak, R.; Deckert, J. Proteins induced by cadmium in soybean cells. *J. Plant Physiol.* **2006**, *163*, 1203–1206.

7

Effect of Aluminum Treatment on Proteomes of Radicles of Seeds Derived from Al-Treated Tomato Plants

Ikenna Okekeogbu [1], Zhujia Ye [1], Sasikiran Reddy Sangireddy [1], Hui Li [1], Sarabjit Bhatti [1], Dafeng Hui [1], Suping Zhou [1,*], Kevin J. Howe [2], Tara Fish [2], Yong Yang [2] and Theodore W. Thannhauser [2,†]

1 Department of Agricultural and Environmental Sciences, College of Agriculture, Human and Natural Sciences, Tennessee State University, 3500 John A Merritt Blvd, Nashville, TN 37209, USA; E-Mails: iokekeog@my.tnstate.edu (I.O.); zhujiaye117@gmail.com (Z.Y.); sangisasi@gmail.com (S.R.S.); hli@my.tnstate.edu (H.L.); sbhatti@Tnstate.edu (S.B.); dhui@Tnstate.edu (D.H.)

2 Plant, Soil and Nutrition Research Unit, USDA-ARS, Tower Rd, Ithaca, NY 14853, USA; E-Mails: kjh46@cornell.edu (K.J.H.); tlf26@cornell.edu (T.F.); yy44@cornell.edu (Y.Y.)

† Correspondence author for proteomics analysis; E-Mail: tt34@cornell.edu.

* Author to whom correspondence should be addressed; E-Mail: zsuping@tnstate.edu

---

Abstract: Aluminum (Al) toxicity is a major constraint to plant growth and crop yield in acid soils. Tomato cultivars are especially susceptible to excessive Al^{3+} accumulated in the root zone. In this study, tomato plants were grown in a hydroponic culture system supplemented with 50 μM $AlK(SO_4)_2$. Seeds harvested from Al-treated plants contained a significantly higher Al content than those grown in the control hydroponic solution. In this study, these Al-enriched tomato seeds (harvested from Al-treated tomato plants) were germinated in 50 μM $AlK(SO_4)_2$ solution in a homopiperazine-1,4-bis(2-ethanesulfonic acid) buffer (pH 4.0), and the control solution which contained the buffer only. Proteomes of radicles were analyzed quantitatively by mass spectrometry employing isobaric tags for relative and absolute quantitation (iTRAQ®). The proteins identified were assigned to molecular functional groups and cellular metabolic pathways using MapMan. Among the proteins whose abundance levels changed significantly were: a number of transcription factors; proteins regulating gene silencing and programmed cell death; proteins in primary and secondary signaling pathways, including phytohormone signaling and proteins for enhancing tolerance to abiotic and biotic stress. Among the metabolic pathways, enzymes in glycolysis and fermentation and sucrolytic pathways were repressed. Secondary metabolic pathways including the mevalonate pathway and lignin biosynthesis were induced. Biological reactions in mitochondria seem to be induced due to an increase in the abundance level of mitochondrial ribosomes and enzymes in the TCA cycle, electron transport chains and ATP synthesis.

Keywords: iTRAQ; orbitrap mass spectrometry; Al-enriched tomato seeds; seed germination; root proteomics; functional pathways

---

1. Introduction

Aluminum is not an essential mineral to plants; Al ions (Al^{3+}), when at an excessive level, are very toxic to seed germination in both tolerant and susceptible plants [1,2]. Of all the Al-induced phytotoxic symptoms, disruption of cell division and growth within the root apex has the most significant impact on plant growth and yield [3,4]. Thus, many studies are focused on the physiological and molecular activities in the root tip zone related to reducing Al^{3+} phytotoxicity. Tolerance to Al is achieved by avoidance mechanisms (e.g., through secretion of organic acids to bind Al^{3+} near the vicinity of root tips) [5–7], and internal resistance, by remodeling cellular processes [8,9] and through apoplastic and symplastic detoxification of internalized Al [10,11].

Seed germination is the process by which an embryo transitions into a complete plant. It begins as the root (radicle) becomes the first embryonic organ to emerge from the seed coat; this is followed by elongation of the hypocotyl and ends in expansion of the cotyledon (s) [12–14]. Radicle emergence involves both cell division and cell enlargement. Radicle growth (mainly involving cell enlargement) is sensitive to metal toxicity [15]. A study using *Arabidopsis thaliana* shows that *de novo* protein synthesis from the pool of stored mRNA is essential for the completion of radicle protrusion; however, the process can proceed even in the absence of transcription (*de novo* mRNA synthesis) [16]. Pre-incubation of wheat (*Triticum aestivum*) seedlings with low doses of Al increased tolerance to subsequent exposure to lethal concentrations. The study also concluded that synthesis of proteins is essential for acquiring tolerance to Al because addition of the protein translation inhibitor, cycloheximide, completely abolished the induced tolerance to Al toxicity [17]. Therefore, proteome changes in the primary root can directly affect the development of tolerance and may represent the key to understanding the molecular mechanism involved.

Tomato (*Solanum lycopersicum*) is among the few species that produce very acidic fruits ($pH < 3.0$ in ripened tomato fruit), thereby providing an environment capable of shifting the equilibrium from the benign Al^{2+} form to the highly toxic Al^{3+}. In this study, tomato plants were grown in a hydroponic system supplemented with Al during the reproductive stages (from flowering until fruit ripening). Tomato seeds produced by these plants were considered to be Al-enriched as they contained a higher Al content than those harvested from plants growing in a solution without added Al. Subsequently, the Al-enriched seeds were germinated in an Al solution, and a proteomics analysis of their radicles was performed to identify proteome changes in response to the Al treatment as a means to identify candidate proteins that could play a key role in acquiring Al tolerance.

2. Experimental

Tomato (*S. lycopersicum* cv. Micro-Tom) plants were grown in a hydroponic culture system. As soon as plants started to set fruits (pea-sized fruits were seen on the first fruit cluster), $AlK(SO_4)_2$ was added up to a final concentration of 7.2 μM of Al^{3+} activity [or 50 μM $AlK(SO_4)_2$]. The pH of the solution was tested daily using pH strips (Fisher Scientific) and the solution was refreshed weekly or when the pH increased to 5.0. Tomato fruits were harvested periodically when the color turned red. To collect tomato seeds, fruits were wrapped in paper towels to squeeze out all the tomato juice. After removal of the gelatinous sack tissues, seeds were soaked in 50% bleach for 5 min followed by three rinses in autoclaved water. Seeds were stored at 4 °C until analysis. These field experiments were performed for two seasons. Mineral analysis of seed tissues (embryo and seed coat separately) found that the Al content of embryo was 10–15 mg per kg dry weight (DW) for seeds derived from Al-treated plants, and it was 6–8 mg per kg DW for those harvested from plants growing in the same hydroponic system but without adding $AlK(SO_4)_2$. In this experiment, it was noticed that control samples including roots, leaves and seeds also contained Al although the content level was much lower than the treated samples. Consistently in the two-season experiments, the Al-treated embryos contained a significantly ($p < 0.01$ using *t*-test) lower amount of boron (B) and iron (Fe) [18].

A seed germination assay was conducted to test the effect of Al treatment on improving Al tolerance of the next-generation offspring. Seeds (Al-enriched) that were harvested from Al-treated plants were germinated on wet filter paper soaked in a 50 mM homopiperazine-1,4-bis(2-ethanesulfonic acid) (Homopipes) buffer containing 50 μM $AlK(SO4)_2$ (pH 4.5–5.0) in the Al-treated group, as opposed to the buffer only in the control treatment. The germination was carried out at 25 ± 2 °C. The lengths of radicles were measured on the third day of germination. There was no significant difference in terms of radicle length between Al-treated (5 ± 1 mm) and untreated groups (4 ± 1 mm). These results indicate that radicle elongation growth was not affected by the presence of Al in the solution for these seeds.

For preparation of this study, seeds were germinated under the same conditions, with three biological replicates for treated and control experiments. Each biological replicate consisted of 100 seeds with 10 seeds wrapped in one filter paper sandwich. Radicles protruding from seed coat were dissected using a sharp blade on the third day after germination. Tissues were frozen in liquid nitrogen and ground into a fine powder immediately after harvest.

2.1. Protein Extraction and Isobaric Tags for Relative and Absolute Quantification Labeling

For protein extraction, tissue powder was washed sequentially in 10% TCA/acetone, 80% methanol/0.1 M ammonium acetate, and 80% acetone with centrifugation to pellet the powder after each wash. The protein was then extracted in a phenol (pH 8.0) and dense SDS buffer [30% sucrose, 2% SDS, 0.1 M Tris-HCl, pH 8.0, 5% beta-mercaptoethanol (v/w)]. After incubation at 4 °C for 2 h, the mixture was centrifuged at 16,000 *g* at 4 °C for 20 min. Protein in the upper phenol phase was precipitated in 0.1 M ammonium acetate in methanol after incubation overnight at −20 °C. Protein pellets were washed in methanol and acetone and were then dissolved in a buffer of 500 mM triethylammonium bicarbonate (TEAB) and 2 M urea, and 0.1% SDS and a proteinase inhibitor cocktail for plant tissue (100 × dilution in the extraction buffer) (Sigma, St. Louis, MO, USA). Protein concentration was determined using a Bradford assay kit (Bio-Rad, Hercules, CA, USA).

One hundred μg of protein from each sample was digested with trypsin and then labeled as previously described [19] following the instructions accompanying the 8-plex iTRAQ® labeling kit (AB SCIEX, Framingham, MA, USA). The treated samples were labeled with tags 113, 114 and 115 and the control samples with 116, 117 and 118 were combined. Unbound tags and SDS were removed through cation exchange cartridge (AB SCIEX), and salts were removed using reverse-phase solid-phase extraction procedure involving 1-cm^3, 50-mg cartridges following the manufacturer's instructions (Sep-Pak C_{18}; Waters, Milford, MA, USA). Peptides were eluted in 500 μL 50% (v/v) acetonitrile with 0.1% TFA and dried under vacuum.

These peptide samples were subjected to a first dimension of high pH Ultra Performance Liquid Chromatography (UPLC) separation using an Acquity UPLC System (Waters) coupled with a robotic fraction collector (Probot; Dionex, Sunnyvale, CA, USA) [19]. One hundred micrograms of the multiplexed sample was injected and fractionated into 48 fractions in a 96-well plate. The 48 fractions were concatenated to yield 16 samples pools by pooling every 16th sample. These were dried at reduced pressure using a CentiVac Concentrator (LabConco, Kansas City, MO, USA). For the low pH 2nd dimension, low pH reverse-phase (RP) chromatography was employed.

Dried samples were reconstituted with 15 μ L of 2% acetonitrile with 0.5% formic acid. Nano-LC separations of tryptic peptides were performed as described previously [20,21]. The eluent from the analytical column was delivered to the LTQ-Orbitrap Elite (Thermo-Fisher Scientific,Waltham, MA, USA) via a "Plug and Play" nano ion source (CorSolutions LLC, Ithaca, NY, USA). The mass spectrometer was externally calibrated across the *m/z* range from 375–1,800 with Ultramark 1621 for the FT mass analyzer, and individual runs were internally calibrated with the background polysiloxane ion at *m/z* 445.1200025 as a lock mass.

The Orbitrap Elite was operated in the positive ion mode with nanosource voltage set at 1.7 kV and source temperature at 250 ℃. A parallel DDA mode was used to obtain one MS survey scan with the FT mass analyzer, followed by isolation and fragmentation of the 15–20 most abundant, multiply-charged precursor ions with a threshold ion count higher than 50,000 in both the LTQ mass analyzer and the high energy collisionally induced dissociation (HCD)-based FT mass analyzer at a resolution of 15,000 (fwhm *m/z* 400). MS survey scans were acquired with resolution set at 60,000 across the survey scan range (*m/z* 375–1800).

Dynamic exclusion was utilized with repeat count set to 1 with a 40 s repeat duration; exclusion list size was set to 500, 20–30 s exclusion duration, and low and high exclusion mass widths set to 1.5. Fragmentation parameters were set with isolation width at 1.5 *m/z*, normalized collision energy at 37%, activation Q at 0.25. Activation time for HCD analysis was 0.1 min. All data were acquired using XCalibur 2.1 (Thermo-Fisher Scientific).

2.2. Data Processing, Database Searching and iTRAQ Quantitation

Proteome Discoverer v 1.4 was used to convert raw spectral data files for each iTRAQ experiment into a merged peak list (mgf format) containing all 2nd dimension fractions for each tomato experiment for subsequent database searching. Mascot Daemon v. 2.3.2 was used to query .mgf files against an iTAG 2.3 tomato protein database [22]. Trypsin was selected as the enzyme with 1 missed cleavage allowed. Methylthiolation of cysteine, oxidation of methionine, and deamidation of asparagine and glutamine were set as variable modifications. Peptide charge was set to 2^+, 3^+, and 4^+. Precursor tolerance was set to 10 ppm, while fragment tolerance was set to 100 mmu. The instrument selected was ESI-FTICR. The iTRAQ quantitation method utilized a weighted protein ratio type, featured outlier removal, and required a minimum of 2 peptides for protein quantitation. Summed normalization was used. For the iTRAQ 8-plex labeling, N-terminal and lysine modification with iTRAQ were set as fixed modifications, and tyrosine labeling was set as a variable modification. Upon completion of searching, each report was opened and results were exported after setting the ion score filter to 0.1, thereby exporting only results with an expectation value below 0.1, specifying unique peptides only. Only the highest scoring matches to a particular peptide sequence, listed under the highest scoring protein containing that match, were considered.

2.3. Protein Quantification, Statistics, and Protein Functional Analysis

For a protein to be included in the quantitative analysis, it was required that at least two unique peptides (with the normalized intensity levels raw intensity >20) were identified in all the six biological samples (three biological replicates each in Al-treated and control groups). The normalized peak intensities of reporter ions of constituent peptides were $\log_2$ transformed. Then, $\log_2$ fold values from all constituent peptides were subjected to *t*-test (general linear model procedure) followed by false discovery rate (FDR) corrections to test the statistical significance of the difference in normalized abundance of each protein between Al-treated and control sample groups [19]. The $\log_2$ transformed abundance ratios were then fit to a normal distribution. Two standard deviations (at a 95% confidence level) of the $\log_2$ fold (from treated to untreated control) in protein abundance were used as the cutoff threshold for significantly changed proteins. Statistical analyses were performed using SAS (version 9.3; SAS Institute, Cary, NC, USA).

MapMan [23] was used to associate the identified tomato proteins with cellular process and metabolic pathways using iTAG 2.3 tomato protein database downloaded from the MapMan website. The putative functions of the identified proteins were also discussed based on relevant information from literature and database searches on tomatoes and other plant species.

3. Results and Discussion

In this study, 3,160 proteins, meeting the quantification analysis criteria of two or more peptides, were identified in all six biological samples. The spectral intensity of each peptide was transformed into log (base 2) values, and principal component analysis (PCA) (Figure 1) separated the treated and control triplicate groups, which indicates that there is a systematic difference in protein composition between the two groups. The low percentage in component 2 compared to component 1 (2.7% *vs.*

94.87%) indicates that differences among proteins (or peptides of the same protein) are much greater which is understandable as the proteome is comprised of proteins of different abundance (high *versus* low content levels). Data of log_2 fold change of proteins from treated to untreated groups fit into a near normal distribution (Figure 2). After *t*-test and FDA corrections, 139 proteins were found to be significantly changed from untreated to treated root samples ($p \leq 0.05$) and the fold change passed the threshold of a two standard deviation (>±0.82). Fifty-two proteins were repressed and 87 proteins were induced, and Al-induced changes in protein abundance were given as the ratio between treated and non-treated control groups which is the antilogarithm of log_2 (fold) (Appendix Table A1).

Figure 1. Principal component distribution (PCA) of proteomes from Al-treated tomato radicle. (Tryptic peptides from six biological samples were labeled with iTRAQ tags (treated samples with tags 113, 114 and 115 and the control samples with 116, 117 and 118). The intensity of reporter ions of peptides from mass spectrometry analysis was log-transformed (base, 2). Protein samples were clustered based on the distribution of log_2 fold change values of all peptides in the six tagged samples. Three control biological replicates: C1, C2, C3; three treated biological replicates: T1, T2, T3).

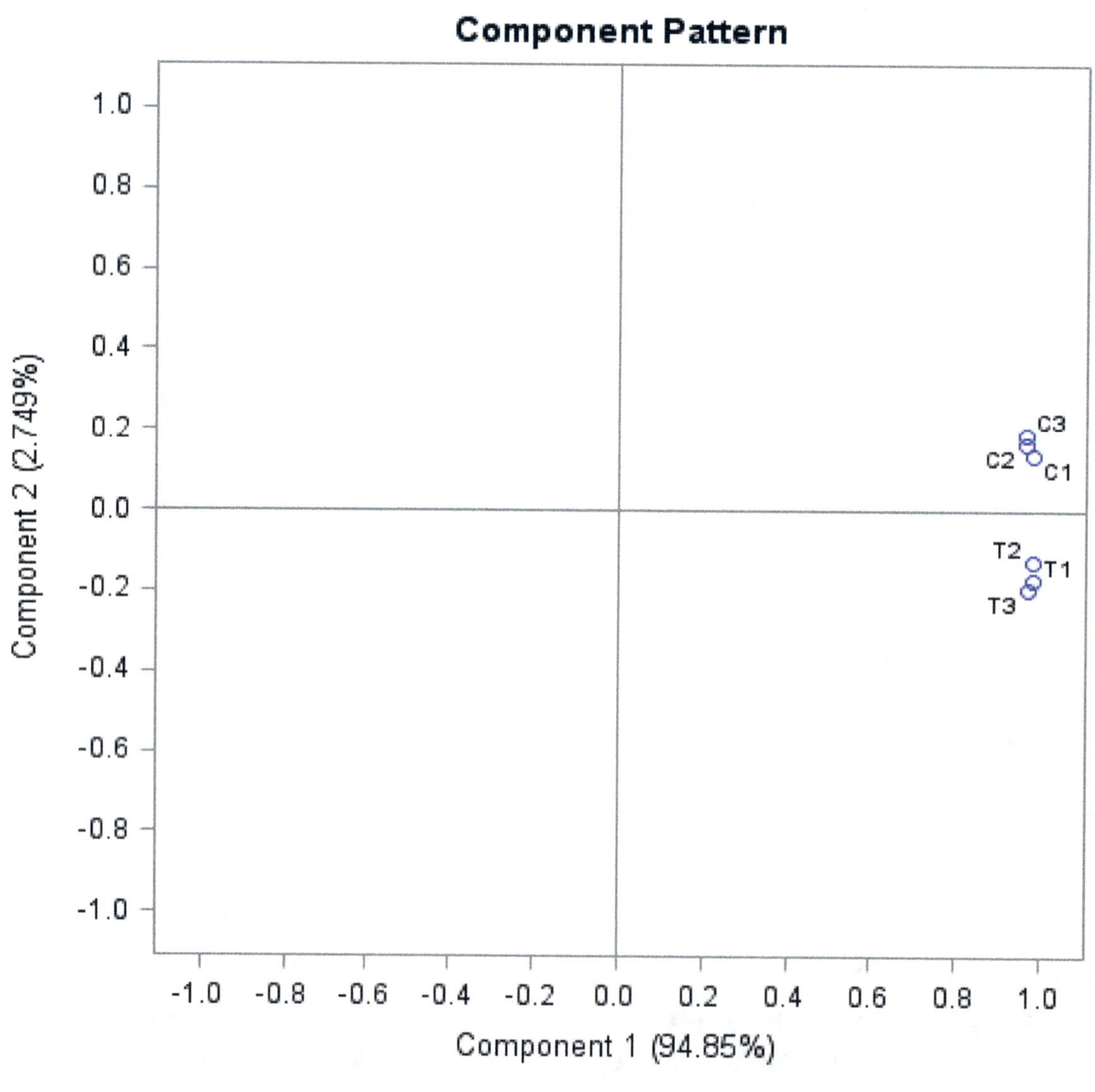

Figure 2. The normal distribution fit of the $\log_2$ fold values of proteins from Al-treated tomato radicles. (Tomato seeds were germinated in 50 mM Homopipes (pH 4.5) buffer supplemented with 50 μM $AlK(SO_4)_2$ and the control solution contained the buffer only. Tryptic peptides were labeled with iTRAQ tags (treated samples with tags 113, 114 and 115 and the control samples with 116, 117 and 118) followed by analysis using mass spectrometry. The reporter ion intensity of all the tags was log-transformed and the $\log_2$ fold changes of protein from Al-treated and untreated tomato samples were plotted against a theoretical normal distribution in SAS program. The purple-colored is the theoretical curve and the blue-colored is the data fit curve).

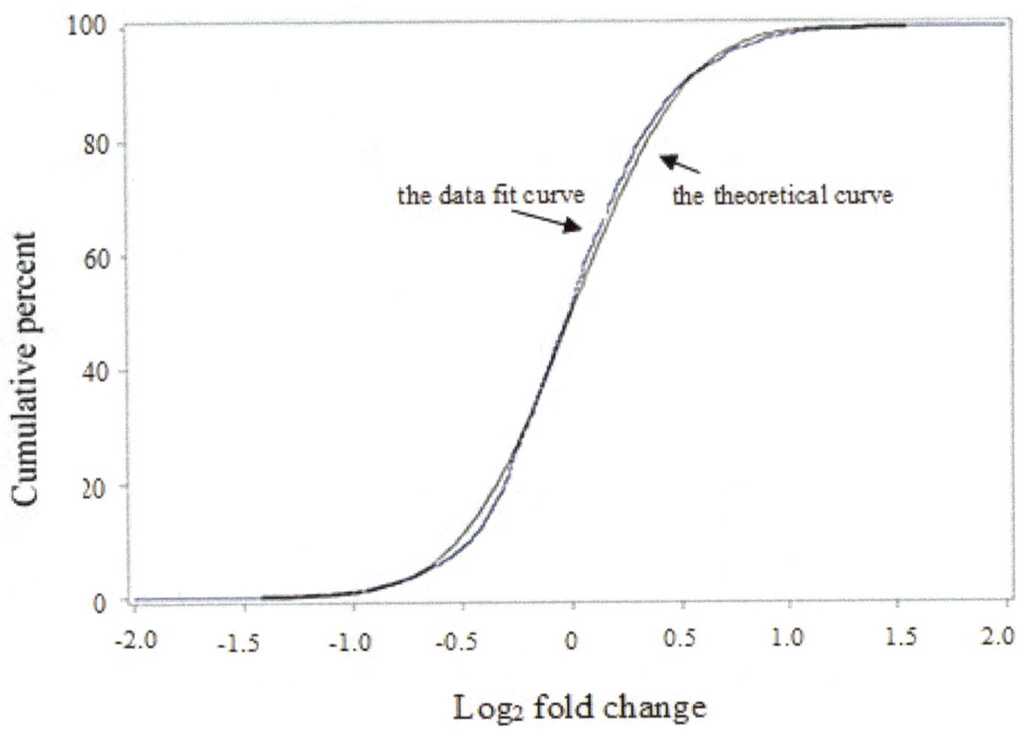

3.1. Al Treatment-Induced Proteome Changes and the Associated Cellular and Molecular Functions

Using MapMan, tomato root proteins were clustered into 20 cellular functional pathways (Figure 3). In each of the functional groups, there were proteins repressed, induced and unchanged (the intensity of the color change corresponded to the $\log_2$ fold change of respective proteins from treated to untreated groups). The majority of the 20 functional pathways contained significantly changed proteins. These results are in agreement with previous findings that many genes (and gene products) located in multiple genome regions and participating in various cellular activities could be involved in the modulation of plant responses to Al stress [24–27].

For those significantly changed proteins, additional manual searches of literature and other plant databases were performed to identify their putative roles in Al and secondary cellular stress. These proteins were divided into eight groups by combining MapMan classification (based on known protein functions) and putative functions derived from other sources.

Figure 3. Cell function overview of proteins from radicles of Al-treated tomato seeds. (The graphic was generated using the Cell-Function Overview and Slyc-iTAG2.3 as the reference database in MapMan. The intensity of the color change corresponds to the scale created based on $\log_2$ fold of protein from Al-treated to untreated groups.)

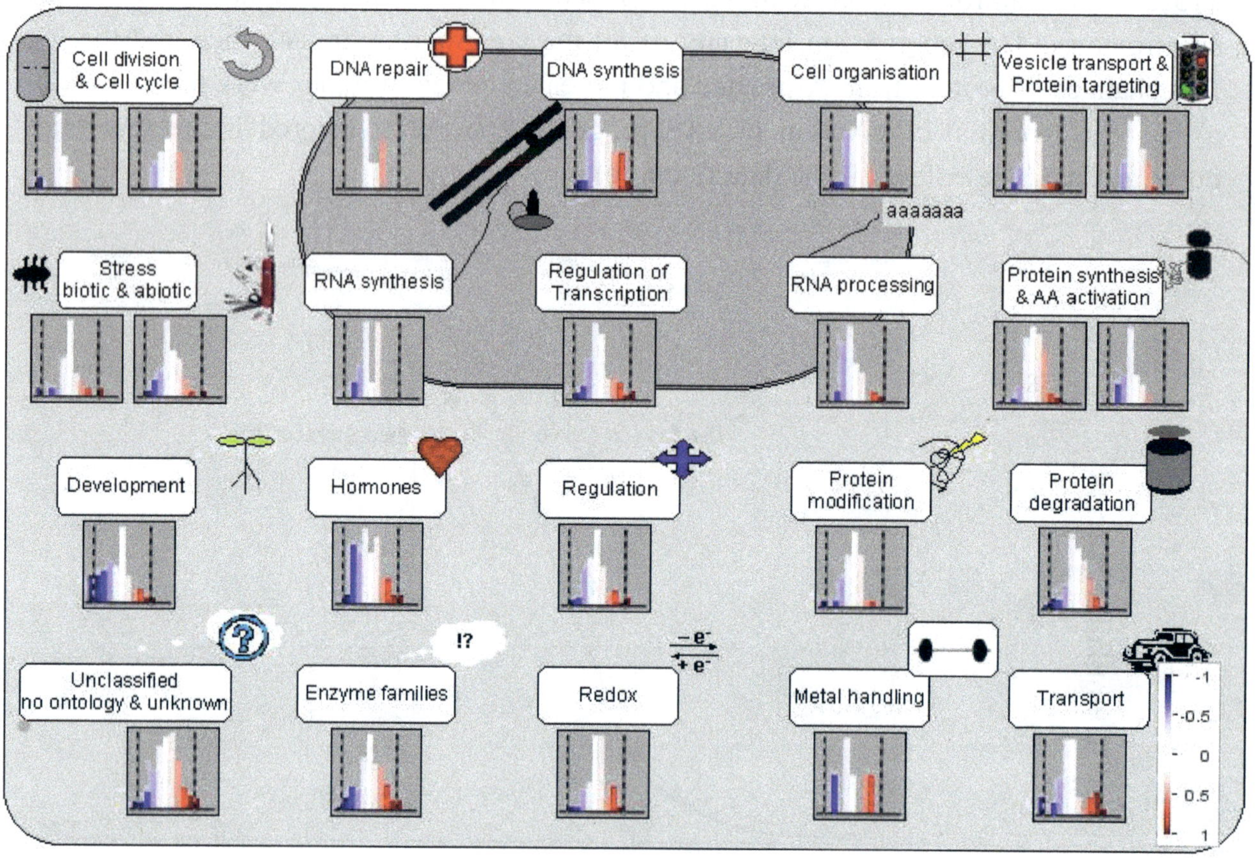

3.1.1. Mobilization of Seed Storage Proteins in Al-Treated Tomato Radicles

Mature seeds contain many species of hydrophilic proteins, such as dehydrins, globulins, and late embryogenesis abundant proteins (LEA). During seed germination, these seed proteins play a key role in maintaining intracellular water balance by controlling water uptake, and they are also recycled to provide a nitrogen source for the germinating embryo. In this study, Al-treated radicles were found to contain a lower level of hydrophilic proteins, such as globulin, vicilin, LEA, seed biotin-containing protein SBP65, dehydrin, and small hydrophilic plant seed protein. These results suggest that the Al treatment induced more active catabolism of hydrophilic proteins or inhibited transport of those proteins from cotyledons to the growing radicles. Both processes can result in a low hydrophilic protein content in the Al-treated radicles.

In addition, two seed oil body-associated proteins, caleosin and oleosin, were also repressed in Al-treated radicles. These lipid body-binding proteins play a key role in the degradation of storage lipid during seed germination [28,29]. A decrease in oleosin content can lead to coalescence of lipid bodies which is harmful to cells, or it may make the radicle more susceptible to dehydration as these proteins also affect tissue tolerance to desiccation [30]. In summary, changes in proteins of this group will make tomato more sensitive to dehydration stress, which can happen under excessive salt and drought stresses. This could be one of the major molecular mechanisms by which Al toxicity can exacerbate the impact of other environmental stress factors during seed germination.

3.1.2. Proteins Involved in Cell Organization, Cell Division and Cell Cycle

In the Al-treated radicles, proteins affecting cell division cycle (protease ftsH homolog) and cell skeleton structure (actin and tubulin) were repressed. On the other hand, proteins in programmed cell death (PCD) or related processes were induced. These are the RPM1 interacting proteins which are essential for hypertensive cell death in reaction to pathogens [31] and the vesicle-associated membrane family protein which has a critical role in regulating execution of PCD by affecting the rate of membrane recycling, especially under oxidative stress [32].

3.1.3. Proteins Involved in Regulation of Transcription

Expression of genetic materials provides the basis for all physiological traits. One of the first critical steps is the regeneration of mature mRNA (gene transcripts). In tomato radicles, a number of transcription factors were affected by Al treatment. Several C2H2 zinc finger family proteins were induced, but CCHC zinc finger, CCCH-type zinc finger and ZF-HD class zinc finger-homeodomain proteins were repressed. Additional induced proteins are Myb transcription factor, BolA-like protein and bZIP transcription factor. DNA silencing and mRNA decay are both important mechanisms in regulating gene expression, especially under stress conditions. Proteins in these categories were induced in the Al-treated tomato radicles, and include DNA (cytosine-5-)-methyltransferase 3, U6 RNA-associated Sm-like proteins and LSm6. Two proteins in the nonsense-mediated mRNA decay (NMD) pathway were also identified; they are eukaryotic translation initiation factor SUI1, and eukaryotic translation initiation factor 4 gamma-MIF4-like. Activation of these mechanisms can help plant cells to rid themselves of aberrant proteins and transcripts, and ensure a "healthy mRNA pool" in the Al-treated tissues.

3.1.4. Proteins Affecting Protein Synthesis and Post-Translational Modification

Plants have three separate sets of genomes in chloroplasts, mitochondria, and cell nucleus. While nuclear genes are translated in the cytoplasm, the mitochondria and chloroplasts each contain its own translation machinery within the respective organelles [33,34]. Changes in ribosomal proteins suggest that protein translation in plastids and mitochondria may be more active as ribosomal proteins annotated to those organelles were induced, which include 30S ribosomal protein S7, ribosomal protein L24 and L27 for chloroplasts, and mitochondrial ribosomal protein L37 and L15 and peptidyl-tRNA hydrolase ICT1. In mitochondria, all the proteins in the electron transport chain and ATP synthase are synthesized by mitochondrial ribosomes [35]. The Al-induced ribosomal protein expression could have some effect on mitochondrial functions (which will be discussed later).

Cytoplasmic ribosomes which are responsible for the translation of nuclear-genome encoded genes revealed more complex changes. Some proteins were induced including ribosomal protein S21e and L32e, whereas others were repressed including ribosomal protein S26e, L15e and L19/L19e. Differential relative abundances of these ribosomal proteins were also found in tomatoes under other abiotic stresses [19,36].

A large number of proteins undergo post-translational modification to generate biologically active forms and/or to be targeted into correct subcellular organelles. One major post-translation modification is protein phosphorylation which is catalyzed by kinase and the dephosphorylation catalyzed by phosphatase. In the Al-treated tomato radicles, a serine/threonine kinase protein was induced, whereas serine/threonine-specific protein phosphatase was repressed. Protein serine/threonine phosphatases are implicated in the regulation of apoptotic pathways [37,38]. A study on wheat showed that the active function of protein kinase seems to be essential in alleviating Al-induced root inhibition as protein phosphorylation was found to be involved in the Al-responsive malate efflux in root tip [39]. Therefore, changes in this pair of enzymes may have some effect on different aspects of cellular processes under Al stress.

3.1.5. Proteins Involved in Protein Degradation and Modification

The ubiquitin-proteasomal degradation pathway is essential for many cellular processes, including the cell cycle, the regulation of gene expression, and responses to oxidative stress. The proteasome complex consists of ubiquitin-activating enzyme (E1), ubiquitin-conjugating enzyme (E2) and ubiquitin ligases (E3). Among the three subunits, E3 controls the specificity of protein degradation. While no significant changes were found in E1, E2, or E3 in the Al-treated tomato radicles, the COP9 signalosome (CSN) subunit 6 was induced. CSN is the protein regulating the function of ubiquitin E3; it is rapidly emerging as a key player in the DNA-damage response, cell-cycle control and gene expression, and plant response to environmental stimuli and stresses [40–42]. This is the first time that this protein has been found to be regulated by Al stress.

3.1.6. Proteins Involved in Hormone Metabolism and Signaling

In the Al-treated radicles, proteins involved in the biosynthesis of three hormones and their signaling pathways were induced. This includes ethylene (1-aminocyclopropane-1-carboxylate oxidase and multiprotein bridging factor 1) and abscisic acid (ABA) signaling pathway (ABA/WDS-induced protein). The jasmonate pathway enzymes were repressed, which include lipoxygenase and allene oxide synthase. Proteins for the biosynthesis of gibberellin were repressed (gibberellin 3 beta-hydroxylase 2–3 and gibberellin 2-oxidase 2), but several gibberellin-regulated proteins were repressed or induced (0.59–2.06-fold).

In general, ABA-related genes are more highly expressed when germination is inhibited and the hormone (ABA) inhibits radicle emergence [43,44]. In contrast, GA-related genes are activated during seed germination [43]. In this study, tomato seeds germinated at the same rate under Al-treated and non-treated conditions, therefore, there was no correlation between this physiological process and changes in these ABA- and GA-related proteins. These results indicate that a more complex hormone signaling and interaction mechanism is involved in radicle growth during tomato seed germination.

3.1.7. Proteins in Signal Transduction

In the Al-treated tomato radicles, mitogen-activated protein kinase (MAPK), a key enzyme in MAPK signaling pathway, was strongly induced. This enzyme plays a key role to communicate an external signal from a receptor on the surface of the cell to the DNA in the nucleus of the cell [45]. Earlier studies using cell suspension cultures of coffee (*Coffea arabica*) found that the MAPK was activated by the oxidative burst induced by Al treatment but this protein is not necessarily associated with Al tolerance [46,47]. Additionally, several calcium-binding proteins that participate in the secondary calcium cell signaling pathways to mediate intracellular stress responses were also induced in the Al-treated tomato radicles. Therefore, the higher abundance of MAPK and Ca-binding proteins in Al-treated tomato radicles may also have a role in enhancing plant tolerance to the secondary cellular stresses induced by Al stress.

More importantly, the rapid alkalinization factor (RALF) proteins were induced in the Al-treated tomato radicles. RALF is a 5-kDa ubiquitous polypeptide initially isolated from tobacco leaves that induces a rapid alkalinization of the culture medium of tobacco suspension-cultured cells and a concomitant activation of an intracellular mitogen-activated protein kinase [48]. A synthetic tomato RALF homolog peptide, when supplied to germinating tomato and *Arabidopsis thaliana* seeds, caused an arrest of root growth and development [49]. This is the first finding that Al induced an increase in the endogenous level of RALF in tomato radicles, which might also have some roles in regulating cell cycle under the stress condition.

3.1.8. Stress Proteins

The induced proteins include the universal stress protein family, major latex-like proteins and germin-like protein and wound/stress protein. Some of these proteins were also induced during tomato seed germination of normal seeds (seeds harvested from non-treated plants) in Al-treated solution [27]. However, it seems that Al-treated radicles contained a lower level of several heat shock proteins, including Hsp40, DnaJ, DnaJ 2, Hsp 70, ClpB chaperone, class IV HSP and class I HSP, as well as a low-temperature-induced 65 kDa protein. These important protection proteins are mostly induced by stress factors, and they may have a different role under Al stress.

Oxidative burst is an important secondary cellular stress induced by Al [50]. Among the antioxidant enzymes, thioredoxin and superoxide dismutase, and germin (oxalate oxidase) were induced in Al-treated radicles. In contrast, germin protein was reduced in tomato "Money Maker" treated with the same type of stress [27]. Previous studies indicate that the Al-induced up-regulation of oxalate oxidase gene in the root tip of wheat helps roots to get rid of Al-damaged cells and maintain a healthy epidermal layer of roots, thus protecting the deeper layer of the meristematic and elongation zone that are essential for root growth [51]. Therefore, the induction of germin and germin-like proteins may enhance Al tolerance which was acquired during exposure to Al during seed germination.

3.1.9. Enzymes in Cellular Metabolism

Protein changes in different metabolic pathways are shown in Figure 4. In the Al-treated tomato radicles, fermentation and glycolysis pathways were repressed, due to the reduced abundance in pyruvate decarboxylase-2, alcohol dehydrogenase, and an additional 10 enyzmes in glycolysis. Phosphate dikinase in glucogenesis pathway was also repressed.

Figure 4. Overview of metabolic pathways in tomato radicles. (The graph was generated using the Metabolism Overview in MapMan [23]. The intensity of the color change corresponds to the scale created based on $\log_2$ fold of respective protein from Al-treated to untreated tomato radicle tissues.)

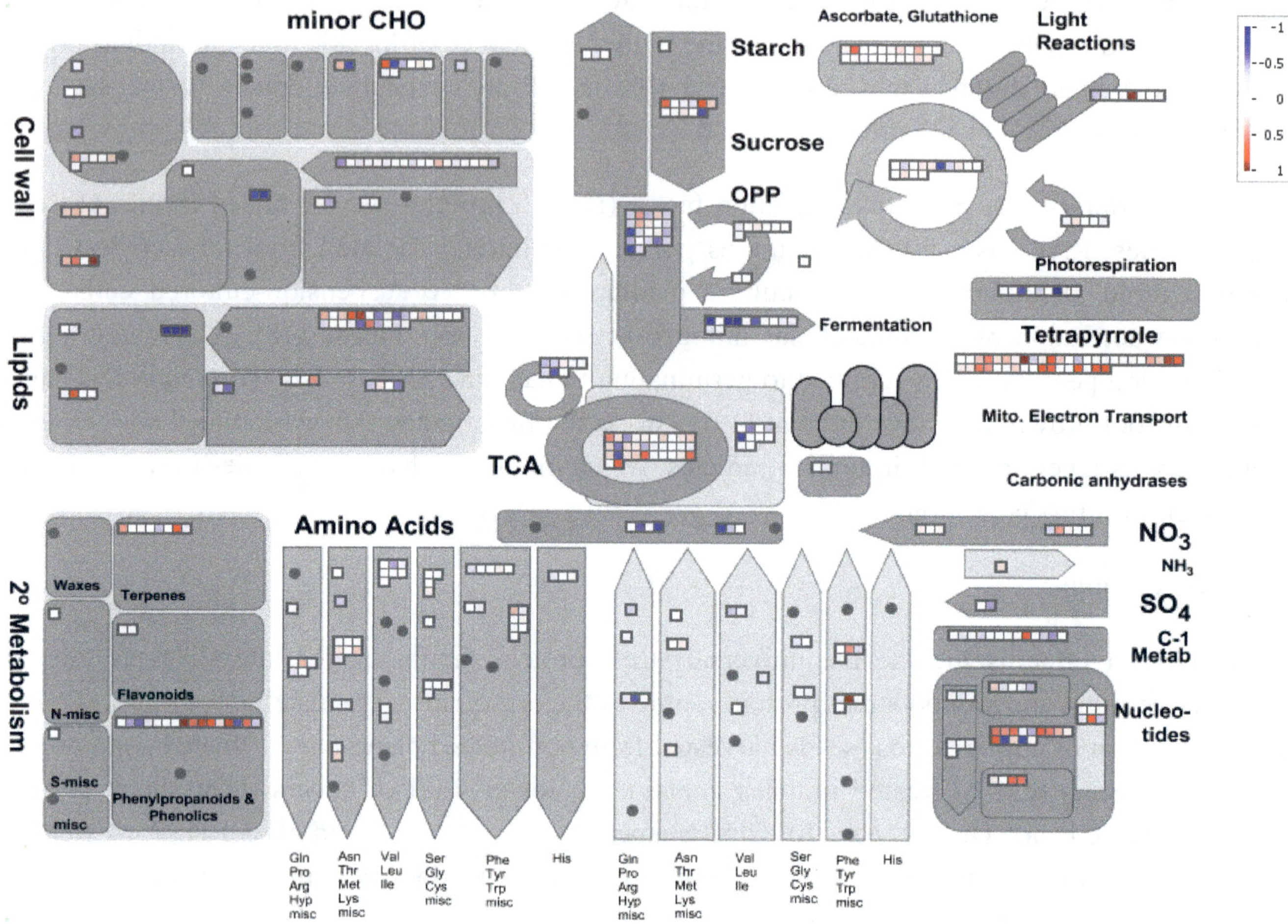

The only pathway that was enhanced in Al-treated radicles is the TCA cycle where malate dehydrogenases were induced. The most systematic changes were found in the mitochondrial electron transfer chain (ETC). ETC consists of complex I, II, III, IV, and ATPases for generation of ATP. The induced proteins are localized in NADH-DH.complex I (NADH ubiquinone dehydrogenase), complex II (succinate dehydrogenase assembly factor 2), complex III (biquinol-cytochrome C reductase complex proteins); a class IA/ IB cytochrome to transfer electrons from complex III to complex IV; complex IV (cytochrome c oxidases), and five ATPase proteins (F0 complex subunit D, F1 complex, OSCP/delta subunit, epsilon subunit and delta/epsilon subunit). Such changes suggest that metabolic pathways from sucrose degradation to glycolysis and fermentation could be repressed, but TCA and ETC in mitochondria were induced in tomato radicles under Al treatment.

Enzymes in glycolysis and fermentation pathways are all encoded by nuclear genes and translated in cytoplasm. As described above, Al induced the repression of several cytosolic ribosomal proteins, which could have affected translation of these proteins. Mitochondial ETC proteins are encoded by mitochondrial genes, and translated within the organelle. Aluminum induced expression of several mitochondrial ribosomal proteins, which may have promoted translation of these proteins in the Al-treated radicles.

Proteins induced in secondary metabolism include isopentenyl-diphosphate delta-isomerase in the mevalonate pathway (MVA), several *O*-methyltransferases, cinnamyl alcohol dehydrogenase-like protein and caffeoyl-CoA O-methyltransferase in the phenylpropanoids and lignin biosynthesis. These pathways are activated by various biotic and abiotic stresses [19,52,53]. In a previous study of Al treatment of tomato roots, none of these proteins were identified [27]. The use of different analytical methods (2D-DIGE in the previous analysis and iTRAQ in the present one) and variation in the seed proteomes (regular seeds compared to Al-enriched from two tomato cultivars) may be factor(s) causing the identification of different proteins in the two experiments.

4. Conclusions

When subjected to excessive Al and other toxic metals, plants have to cope with the direct ion toxicity and the induced secondary cellular stresses, such as accumulation of reactive oxygen species [50,54] and toxic metabolic aldehydes [55] among harmful biomolecules. Correspondingly, multiple and various cellular pathways are affected during the process [5,56], with concomitant alteration in expression of proteins in multiple functional groups. As shown in this proteomics study, proteins in nearly all of the 20 functional categories displayed significant changes in abundance in the Al treated condition. The Al-induced changes in the proteomes of radicles generated from Al-enriched tomato seeds can be summarized as follows:

1. The Al-treated radicles contained lower abundance of hydrophilic (seed) proteins and oil body membrane proteins, which could reduce the tolerance to dehydration. This could cause the tomato seedlings to be more susceptible to drought and salt and other factors;
2. Mitochondria function was enhanced because of the active protein translational machinery (induction in ribosomal proteins) and TCA–oxidative phosphorylation cycle;
3. The identified proteins include regulatory proteins for gene expression, signaling pathways, cell cycle and programmed cell death.

Aluminum (Al) is ubiquitous in soil being the most abundant metal in the earth's crust (>8% by weight). When solubilized at pH values below 5.0, it is highly toxic to plants as Al^{3+}. Approximately 50% of the world's potentially arable lands are acidic [57]. A large proportion of the acid soils occur in developing countries in the tropics and subtropics and it has been estimated that the humid tropics account for 60% of the acid soils in the world. Thus, acid soils limit the growth of crops in many developing countries where food production is critical. Acid soils also have a significant impact on U.S. agriculture as approximately 135 million hectares of land in the U.S. are highly acidic. Furthermore, intensive agricultural practices used in the U.S. and in other developed countries, including the widespread use of N fertilization with anhydrous ammonia, can cause significant acidification of surface soils [58], thereby exacerbating an already thorny problem. Thus, there is a genuine need to better understand Al tolerance mechanisms and the genes/proteins that define them, to sustain and enhance crop production on acid soils. The knowledge gained from this and similar studies will provide the scientific underpinning for novel strategies to overcome the challenge of Al stress, and to sustain agricultural production on acid soils.

Acknowledgments

The authors wish to thank Wei Chen and Sheng Zhang of the Proteomics and Mass Spectrometry Facility of the Cornell University Institute of Biotechnology for expert technical assistance and helpful discussions. The authors also want to acknowledge Michael A. Rutzke and Shree K. Giri for providing mineral analysis of tomato tissues. This project was supported by the Agriculture and Food Research Initiative competitive Grant No. 2010-65114-20405 from the USDA National Institute of Food, and Agriculture, the Evans-Allen Research Funds and ARS CRIS Projects 1907-21000-036/037-00D. ARS disclaimer: "Mention of trade names or commercial products in this publication is solely for the purpose of providing specific information and does not imply recommendation or endorsement by the U.S. Department of Agriculture."

Author Contributions

Tomato plant treatments and protein preparation and iTRAQ labeling were performed by Ikenna Okekeogbu, Zhujia Ye, Sasikiran Reddy Sangireddy and Hui Li with the assistance from Sarabjit Bhatti. Statistical analysis was performed by Suping Zhou and Dafeng Hui. Tara Fish assisted in the protein extractions, protein quantifications and labeling experiments as well as being responsible for various analytical experiments to guarantee the quality of the analysis. Tara Fish also made significant contributions to the manuscript in revision. Yong Yang and Kevin Howe carried out all of the MS analysis, evaluated the data, compiled the experimental results and contributed to the revising of the manuscript. Theodore W. Thannhauser, together with Suping Zhou, developed the experimental design. Furthermore, Theodore W. Thannhauser provided critical oversight for all the MS analysis, conducted an independent (and concurring) statistical analysis and contributed significantly to the drafting and revising of the manuscript.

References

1. Nasr, N. Germination and seedling growth of maize (*Zea mays* L.) seeds in toxicity of aluminum and nickel. *Merit Res. J. Environ. Sci. Toxic.* **2013**, *1*, 110–113.
2. Zhu, L.; Wang, J.; Fang, X.; Wang, Y.; Hao, J.; Weiwei, M.; Jiao, T. Effect of seed soaking with aluminum on seed germination and seedling physiology of *Platycodon grandiflorum. Zhongguo Zhong Yao Za Zhi* **2010**, *35*, 3255–3259.
3. Liang, C.; Piñeros, M.A.; Tian, J.; Yao, Z.F.; Sun, L.L.; Liu, J.P.; Shaff, J.; Coluccio, A.; Kochian, L.V.; Liao, H. Low pH, aluminum, and phosphorus coordinately regulate malate exudation through GmALMT1 to improve soybean adaptation to acid soils. *Plant Physiol.* **2013**, *161*, 1347–1361.

4. Liao, H.; Wan, H.; Shaff, J.; Wang, X.; Yan, X.; Kochian, L.V. Phosphorus and aluminum interactions in soybean in relation to aluminum tolerance: Exudation of specific organic acids from different regions of the intact root system. *Plant Physiol.* **2006**, *141*, 674–684.
5. Kochian, L.V. Cellular mechanisms of aluminum toxicity and resistance in plants. *Annu. Rev. Plant Physiol. Plant Mol. Biol.* **1995**, *46*, 237–260.
6. Kochian, L.V.; Hoekenga, O.A.; Piñeros, M.A. How do crop plants tolerate acid soils? Mechanisms of aluminum tolerance and phosphorous efficiency. *Annu. Rev. Plant Biol.* **2004**, *55*, 459–493.
7. Yang, L.T.; Qi, Y.P.; Jiang, H.X.; Chen, L.S. Roles of organic acid anion secretion in aluminum tolerance of higher plants. *Biomed. Res. Int.* **2013**, *2013*, e173682, doi:10.1155/2013/173682.
8. Eticha, D.; Stass, A.; Horst, W.J. Cell-wall pectin and its degree of methylation in the maize root-apex: Significance for genotypic differences in aluminium resistance. *Plant Cell. Environ.* **2005**, *28*, 1410–1420.
9. Fry, S.C.; Smith, R.C.; Renwick, K.F.; Martin, D.J.; Hodge, S.K.; Matthews, K.J. Xyloglucan endotransglycosylase, a new wall-loosening enzyme activity from plants. *Biochem. J.* **1992**, *282*, 821–828.
10. Huang, C.F.; Yamaji, N.; Chen, Z.; Ma, J.F. A tonoplast-localized half-size ABC transporter is required for internal detoxification of aluminum in rice. *Plant J.* **2012**, *69*, 857–867.
11. Zhu, X.F.; Lei, G.J.; Wang, Z.W.; Shi, Y.Z.; Braam, J.; Li, G.X.; Zheng, S.J. Coordination between apoplastic and symplastic detoxification confers plant aluminum resistance. *Plant Physiol.* **2013**, *162*, 1947–1955.
12. Bewley, J.D.; Black, M. *Seeds, Physiology of Development and Germination*, 2nd ed.; Plenum Press: New York, NY, USA, 1994.
13. Bewley, J.D. Seed germination and Dormancy. *Plant Cell* **1997**, *9*, 1055–1066.
14. Koornneef, M.; Bentsink, L.; Karssen, C.M. Seed dormancy and germination. *Curr. Opin. Plant Biol.* **2002**, *5*, 33–36.
15. Di Salvatore, M.; Carafa, A.M.; Carratù, G. Assessment of heavy metals phytotoxicity using seed germination and root elongation tests: A comparison of two growth substrates. *Chemosphere* **2008**, *73*, 1461–1464.
16. Rajjou, L.; Gallardo, K.; Debeaujon, I.; Vandekerckhove, J.; Job, C.; Job, D. The effect of alpha-amanitin on the Arabidopsis seed proteome highlights the distinct roles of stored and neosynthesized mRNAs during germination. *Plant Physiol.* **2004**, *134*, 1598–1613.
17. Aniol, A. Induction of aluminum tolerance in wheat seedlings by low doses of aluminum in the nutrient solution. *Plant Physiol.* **1984**, *76*, 551–555.
18. Zhou, S.P.; Ikenna, O.; Rutzke, M.A.; Giri, S.K. Tennessee State University: Nashville, TN, USA. Unpublished data, 2014.
19. Zhou, S.; Palmer, M.; Zhou, J.; Bhatti, S.; Howe, K.J.; Fish, T.; Thannhauser, T.W. Differential root proteome expression in tomato genotypes with contrasting drought tolerance exposed to dehydration. *J. Am. Soc. Hortic. Sci.* **2013**, *138*, 1–11.
20. Thannhauser, T.W.; Shen, M.; Sherwood, R.; Howe, K.; Fish, T.; Yang, Y.; Chen, W.; Zhang, S. A workflow for large-scale empirical identification of cell wall N-linked glycoproteins of tomato (*Solanum lycopersicum*) fruit by tandem mass spectrometry. *Electrophoresis* **2013**, *34*, 2417–2431.

21. Yang, Y.; Qiang, X.; Owsiany, K.; Zhang, S.; Thannhauser, T.W.; Li, L. Evaluation of different multidimensional LC-MS/MS pipelines for iTRAQ-based proteomic analysis of potato tubers in response to cold storage. *J. Proteome Res.* **2011**, *10*, 4647–4660.
22. International Tomato Genome Sequencing Project. Available online: http://solgenomics.net/tomato/ (accessed on 16 September 2011).
23. MapMan, version 3.1.0. Available online: http://mapman.gabipd.org/web/guest/mapman (accessed on 30 January 2014).
24. Delhaize, E.; Ma, J.F.; Ryan, P.R. Transcriptional regulation of aluminium tolerance genes. *Trends Plant Sci.* **2012**, *17*, 341–348.
25. Famoso, A.N.; Zhao, K.; Clark, R.T.; Tung, C.W.; Wright, M.H.; Bustamante, C.; Kochian, L.V.; McCouch, S.R. Genetic Architecture of Aluminum Tolerance in Rice (Oryza sativa) Determinedthrough Genome-Wide Association Analysis and QTL Mapping. *PLoS Genet.* **2011**, *7*, e1002221.
26. Ryan, P.R.; Tyerman, S.D.; Sasaki, T.; Furuichi, T.; Yamamoto, Y.; Zhang, W.H.; Delhaize, E. The identification of aluminium-resistance genes provides opportunities for enhancing crop production on acid soils. *J. Exp. Bot.* **2011**, *62*, 9–20.
27. Zhou, S.; Sauvé, R.; Thannhauser, T.W. Proteome changes induced by aluminium stress in tomato roots. *J. Exp. Bot.* **2009**, *60*, 1849–1857.
28. Kim, Y.Y.; Jung, K.W.; Yoo, K.S.; Jeung, J.U.; Shin, J.S. A stress-responsive caleosin-like protein, AtCLO4, acts as a negative regulator of ABA responses in *Arabidopsis*. *Plant Cell Physiol.* **2011**, *52*, 874–884.
29. Poxleitner, M.; Rogers, S.W.; Samuels, A.L.; Browse, J.; Rogers, J.C. A role for caleosin in degradation of oil-body storage lipid during seed germination. *Plant J.* **2006**, *47*, 917–933.
30. Murphy, D.J.; Hernandez-Pinzon, I.; Patel, K. Role of lipid bodies and lipid-body proteins in seeds and others tissues. *J. Plant Physiol.* **2001**, *158*, 471–478.
31. Mackey, D.; Holt, B.F., III; Wiig, A.; Dangl, J.L. RIN4 interacts with Pseudomonas syringae type III effector molecules and is required for RPM1-mediated resistance in *Arabidopsis*. *Cell* **2002**, *22*, 743–754.
32. Levine, A.; Belenghi, B.; Damari-Weisler, H.; Granot, D. Vesicle-associated membrane protein of Arabidopsis suppresses Bax-induced apoptosis in yeast downstream of oxidative burst. *J. Biol. Chem.* **2001**, *276*, 46284–46289.
33. Harris, E.H.; Boynton, J.E.; Gillham, N.W. Chloroplast ribosomes and protein synthesis. *Microbiol. Rev.* **1994**, *58*, 700–754.
34. Pietromonaco, S.F.; Denslow, N.D.; O'Brien, T.W. Proteins of mammalian mitochondrial ribosomes. *Biochimie* **1991**, *73*, 827–835.
35. Spremulli, L. Protein synthesis of mammalian mitochondria. Available online: http://www.chem.unc.edu/people/faculty/spremulli/ (accessed on 10 February 2014).
36. Zhou, S.; Sauvé, R.J. ; Liu, Z.; Reddy, S.; Bhatti, S.; Hucko, S.D.; Fish, T.; Thannhauser, T.W. Identification of salt-induced changes in leaf and root proteomes of the wild tomato, *Solanum chilense*. *J. Am. Soc. Hortic. Sci.* **2011**, *136*, 288–302.
37. Berndt, N. Roles and regulation of serine/threonine-specific protein phosphatases in the cell cycle. *Prog. Cell Cycle Res.* **2003**, *5*, 497–510.

38. Garcia, A.; Cayla, X.; Guergnon, J.; Dessauge, F.; Hospital, V.; Rebollo, M.P.; Fleischer, A.; Rebollo, A. Serine/threonine protein phosphatases PP1 and PP2A are key players in apoptosis. *Biochimie* **2003**, *85*, 721–726.
39. Osawa, H.; Matsumoto, H. Possible involvement of protein phosphorylation in aluminum-responsive malate efflux from wheat root apex. *Plant Physiol.* **2001**, *126*, 411–420.
40. Alcaide-Loridan, C.; Jupin, I. Ubiquitin and Plant Viruses, Let's Play Together! *Plant Physiol.* **2012**, *160*, 72–82.
41. Nezames, C.D.; Deng, X.Y. The COP9 Signalosome: Its regulation of Cullin-Based E3 Ubiquitin ligases and role in photomorphogenesis. *Plant Physiol.* **2012**, *160*, 38–46.
42. Wei, N.; Serino, G.; Deng, X.W. The COP9 signalosome: More than a protease. *Trends Biochem. Sci.* **2008**, *33*, 592–600.
43. Argyris, J.; Dahal, P.; Hayashi, E.; Still, D.W.; Bradford, K.J. Genetic Variation for lettuce seed thermoinhibition is associated with temperature-sensitive expression of abscisic acid, gibberellin, and ethylene biosynthesis, metabolism, and response genes. *Plant Physiol.* **2008**, *148*, 926–947.
44. Finkelstein, R.R.; Lynch, T.J. Abscisic acid inhibition of radicle emergence but not seedling growth is suppressed by sugars. *Plant Physiol.* **2000**, *122*, 1179–1186.
45. Orton, R.J.; Sturm, O.E.; Vyshemirsky, V.; Calder, M.; Gilbert, D.R.; Kolch, W. Computational modelling of the receptor-tyrosine-kinase-activated MAPK pathway. *Biochem. J.* **2005**, *392*, 249–261.
46. Arroyo-Serralta, G.A.; Ku´-Gonza´lez, A.; Herna´ndez-Sotomayor, S.M.T.; Zu´n˜iga-Aguilar, J.J. Exposure to toxic concentrations of Aluminum activates a MAPK-like protein in cell suspension cultures of *Coffea arabica*. *Plant Physiol. Biochem.* **2005**, *43*, 27–35.
47. Ramı´rez-Benı´tez, J.E.; Chee-Gonza´lez, L.; Herna´ndez-Sotomayor, S.M.T. Aluminum induces changes in organic acids metabolism in *Coffea arabica* suspension cells with differential Al-tolerance. *J. Inorg. Biochem.* **2008**, *102*, 1631–1637.
48. Schaller, A.; Oecking, C. Modulation of plasma membrane H^+-ATPase activity differentially activates wound and pathogen defense responses in tomato plants. *Plant Cell* **1999**, *11*, 263–272.
49. Pearce, G.; Moura, D.S.; Stratmann, J.; Ryan, C.A., Jr. RALF, a 5-kDa ubiquitous polypeptide in plants, arrests root growth and development. *Proc. Natl. Acad. Sci. USA* **2001**, *98*, 12843–12847.
50. Richards, K.D.; Schott, E.J.; Sharma, Y.K.; Davis, K.R.; Gardner, R.C. Aluminum induces oxidative stress genes in *Arabidopsis thaliana*. *Plant Physiol.* **1998**, *116*, 409–418.
51. Delisle, G.; Champoux, M.; Houde, M. Characterization of oxalate oxidase and cell death in Al-sensitive and tolerant wheat roots. *Plant Cell Physiol.* **2001**, *42*, 324–333.
52. Soto, G.; Stritzler, M.; Lisi, C.; Alleva, K.; Pagano, M.E.; Ardila, F.; Mozzicafreddo, M.; Cuccioloni, M.; Angeletti, M.; Ayub, N.D. Acetoacetyl-CoA thiolase regulates the mevalonate pathway during abiotic stress adaptation. *J. Exp. Bot.* **2011**, *62*, 5699–5711.
53. Nveawiah-Yoho, P.; Zhou, J.; Palmer, M.; Sauve, R.; Zhou, S.; Howe, K.J.; Fish, T.; Thannhauser, T.W. Identification of Proteins for Salt Tolerance Using a Comparative Proteomics Analysis of Tomato Accessions with Contrasting Salt Tolerance. *J. Am. Soc. Hortic. Sci.* **2013**, *138*, 382–394.

54. Yamamoto, Y.; Kobayashi, Y.; Devi, S.R.; Rikiishi, S.; Matsumoto, H. Aluminum toxicity is associated with mitochondrial dysfunction and the production of reactive oxygen species in plant cells. *Plant Physiol.* **2002**, *128*, 63–72.
55. Yin, L.; Mano, J.; Wang, S.; Tsuji, W.; Tanaka, K. The involvement of lipid peroxide-derived aldehydes in aluminum toxicity of tobacco roots. *Plant Physiol.* **2010**, *152*, 1406–1417.
56. Von Uexküll, H.R.; Mutert, E. Global extent, development and economic impact of acid soils. In *Plant-Soil Interactions at Low pH: Principles and Management*; Date, R.A., Grundon, N.J., Raymet, G.E., Probert, M.E., Eds.; Kluwer Academic Publishers: Dordrecht, The Netherlands, 1995; pp. 5–19.
57. Carver, B.F.; Ownby, J.D. Acid soil tolerance in wheat. *Adv. Agron.* **1995**, *54*, 117–173.
58. Sawyer, S. Anhydrous ammonia application and dry Soils. Available online: http://www.extension.iastate.edu/CropNews/2011/1028sawyer.htm (accessed on 1 September 2012).

Appendix Table A1. Induction of significantly changed proteins in radicles of seeds derived from aluminum-treated tomato plants [z].

Protein accessions [y]	Protein name [x]	log2 fold [w] (Treated/ control)	Ratio [v] Treated/ control)
Solyc03g019820.2.1	Aquaporin	−1.88	0.27
Solyc06g034040.1.1	Oleosin	−1.73	0.30
Solyc06g072130.2.1	Aquaporin	−1.64	0.32
Solyc02g086490.2.1	Oleosin	−1.61	0.33
Solyc02g084840.2.1	Dehydrin DHN1	−1.53	0.35
Solyc03g112440.1.1	Oleosin	−1.52	0.35
Solyc06g072670.2.1	Short-chain dehydrogenase/reductase SDR	−1.47	0.36
Solyc06g053740.2.1	Ubiquitin carboxyl-terminal hydrolase	−1.42	0.37
Solyc01g109920.2.1	Dehydrin	−1.35	0.39
Solyc06g065050.1.1	Transmembrane protein 205	−1.33	0.40
Solyc02g077240.2.1	Pyruvate decarboxylase	−1.30	0.41
Solyc12g010920.1.1	Oleosin	−1.28	0.41
Solyc09g082330.1.1	7S vicilin	−1.24	0.42
Solyc12g096930.1.1	Caleosin	−1.18	0.44
Solyc10g008040.2.1	Seed biotin-containing protein SBP65	−1.17	0.44
Solyc11g067250.1.1	Poly (AHRD V1 ***- B9SCR8_RICCO)	−1.11	0.46
Solyc02g085590.2.1	Vicilin	−1.10	0.47
Solyc11g072380.1.1	Vicilin-like protein	−1.09	0.47
Solyc06g075270.2.1	Convicilin	−1.07	0.48
Solyc01g100390.2.1	Pyrophosphate-energized proton pump	−1.06	0.48
Solyc06g009210.2.1	Ribosomal protein L19	−1.03	0.49
Solyc09g025210.2.1	Legumin 11S-globulin	−1.03	0.49
Solyc05g053140.2.1	26S proteasome non-ATPase regulatory subunit 13	−1.02	0.49
Solyc10g076510.1.1	Pyruvate decarboxylase	−1.01	0.50
Solyc05g053120.1.1	Glucosyltransferase	−0.99	0.50
Solyc08g014000.2.1	Lipoxygenase	−0.98	0.51
Solyc08g078850.2.1	L-lactate dehydrogenase	−0.95	0.52
Solyc01g009660.1.1	Low-temperature-induced 65 kDa protein	−0.94	0.52
Solyc01g098850.2.1	Short-chain dehydrogenase/reductase family protein	−0.94	0.52
Solyc06g076640.2.1	Tubulin beta chain	−0.94	0.52
Solyc03g116590.2.1	Embryo-specific 3	−0.94	0.52
Solyc06g059740.2.1	Alcohol dehydrogenase 2	−0.93	0.53
Solyc11g042800.1.1	Late embryogenesis abundant protein	−0.93	0.53
Solyc03g025810.2.1	Low-temperature-induced 65 kDa protein	−0.92	0.53
Solyc03g083970.2.1	IQ calmodulin-binding motif family protein	−0.91	0.53
Solyc08g013860.2.1	NAD-dependent malic enzyme 2	−0.90	0.53
Solyc10g078770.1.1	Seed maturation protein LEA 4	−0.90	0.53
Solyc09g090150.2.1	Legumin 11S-globulin	−0.89	0.54
Solyc03g112590.2.1	Cell division protease ftsH homolog	−0.89	0.54

Table A1. *Cont.*

Protein accessions [y]	Protein name [x]	log2 fold [w] (Treated/ control)	Ratio [v] Treated/ control)
Solyc04g064710.2.1	Alcohol dehydrogenase 2	−0.89	0.54
Solyc00g297330.1.1	Unknown Protein	−0.88	0.54
Solyc07g053360.2.1	Seed biotin-containing protein SBP65	−0.87	0.55
Solyc08g080480.2.1	Unknown Protein	−0.87	0.55
Solyc06g074750.1.1	Histone H2B	−0.86	0.55
Solyc12g098940.1.1	Ubiquitin	−0.85	0.55
Solyc07g032740.2.1	Aspartate aminotransferase	−0.85	0.56
Solyc01g007940.2.1	Alanine aminotransferase 2	−0.83	0.56
Solyc09g065470.2.1	Vicilin	−0.83	0.56
Solyc09g015070.2.1	Reductase 1	−0.83	0.56
Solyc12g014380.1.1	Glucose-6-phosphate isomerase 1	−0.81	0.57
Solyc07g005390.2.1	Aldehyde dehydrogenase	−0.81	0.57
Solyc09g082340.2.1	Vicilin-like protein	−0.80	0.57
Solyc01g107910.2.1	Caffeoyl CoA 3-O-methyltransferase	0.80	1.74
Solyc00g009020.2.1	Mitochondrial ATP synthase	0.80	1.74
Solyc06g063220.2.1	ATP synthase subunit epsilon mitochondrial	0.80	1.74
Solyc02g082090.2.1	Peroxidase	0.80	1.74
Solyc01g102830.2.1	Unknown Protein	0.81	1.75
Solyc00g147570.2.1	Gelsolin	0.81	1.75
Solyc01g080510.2.1	Os05g0406000 protein	0.81	1.75
Solyc08g068220.2.1	50S ribosomal protein L27	0.81	1.75
Solyc03g078000.2.1	High-affinity fructose transporter ght6	0.81	1.76
Solyc03g096840.2.1	Seed specific protein Bn15D1B	0.82	1.76
Solyc06g075810.2.1	NADH dehydrogenase	0.82	1.77
Solyc06g007630.1.1	Ferredoxin	0.82	1.77
Solyc05g007800.2.1	Negatively light-regulated protein	0.83	1.77
Solyc11g072450.1.1	Mitochondrial F0 ATP synthase D chain	0.83	1.78
Solyc10g078450.1.1	U6 snRNA-associated Sm-like protein LSm6	0.83	1.78
Solyc10g011760.2.1	Aldose 1-epimerase family protein	0.84	1.79
Solyc11g066390.1.1	Superoxide dismutase	0.84	1.79
Solyc03g078670.1.1	Unknown Protein	0.85	1.80
Solyc05g053960.2.1	Cysteine-rich extensin-like protein-2	0.85	1.81
Solyc03g097360.2.1	BolA-like	0.85	1.81
Solyc05g056020.2.1	V-type proton ATPase subunit G 2	0.86	1.81
Solyc07g063630.2.1	Vesicle-associated membrane family protein	0.86	1.82
Solyc04g082590.2.1	Canopy homolog 2	0.86	1.82
Solyc02g079750.2.1	Flavoprotein wrbA	0.87	1.82
Solyc02g078540.2.1	Unknown Protein	0.87	1.82
Solyc11g065270.1.1	CHCH domain containing protein	0.87	1.83
Solyc01g007670.2.1	30S ribosomal protein S7 chloroplastic	0.87	1.83
Solyc07g021500.1.1	Unknown Protein	0.87	1.83

Table A1. *Cont.*

Protein accessions [y]	Protein name [x]	log2 fold [w] (Treated/ control)	Ratio [v] Treated/ control)
Solyc06g083820.2.1	60 ribosomal protein L14	0.88	1.84
Solyc00g072400.2.1	Peroxidase 1	0.88	1.84
Solyc08g006900.2.1	Ribosomal protein L32	0.88	1.84
Solyc10g007350.2.1	Multiprotein bridging factor 1	0.88	1.84
Solyc07g055250.2.1	Cell wall-associated hydrolase	0.89	1.85
Solyc08g075830.2.1	Peroxidase 27	0.89	1.85
Solyc05g041610.1.1	Caffeoyl-CoA O-methyltransferase	0.89	1.86
Solyc07g008350.2.1	Porin/voltage-dependent anion-selective channel protein	0.89	1.86
Solyc11g011340.1.1	Alcohol dehydrogenase	0.90	1.86
Solyc12g094700.1.1	Cathepsin B-like cysteine proteinase	0.90	1.87
Solyc01g049960.2.1	Unknown Protein	0.92	1.89
Solyc03g114970.2.1	Nitrilase associated protein-like	0.92	1.89
Solyc09g082710.2.1	Histone H2A	0.92	1.89
Solyc08g016420.2.1	Prefoldin subunit 6	0.92	1.90
Solyc03g025850.2.1	Remorin 1	0.93	1.91
Solyc01g103220.2.1	Cytochrome c	0.94	1.92
Solyc07g065640.2.1	RPM1 interacting protein 4 transcript 2	0.94	1.92
Solyc05g056290.2.1	Acetyl-CoA carboxylase biotin carboxyl carrier protein	0.95	1.93
Solyc04g049330.2.1	V-type proton ATPase subunit G 1	0.96	1.94
Solyc01g091130.2.1	Nitroreductase	0.96	1.95
Solyc01g095150.2.1	Late embryogenesis abundant protein	0.97	1.96
Solyc07g005240.2.1	FAD-dependent oxidoreductase family protein	0.97	1.97
Solyc01g090360.2.1	Non-specific lipid-transfer protein	0.98	1.97
Solyc01g095050.2.1	Negatively light-regulated protein	0.98	1.98
Solyc04g082010.1.1	Plastocyanin	1.00	1.99
Solyc11g008990.1.1	Phage shock protein A PspA	1.00	2.00
Solyc04g007750.2.1	Major latex-like protein	1.01	2.01
Solyc03g113730.2.1	B12D protein	1.03	2.04
Solyc08g013930.2.1	Peroxidase family protein	1.04	2.05
Solyc01g088140.2.1	Unknown Protein	1.04	2.05
Solyc02g085230.2.1	Nucleolar protein 6	1.04	2.06
Solyc06g036380.1.1	Ulp1 protease family C-terminal catalytic domain containing protein	1.04	2.06
Solyc12g019040.1.1	Exostosin family protein	1.04	2.06
Solyc03g116060.2.1	Gibberellin-regulated protein	1.04	2.06
Solyc06g054520.1.1	3-hydroxyisobutyryl-CoA hydrolase	1.05	2.07
Solyc02g043900.1.1	Unknown Protein	1.06	2.08
Solyc07g041490.1.1	Stress responsive alpha-beta barrel domain protein	1.07	2.09
Solyc04g071580.2.1	Unknown Protein	1.08	2.12

Table A1. *Cont.*

Protein accessions [y]	Protein name [x]	log2 fold [w] (Treated/ control)	Ratio [v] Treated/ control)
Solyc08g008330.2.1	Unknown Protein	1.09	2.13
Solyc09g074890.1.1	Rapid alkalinization factor 1	1.11	2.15
Solyc04g024840.2.1	GDSL esterase/lipase 1	1.11	2.16
Solyc04g074900.2.1	40S ribosomal protein S21	1.12	2.17
Solyc06g054250.2.1	5&apos-nucleotidase surE	1.14	2.21
Solyc02g092270.2.1	NADH dehydrogenase	1.14	2.21
Solyc12g008950.1.1	At1g17490/F1L3_4	1.15	2.22
Solyc10g076240.1.1	Peroxidase 1	1.21	2.31
Solyc03g113580.1.1	Germin-like protein	1.26	2.40
Solyc03g118110.2.1	Succinate dehydrogenase assembly factor 2, mitochondrial	1.30	2.45
Solyc07g054960.1.1	Myb-related transcription factor	1.35	2.55
Solyc10g005660.2.1	COP9 signalosome subunit 6	1.35	2.55
Solyc11g010160.1.1	Cc-nbs-lrr, resistance protein	1.35	2.55
Solyc06g062770.2.1	At1g17490/F1L3_4	1.36	2.56
Solyc03g117810.2.1	Phosphate import ATP-binding protein pstB 1	1.36	2.57
Solyc11g066270.1.1	Xyloglucan endotransglucosylase/hydrolase 9	1.37	2.59
Solyc05g007090.2.1	Zinc knuckle	1.37	2.59
Solyc01g107990.2.1	MAP protein kinase-like protein	1.42	2.68
Solyc00g015000.1.1	DNA (Cytosine-5-)-methyltransferase 3	1.55	2.93
Solyc02g093230.2.1	Caffeoyl-CoA *O*-methyltransferase	1.55	2.93
Solyc04g028490.1.1	Ulp1 protease family *C*-terminal catalytic domain containing protein	1.55	2.93

[z] Tomato proteins identified as significantly induced or repressed in tomato radicles from seeds germinated in 50 μM AlK $(SO_4)_2$ in 50 mM Homopipes buffer (pH, 4.5) (treated) and those in the buffer only (control). Tomato seeds were harvested from plants grown in a hydroponic solution supplemented with 50 μM AlK $(SO4)_2$. [y] Protein accession number is from the ITAG Protein database (release 2.3 on 26 April 2011; Sol Genomics Network, Boyce Thompson Institute, Ithaca, NY, USA). [X] protein name annotated in ITAG2.3database. [w] The log2 ratio of each protein between treated and control samples measured by the intensity of its constituent peptides. All the listed proteins have passed the *t* test [general linear model (GLM)] with false discovery rate (FDR) corrections ($p \leq 0.05$), and with a log2 fold change greater than 0.80 (±) which equals to two standard deviations of the near-normal distribution of log2 fold for all proteins identified in the experiment. Statistical analyses were performed using SAS (Version 9.3; SAS Institute Inc. Cary, NC, USA). [V] The ratio of protein abundance between treated and control samples, which is antilogarithm of the log2 ratio.

8

Enhanced Synthesis of Antioxidant Enzymes, Defense Proteins and Leghemoglobin in Rhizobium-Free Cowpea Roots after Challenging with *Meloydogine incognita*

Jose T. A. Oliveira [1,*], Jose H. Araujo-Filho [2], Thalles B. Grangeiro [3], Darcy M. F. Gondim [1,4], Jeferson Segalin [5], Paulo M. Pinto [6], Celia R. R. S. Carlini [5], Fredy D. A. Silva [1], Marina D. P. Lobo [3], Jose H. Costa [1] and Ilka M. Vasconcelos [1]

1 Department of Biochemistry and Molecular Biology, Federal University of Ceara, Fortaleza 60451-970, Brazil; E-Mails: darcymayra@hotmail.com (D.M.F.G.); fredydavi@hotmail.com (F.D.A.S.); costajhe@yahoo.com.br (J.H.C.); imvasco@ufc.br (I.M.V.)

2 Department of Biological Sciences, State University of Rio Grande do Norte, Rio Grande do Norte, Mossoro 59610-210, Brazil; E-Mail: heliofilho10@uol.com.br

3 Department of Biology, Federal University of Ceara, Fortaleza 60451-970, Brazil; E-Mails: thalles@ufc.br (T.B.G.); marinadplobo@gmail.com (M.D.P.L.)

4 University of Fortaleza (UNIFOR), Ceara, Fortaleza 60451-970, Brazil

5 Department of Biophysics and Center of Biotechnology, Federal University of Rio Grande do Sul, Rio Grande do Sul, Porto Alegre 91501-970, Brazil; E-Mails: jsegalin@inmetro.rs.gov.br (J.S.); ccarlini@ufrgs.br (C.R.R.S.C.)

6 Institute of Biotechnology, University of Caxias do Sul, Caxias do Sul, RS, Rio Grande do Sul, Caxias do Sul 95070-560, Brazil; E-Mail: paulomarcospinto@gmail.com

* Author to whom correspondence should be addressed; E-Mail: jtaolive@ufc.br

External Editors: Dr. Setsuko Komatsu and Dr. Zahed Hossain

Abstract: The root knot nematodes (RKN), *Meloydogine* spp., particularly *Meloidogyne incognita* and *Meloidogyne javanica* species, parasitize several plant species and are responsible for large annual yield losses all over the world. Only a few available chemical nematicides are still authorized for RKN control owing to environmental and health reasons. Thus, plant resistance is currently considered the method of choice for controlling RKN, and research performed on the molecular interactions between plants and nematodes to identify genes of interest is of paramount importance. The present work aimed to identify the differential accumulation of root proteins of a resistant cowpea genotype (CE-31) inoculated with *M. incognita* (Race 3) in comparison with mock-inoculated control, using 2D electrophoresis assay, mass spectrometry identification and gene expression analyses by RT-PCR. The results showed that at least 22 proteins were differentially represented in response to RKN challenge of cowpea roots mainly within 4–6 days after inoculation. Amongst the up-represented proteins were SOD, APX, PR-1, β-1,3-glucanase, chitinases, cysteine protease, secondary metabolism enzymes, key enzymes involved in ethylene biosynthesis, proteins involved in MAPK pathway signaling and, surprisingly, leghemoglobin in non-rhizobium-bacterized cowpea. These findings show that an important rearrangement in the resistant cowpea root proteome occurred following challenge with *M. incognita.*

Keywords: cowpea; defense proteins; *Meloidogyne incognita*; plant proteomics

---

1. Introduction

The root knot nematodes (RKN, *Meloidogyne* spp.) are among the most damaging plant parasites, as they establish feeding sites in the roots of major crops, preventing the normal uptake of water and nutrients. They are responsible for large annual yield losses all over the world [1], and their economic importance is increasing, as only a few available chemical nematicides are still authorized for RKN control, owing to environmental and health reasons. Thus, plant resistance is currently considered the method of choice for controlling root-knot nematodes, and research performed on the molecular interactions between plants and nematodes to identify genes of interest is of paramount importance.

RKNs are obligate biotrophic pathogens that establish and maintain permanent feeding cells, the giant cells, inside the root system of host plants, from which they draw off nutrients to complete their life cycle. The giant cells result from repeated rounds of karyokinesis without cell division. Hyperplasia and hypertrophy of the cells surrounding the feeding sites lead to the formation of tumors designated as root galls, the primary visible symptom of infection [1]. These symptoms occur in susceptible plants, presumably because they do not perceive the enemy nor activate their defense mechanisms efficiently. Resistant plants can trigger plant immune responses, as they possess the pattern recognition receptors (PRR) that recognize conserved pathogen-derived molecules, the pathogen- or microbe-associated molecular patterns (PAMPs/MAMPs) and/or possess R proteins (NB-LRR proteins) composed of a central nucleotide-binding site (NBS) and a C-terminal leucine-rich repeat (LRR), which detect pathogen effectors [2]. Most of the several plant proteins conferring resistance to nematodes have been identified as NB-LRR proteins [3]. After recognition, transcriptional reprogramming of the plant is induced by the nematode, both locally and systemically throughout the plant [4]. In various incompatible relationships between pathogens and the resistant plant, one of the first events observed after recognition is the oxidative burst, during which a rapid generation of reactive oxygen species (ROS), such as superoxide anion (O_2^-), hydrogen peroxide (H_2O_2) and also nitric oxide (NO), occur locally in the site of attempted infection. ROS generation is often associated with the hypersensitive response (HR), a programmed cell death (PCD) process that occurs around the infection site [5] as a plant attempts to hamper the pathogen invasion. An excess of ROS generated during HR causes considerable cell damage, but plants can activate various mechanisms for the efficient scavenging of these transient augmentations in ROS. These include the non-enzymatic antioxidant systems, such as ascorbate and glutathione, and the enzymatic ROS-scavenging mechanisms in which catalase, peroxidase, ascorbate peroxidase, superoxide dismutase, glutathione peroxidases and peroxiredoxins participate. Transiently elevated ROS levels have also been considered as second messengers in plant, as they are perceived by different receptors, proteins or enzymes and seem to be involved with the regulation of phytohormones, such as ethylene (ET), salicylic acid (SA) and jasmonic acid (JA), which play important roles in plant-pathogen interactions [6].

After HR, a second kind of induced response against pathogen attack, the systemic acquired response (SAR), takes place, in which various defense genes are over- or down-regulated, mainly by intervention of SA, JA and ET [6]. Das *et al.* [7] showed that 552 genes were significantly differentially expressed between the *M. incognita*-infected and non-infected resistant cowpea CB46 plants and amongst the upregulated genes, there were those involved in metabolism (42.8%), genes coding for proteins with binding functions (25.3%) and genes involved in the interaction with the environment (15.8%), whereas those gene downregulated the code for proteins with binding functions (34.7%), metabolism (29.6%) and protein fate (20.3%).

The cowpea (*Vigna unguiculata* (L.) Walp.) legume is an important crop used as food mostly in tropical and semi-arid regions of the world. The dried seeds, leaves, immature seeds and fresh green pods are all consumed. However, the cowpea seeds represent the major form of utilization, because of their nutritional profile, particularly protein (20.3%–29.3%) and carbohydrate (55.6%–74.5%) contents [8]. The resistance of cowpeas to *M. incognita* resides on a single gene or locus, designated Rk, with alleles rk, rki, Rk, Rk2 and Rk3, which effectively inhibit the reproduction of *M. incognita* [9]. The cowpea genotype CE-31 is highly resistant to *Meloydogine incognita* Race 3 [10].

The aim of this work was to analyze the differential accumulation of proteins in the roots of the resistant cowpea genotype CE-31 inoculated with *M. incognita* (Race 3) and non-inoculated control, using a 2D electrophoresis assay associated with mass spectrometry identification and gene expression analyses by reverse transcription-polymerase chain reaction (RT-PCR).

2. Experimental Section

2.1. Nematode Inoculums

The root knot nematode (RKN) inoculum was obtained from a population of *Meloidogyne incognita* (Race 3) isolated from susceptible cowpea plants (cv. Vita-3), growing in 1.5-L plastic pots containing exhaustive tap water washed river bottom sand that was previously mixed with humus (5:1, m/m) and autoclaved (121 °C, 30 min, 1.5 kgf/cm^2 (a kilogram-force per square centimeter)). Plants were maintained in a greenhouse, where the average temperature varied from 25 °C (night) to 35 °C (day), relative humidity (RH) from 55% (day) to 80% (night) and natural light *ca.* 700 μmol·m^{-2}·s^{-1} of photosynthetically active radiation (PAR) at the plant canopy. Irrigation was done daily with distilled water for up to 4 days after sowing, followed by irrigation (100 mL/pot) with 5-times diluted nutritive Hoagland and Arnon solution, as previously described [11]. Egg masses from *M. incognita* were isolated from galled cowpea roots using a stylet under a stereoscopic microscope (ausJENA, Jena, Germany). The egg masses were sterilized by immersion in sodium hypochlorite (0.05%, v/v active chloride) for 3 min, followed by three washings with sterile Milli-Q grade water [12]. Next, they were placed in sterile Milli-Q grade water in a Petri dish, and the infective, motile second-stage juveniles (J2) were allowed to hatch at around 26 °C in the dark. The J2 hatched within 24 h were discarded, and those of 48 and 72 h were collected every 12 h, concentrated using a 30-µm pore size nylon sieve and resuspended in sterile Milli-Q grade water to a 2000 J2/mL population that was used as the inoculum within 1 to 3 days of collection [13].

2.2. Plant Material and Nematode Inoculation

The cowpea seeds from cv. CE-31, highly resistant to *M. incognita* (Race 3) [10], were surface sterilized with sodium hypochlorite (0.05% active chloride) for 5 min, washed exhaustively with sterile Milli-Q grade water and germinated between two moist filter papers (Germitest™), which were placed in a plastic tray in the dark. The filter papers were watered with sterile Milli-Q grade water twice a day. Three days later, the seeds that had germinated were selected and transplanted to 1.5-L plastic jars (five per jar) containing river bottom sand thoroughly washed with tap water and autoclaved (121 °C, 30 min, 1.5 kgf/cm^2). The jars were kept inside a four-legged aluminum framework covered with an air permeable and transparent nylon net to protect the seedlings from dust and insects and maintained in a greenhouse under the same conditions of irrigation, temperature, photoperiod and PAR, as described above. Twelve days after seedling transplantation to the jars (15 days after planting), two plants were removed from each jar, and the three remaining ones inoculated with 2000 *M. incognita* J2 suspended in 2 mL of sterile water. The J2 suspension was placed in a 2-cm deep hole in the soil neighboring the main root axis of each plant and the hole filled with river sand. Controls were mock-inoculated with sterile water. The jars were arranged in a completely randomized-block design experiment, with twelve plants and three repetitions for each studied time point. Twelve plants were harvested at each time point (0, 12, and 24 h and 2, 4, 6, 8 and 10 days after inoculation). The plants were uprooted and the roots washed free of soil with distilled water, dried between two layers of paper towel, frozen in liquid nitrogen, powdered and stored at −80 °C for posterior use.

2.3. Extraction of Proteins from the Cowpea Roots

The frozen root powder (4.0 g) was resuspended in 8.0 mL of 100 mM Tris-HCl buffer, pH 8.0, containing 20% (v/v) glycerol, 3% (v/v) PEG, polyvinylpolypyrrolidone (PVPP) (1:2, m/v), 10.0 mM EDTA, 1.0 mM DTT (Dithiothreitol) and 1.0 mM PMSF (phenylmethylsulfonyl fluoride). After 2 h under gentle agitation at 4 °C, the suspension was centrifuged at 20,000× *g* for 20 min, 4 °C, the supernatant collected and the extracted proteins precipitated overnight at −80 °C with 30% (v/v) TCA (Trichloroacetic acid) in acetone. The proteins precipitated were centrifuged at 5000× *g* for 10 min, 4 °C, and the pellet washed twice in methanol, twice in acetone, vacuum dried and resuspended in 7 M urea/2 M thiourea. The final homogenate was centrifuged as above, the supernatant collected and the protein concentration measured using BSA as the standard [14]. To detect any possible contamination of the RKN-inoculated root protein samples with proteins originating from the nematode itself, the same above procedure to extract and process the root proteins was used to extract the nematode proteins from a mixture of 2000 J2 + 50 females + 50 eggs masses, and 2D electrophoresis gels were produced. The amount of J2, female and egg masses tested was about 50-times higher than those found, on average, in the roots of the resistant genotype, CE-31, infected with the *M. incognita* nematode [10].

2.4. Two-Dimensional Polyacrylamide Gel Electrophoresis (2D-SDS-PAGE)

Root proteins (200 μg) in 250 μL of 7 M urea, 2 M thiourea, 0.065 M DTT, 0.5% (m/v) CHAPS (3-[(3-cholamidopropyl)dimethylammonio]-1-propanesulfonate), 0.5% (m/v) IPG buffer and 0.002% (m/v) bromophenol blue were loaded onto 13-cm IPG immobilized pH (4–7) gradient strips (GE Healthcare, Uppsala, Sweden). The strips were rehydrated for 10 h (overnight) at 25 °C and isoelectric focused in an Ettan IPGphor II system (Amersham Biosciences, Piscataway, NJ, USA) programmed as follows: 200 V for 1 h, 500 V for 2 h, 5000 V for 2.5 h and, finally, 10,000 V for 1 h to achieve a total of 31.4 kVh. After isoelectric focusing, the strips were equilibrated for 15 min by shaking in 1.5 M Tris-HCL buffer, pH 8.8, containing 6 M urea, 30% (v/v) glycerol, 2% (m/v) SDS, 2% (m/v) DTT and trace amounts of bromophenol blue.

These same strips were re-equilibrated for a further 15 min in the above buffer, except that it contained 2.5% (m/v) iodoacetamide instead of DTT. For 2D-electrophoresis [15], the strip was fitted on a 15% (m/v) acrylamide gel (150 mm × 180 mm × 1.5 mm) sealed with 0.5% (w/v) agarose prepared in the electrode buffer. Electrophoresis was carried out at 40 mA/gel, in a vertical electrophoresis SE 600 unit (18 × 16 cm, GE-Healthcare; Amersham Bioscience) coupled to a circulating bath (MultTemp II, Pharmacia, LKB, Uppsala, Sweden) set at 5 °C. Protein spots were detected by colloidal Coomassie blue stain [16] and scanned at 300 dpi (ImageScanner Amersham Bioscience). Images (.tiff) were analyzed by the ImageMaster 2-D Platinum version 6.0 software, (GE-Healthcare; Amersham Bioscience). Three 2D patterns of proteins from RKN-inoculated and mock-inoculated control were produced for each time point after inoculation and compared in order to identify common, distinct, as well as differentially represented proteins. Only the protein spots that were at least two-fold up- or down-regulated after RKN inoculation, compared with the corresponding control (non-inoculated), were excised from 2D gels for protein identification after mass spectrometry.

2.5. In-Gel Digestion

Upregulated protein spots were manually excised from 2D gels run with RKN-inoculated cowpea root samples, whereas those downregulated were excised from control (non-inoculated) gels where they were more prominent. Each protein spot was individually transferred to 0.6 mL tubes, fragmented to about 1 mm^3 pieces, washed twice with ultrapure water and destained three-times with a 1:1 mixture of 25 mM ammonium bicarbonate and 50% (v/v) acetonitrile (ACN), pH 8.0, followed by two dehydration steps with 100% ACN for 5 min each.

After being dried under vacuum, the fragments were rehydrated with 20 μL of the digestion solution consisting of 25 mM ammonium bicarbonate, 1 mM $CaCl_2$ and 0.2 μg sequencing-grade modified trypsin (Promega, Madison, WI, USA). The reaction was done in a water bath at 37 °C for 16 h [17]. The resulting tryptic fragments were recovered by diffusion into a solution composed of 50% (v/v) ACN and 5% (v/v) trifluoroacetic acid (TFA) in 50 mM ammonium bicarbonate for 30 min, in three washes, transferred to micro tubes and dried under vacuum.

2.6. Mass Spectrometry Analysis

Prior to analysis, the dried peptides were dissolved in 10 μL of 0.1% formic acid. MS/MS analyses were performed by an electrospray ionization (ESI) quadrupole time-of-flight (Q-TOF) Micro™ mass spectrometer coupled to a nanoACQUITY® UltraPerformance liquid chromatography system (Waters, Milford, TX, USA). The peptides were loaded on a nanoeasy-C18 (75 μm ID) capillary column equilibrated with 98% Solution A (0.1% formic acid/water) and 2% B (ACN/0.1% formic acid). Elution was done with the gradient schedule: 2%–60% B for 13 min; 60%–95% B for 6 min; 95%–2% B for 11 min. Data were acquired in data-dependent mode (DDA), and multiple charged peptide ions (+2 and +3) were automatically mass selected and dissociated in MS/MS experiments. The ionization conditions and liquid chromatography were: 0.6 μL/min flow; 3.5 kV nanoflow capillary voltage; 100 °C block temperature; 50 V as the cone voltage.

The mass spectra were acquired and processed using the Mascot Distiller software (Matrix Science, London, UK), and the Mascot Generic Format (MGF) files generated were searched against the non-redundant protein sequence databases from NCBI (National Center for Biotechnology Information), using the MASCOT v. 2.2 software (Matrix Science, London, UK, www.matrixscience.com). Searches were performed using the following criteria: Viridiplantae as the taxonomic category, tolerance of one missed cleavage; cysteine carbamidomethylation; methionine oxidation; and 0.2 Da for peptide mass tolerance. The limit of significance was fixed at $p < 0.05$ and identification required that each protein contained at least one peptide with an expected value <0.05. The statistical test (ANOVA) was performed automatically. The proteins were categorized based on Bevan *et al.* [18].

2.7. Gene Expression Analysis

Gene expression analysis was accomplished by reverse transcription of mRNA templates coupled to *in vitro* amplification by polymerase chain reaction (RT-PCR). To this end, total RNA was extracted from RKN-inoculated and non-inoculated cowpea fresh roots at 4, 5 and 6 days after inoculation (DAI) by the Tris-lithium chloride procedure adapted from Chang *et al.* [19]. Two grams of fresh roots were ground with liquid nitrogen in a mortar and pestle and the fine powder transferred to Falcon tubes (15 mL) to which 6 mL of the extraction buffer (100 mM Tris-HCl pH 8.0 buffer, containing 2% CTAB (m/v), 2 M NaCl and 25 mM EDTA) were slowly added and gently mixed by inversion. The suspension was incubated at 25 °C for 1 h under gentle inversion and, next, an equal volume of chloroform/isoamyl alcohol solution (24:1, v/v) was added. The mixture was incubated at 25 °C for 20 min under gentle agitation and centrifuged (6500× *g*, 10 min, 25 °C).

The upper phase was transferred to a new Falcon tube, to which 7.0 mL of chloroform/isoamyl alcohol were added, mixed for 5 min and centrifuged under the above conditions. The supernatant was transferred to a new Falcon tube, and lithium chloride (10 M, 1/3 of total volume) was immediately added. The nucleic acids were left to precipitate overnight at 4 °C. Next, the suspension was centrifuged (8000× *g*, 45 min, 4 °C), the supernatant discarded, the precipitate washed two times with 70% ethanol and collected by centrifugation (8000× *g*, 10 min, 4 °C). The final pellet was left to dry at 25 °C and resuspended with 250 μL of diethyl pyrocarbonate (DEPC)-treated water. The integrity of the RNA samples was checked by 1% (m/v) agarose gel electrophoresis, and the yield was estimated by measuring the absorbance at 260 nm [20].

Prior to cDNA synthesis, residual DNA was removed with RQ1 RNAase-free DNAse I (Promega) and the RNA purified with the RNeasy mini Kit (Qiagen, Hilden, Germany), according to the manufacturer's instructions. First strand cDNA was synthesized from the mRNA present in the purified total RNA using oligo(dT)$_{18}$ (Fermentas Life Sciences, Burlington, ON, Canada) and the ImProm-II™ Reverse Transcriptase (Promega), according to the supplier's recommendations. The first-strand cDNA products were then amplified by PCR using oligonucleotide primers (Table 1) targeting some genes whose products showed a decreased or increased amount in RKN-inoculated roots in comparison to roots of non-inoculated plants, as detected by 2D-PAGE. Moreover, although chitinases were not identified among the selected protein spots that showed differential response in RKN-inoculated cowpea roots, primers targeting genes encoding chitinases were also designed and included in the gene expression analysis, owing to the well-known role of these proteins in plant defense. In addition, specific oligonucleotide primers targeting conserved regions of *nodC* [21], one of the genes responsible for the synthesis of the Nod-factors core structure and present in all nodulating rhizobia, were also used. This aimed at ensuring that root samples used in this study were not infected by *Rhizobium* spp. Cowpea genomic DNA was isolated by a CTAB-based protocol [22] and used as a template in pilot PCR amplifications to select the optimal annealing temperature for each pair of primers.

The reactions were performed in a final volume of 10 μL containing first-strand cDNA (750 ng), 1× GoTaq reaction buffer (Promega), 1.5 mM $MgCl_2$, 200 μM of each dNTP, 0.5 μM of each primer and 1.25 U of GoTaq DNA Polymerase (Promega). Amplifications were performed in a PTC-200 thermocycler (MJ Research, Waltham, MA, USA) using the following cycling parameters: an initial denaturation step (95 °C for 2 min) followed by cycles of denaturation (95 °C, 1 min), annealing (1 min) and extension (72 °C, 3 min). The number of cycles and the annealing temperatures varied according to the target transcript (Table 1). After the last cycle, the reactions were further incubated for 10 min at 72 °C. The PCR products were analyzed by 1% (m/v) agarose gel electrophoresis, and the DNA bands were stained with 0.5 μg/mL ethidium bromide and visualized under UV light.

3. Results

3.1. Proteomic Analysis of RKN-Inoculated and Non-Inoculated Cowpea Roots

Eleven solutions/buffers for the extraction of cowpea root proteins (Table 2) were individually mixed with the cowpea genotype CE-31 root powder (1:2, m/v) to test which one better extracted the proteins for proteome analysis. Dependent on the buffer system used, the protein concentration varied from 0.11 to 0.38 mg/mL with significant differences between some of the buffers tested. As in preliminary tests, we have carried out 2D electrophoresis runs loading 100, 150, 200, 250, 300, 350 and 400 μg root protein/gel in order to choose which buffer and concentration gave the best resolution of the protein spots. Repeated protein extraction procedures for every buffer were done, and the soluble proteins were obtained, concentrated by precipitation with TCA in acetone, as described in Section 2.3, and recovered to reach the desired amounts. Buffer 9 (Table 2) extracted the highest quantity of proteins, and the best concentration to load the 2D-gels here discussed was 200 μg protein, as below this value, most of the spots were not visible; at higher protein concentrations, the gels were overloaded and lost resolution. Moreover, the Buffer 9 system apparently removes unwanted compounds, as it produced stained gels with low backgrounds that did not interfere with image acquisition (ImageScanner Amersham Bioscience), as described in Section 2.4.

Table 1. Nucleotide sequences of the primers [1] used in RT-PCR.

Target Gene (GenBank Accession Number)	Oligonucleotide Sequences [2]	Position [3]	Amplicon Size (bp)	Annealing Temperature (°C)/Cycle	Reference
Asparaginyl endopeptidase (D89971)	5'-AACGGCTATTGGAACTAC-3' (f) 5'-GAGATCAGCATCCCTTTG-3' (r)	217–234 1074–1092	876	52.5/35	This work
ACC synthase (Z12135)	5'-CAAATGGGTCTTGCTGAGAAT-3' (f) 5'-TCTCAGCCTCTCCCTGTT-3' (r)	70–90 910–927 [4]	858	58.9/20	This work
ARG 10 (AB012110)	5'-CGAACACCATCGCCAAAG-3' (f) 5'-AGGGAAAGAAGCAAGCGA-3' (r)	104–121 584–601	498	56.4/35	This work
Chalcone-flavanone isomerase (AB073787)	5'-GAGAGGGGTTGACGATT-3' (f) 5'-GCCAATCATCGTCTCCAA-3' (r)	112–129 571–588	477	54.2/35	This work
Cysteinyl endopeptidase (U49445)	5'-TACGAGAGATGGAGGAGT-3' (f) 5'-TCCGACAATTGCTACACC-3' (r)	119–136 866–883	765	51.2/25	This work
Leghemoglobin (U33205)	5'-ATGGTTGCTTTCTCTGACAAG-3' (f) 5'-TTCATCACTCCATTTGTCTCC-3' (r)	46–66 400–420	375	64.7/35	This work
CuZn-superoxide dismutase (AJ278668)	5'-AAAGCGGTGGCGGTGCTGAAA-3' (f) 5'-GCTCAGTTCATGGCCGCCTTT-3' (r)	195–215 567–587	393	61.0/35	This work
nodC (AE006469)	5'-TGATYGAYATGGARTAYTGGCT-3' (f) 5'-CGYGACARCCARTCGCTRTTG-3' (r)	545–566 1164–1184	640	55.6/35	Sarita *et al.* 2005 [21]
Chitinase I (X88800)	5'-AGGATGATATGGAGCGTAGC-3' (f) 5'-GACAGGGTGAGATGTAGATC-3' (r)	14–33 966–985	972	58.0/28	This work
Chitinase IIIa (X88802)	5'-CTATCAACAACTGCAACGTG-3' (f) 5'-ATTTGGAAGAACCCTTGATG-3' (r)	152–171 708–727	576	55.0/28	This work
Chitinase IIIb (X88801)	5'-ACGTCAACATAGCTTTCCTC-3' (f) 5'-CTTCCCAGCAGGTACTGTAC-3' (r)	175–194 718–737	563	55.0/28	This work
Chitinase IV (X88803)	5'-GCTCAGAACTGTGGTTGTGC-3' (f) 5'-TAGCAAGTAAGATTATCACC-3' (r)	8–27 731–750	743	57.0/28	This work
Actin (AF143208)	5'-GCGTGATCTCACTGATGC-3' (f) 5'-TCGCAATCCACATCTGTTGG-3' (r)	669–686 1179–1198	530	59.0/35	Costa *et al.*, 2004 [23]

[1] The properties of each primer, including melting temperature, percent G + C content and PCR suitability, were determined using the PCR Primer Stats tool of the Sequence Manipulation Suite version 2 program (http://www.bioinformatics.org/sms2/index.html). [2] The forward and reverse primers are indicated by f and r, respectively (shown in parenthesis after each primer sequence). [3] Primer-binding sites in the corresponding target sequences deposited in the GenBank nucleotide database. [4] The numbers refer to the coding sequence extracted from Z12135.

Table 2. Solutions and buffers used to extract the root proteins * from cowpea genotype CE-31 for two-dimensional polyacrylamide gel electrophoresis (2D-SDS-PAGE) analysis.

(1) 50 mM pyridine + 10 mM thiourea + 1% (m/v) SDS + 100 M HCl, pH 5.0
(2) 50 mM pyridine + 10 M thiourea + 1% (m/v) SDS + 100 M HCl, pH 5.0, + 1:2 (m/v) polyvinylpolypyrrolidone (PVPP)
(3) 20 mM Tris-HCl pH 6.0, containing 20% (v/v) glycerol + 3% (v/v) *polyethylene glycol* (*PEG*)
(4) 20 mM Tris-HCl pH 6.0, containing 20% (v/v) glycerol, 3% (v/v) PEG, 1:2 (m/v) PVPP
(5) 40 mM Tris-HCl pH 7.0, containing 250 mM sucrose, 1% (v/v) triton X-100, 10 mM ethylenediaminetetraacetic acid (*EDTA*), 1.0 mM DTT, 1.0 mM phenylmethylsulfonyl fluoride (PMSF)
(6) 40 mM Tris-HCl pH 7.0, containing 250 mM sucrose, 1% (v/v) Triton X-100, 10 mM EDTA, 1.0 mM dithiothreitol (DTT), 1.0 mM PMSF, 1:2 (m/v) PVPP
(7) 100 mM Tris-HCl pH 8.0, containing 20% (v/v) glycerol, 3% (v/v) PEG
(8) 100 mM Tris-HCl pH 8.0, containing 20% (v/v) glycerol, 3% (v/v) PEG, 1:2 (m/v) PVPP.
(9) 100 mM Tris-HCl pH 8.0, containing 20% (v/v) glycerol, 3% (v/v) PEG, 1:2 (m/v) PVPP, 10 mM EDTA, 1.0 mM DTT, 1.0 mM PMSF
(10) 100 mM Tris-HCl pH 9.0, containing 0.01 M EDTA, 1% (v/v) Triton X-100
(11) 100 mM Tris-HCl pH 9.0, containing 0.01 M EDTA, 1% (v/v) Triton X-100, 1:2 (m/v) PVPP

* The proportion of cowpea genotype CE-31 root powder to solutions/buffers was 1:2 (m/v).

The 2D-gels obtained from three independent experiments for cowpea root samples collected at 0, 12, 24, 48 h after inoculation (HAI) and 4, 6, 8 and 10 days after inoculation (DAI) were matched. Only the spots present in all gels developed for each time point within the pI range of 4–7 were considered (Figure 1). In the RKN-infected cowpea roots compared with mock-inoculated plants (control), the protein spot alterations were more prominent between 4 and 6 DAI. About 339, 370 and 368 protein spots were detected in the control plants at 4, 5 and 6 DAI, respectively, whereas for the RKN-inoculated plants, they were around 347, 370 and 368. Taking into consideration only the protein spots that were two-fold or more altered ($p \leq 0.05$), a total of 32 proteins were up- (26 spots) or down-represented (six spots) in the RKN-inoculated compared with the non-inoculated roots of the resistant cowpea cv. CE-31. Out of these 32 proteins spots, 22 (17 up- and five down-represented) (Figures 1 and 2) were excised and further analyzed by mass spectrometry.

The remaining 10 protein spots (nine up- and one down-represented), although significantly ($p <$ 0.05) reprogrammed in the RKN-inoculated roots in comparison with control plants, were less abundant proteins only visualized at a certain zoom level by the ImageMaster 2-D Platinum version 6.0 software and, thus, were not excised due to technical difficulties, nor analyzed by mass spectrometry for identification. To exclude the possibility of misinterpretation because of the possible presence of RKN-derived proteins together with those originating from RKN-infected cowpea roots, 2D gels were also run only with the nematode-derived proteins under the same conditions of cowpea root sample and revealed using colloidal Coomassie or the silver nitrate staining method. However, no protein spot could be visualized, making it evident that all protein spots found in 2D gels were exclusively derived from the cowpea roots. The 22 root proteins excised from the gels were trypsinized and subjected to mass spectrometry analysis and their sequences compared for similar protein sequences deposited in the NCBI database. However, only 17 out of 22 spots were identified. Table 3 shows the identification of these proteins grouped according to their biological functions, as well as pI, molecular weight, fragment amino acid sequence, statistical scores and the percentages of coverage of their sequences. Fifteen (two downregulated and 13 upregulated) similar sequences were found. The upregulated proteins were: aminocyclopropane-1-carboxylic acid synthase (ACC synthase, Spot 2), 1-aminocyclopropane-1-carboxylate oxidase (ACC oxidase, Spot 3) cysteinyl endopeptidase (Spot 4), chalcone-flavone isomerase (Spot 6), ascorbate peroxidase (Spot 11), auxin-induced protein (Spot 12), superoxide dismutase CuZn-dependent (Spot 15), Class I heat shock protein (Spot 16), PR-1 (Spot 17), PR-3 (Spot 18), PR-2 (Spot 19), leghemoglobin (Spot 21) and nucleoside diphosphate kinase (Spot 22). The two downregulated proteins were identified as asparaginyl endopeptidase (Spot 1) and ARG 10 (Spot 10). All of the identified proteins showed score values significantly above the minimum threshold of reliability calculated automatically by the Mascot program for mass spectra analysis. These identified proteins showed similar mass spectra with those of plant species within the Fabaceae family to which *Vigna unguiculata* (cowpea) belongs. Six of them (Spots 1, 2, 3, 4, 10 and 12) showed peptide fragments similar to those of two other plant species belonging to the *Vigna* genus (*V. mungo*, *V. radiata*), and the remaining three proteins (Spots 11, 18, and 21) matched with peptides fragments of *V. unguiculata*. These findings show that the data of our study are consistent. Two protein spots (7 and 9) were matched with the protein sequences deduced *in silico*, but of unknown biological functions. Spots 5, 8, 13, 14 and 20 were not successfully identified.

Figure 1. Protein profiles for control and *M. incognita*-infected cowpea genotype CE-31 roots at 4, 5 and 6 days after inoculation (DAI). Root proteins (200 μg) were extracted and separated in the first dimension by isoelectric focusing (pI 4–7) and in the second dimension by SDS-PAGE. Protein spots were detected by colloidal Coomassie blue stain [16]. Proteins that had differential accumulation in *M. incognita*-infected plants 4, 5 and 6 DAI in relation to the respective controls (mock-inoculated) are circled. Arrows in the gels of control plants indicate proteins that were down-represented after challenging with *M. incognita*.

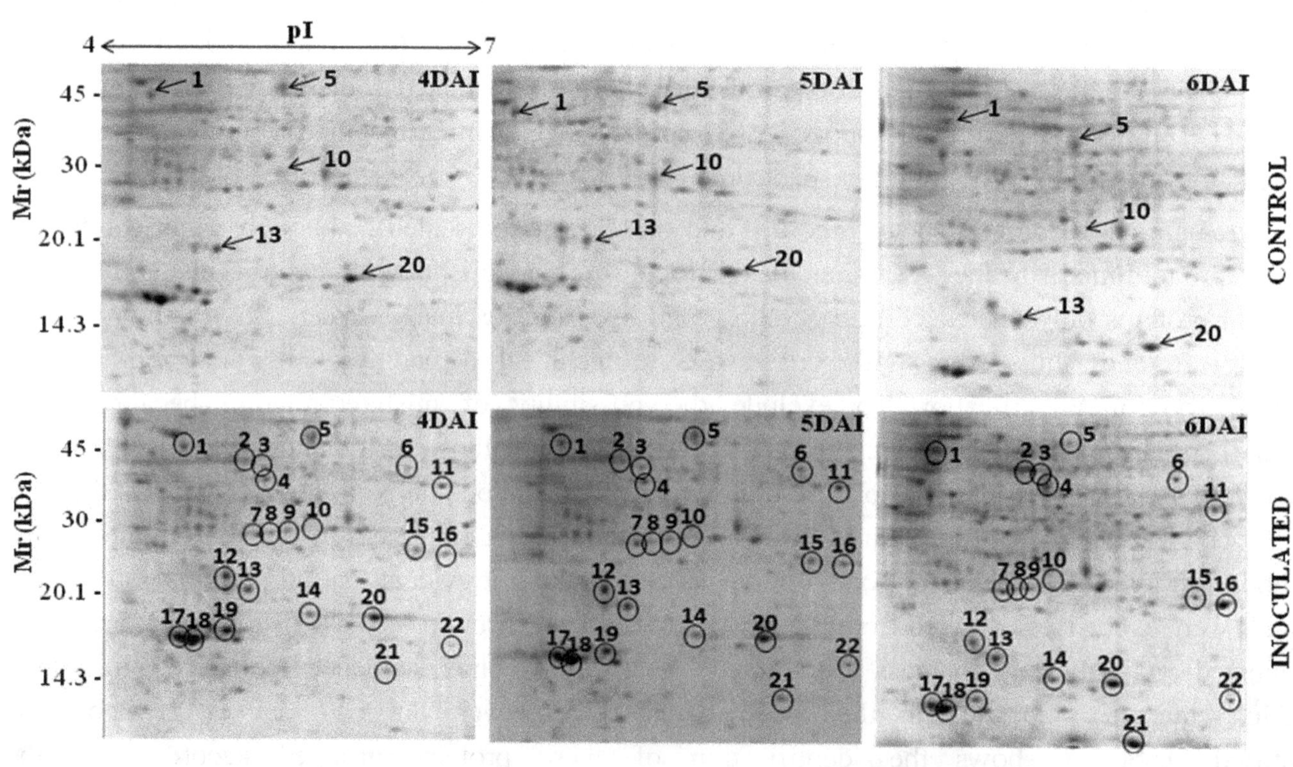

Figure 2. Enlarged views of the up- and down-represented proteins that were identified in the roots of the cowpea genotype CE-31 inoculated with *M. incognita* (Race 3) and the non-inoculated control. Numbers at the left correspond to those protein spots denoted in Figure 1. Arrows are placed on Spots 17 and 18 to indicate that they appear in pairs, whose levels changed in response to *M. incognita* infection. Numbers at the right of every double column denote the mean of the protein fold change measured as the difference in intensity for each spot between root knot nematode (RKN)-infected and control plants from three gels using different biological samples. The asterisk on the numbers denotes significant difference ($p \leq 0.05$) after application of the t-test.

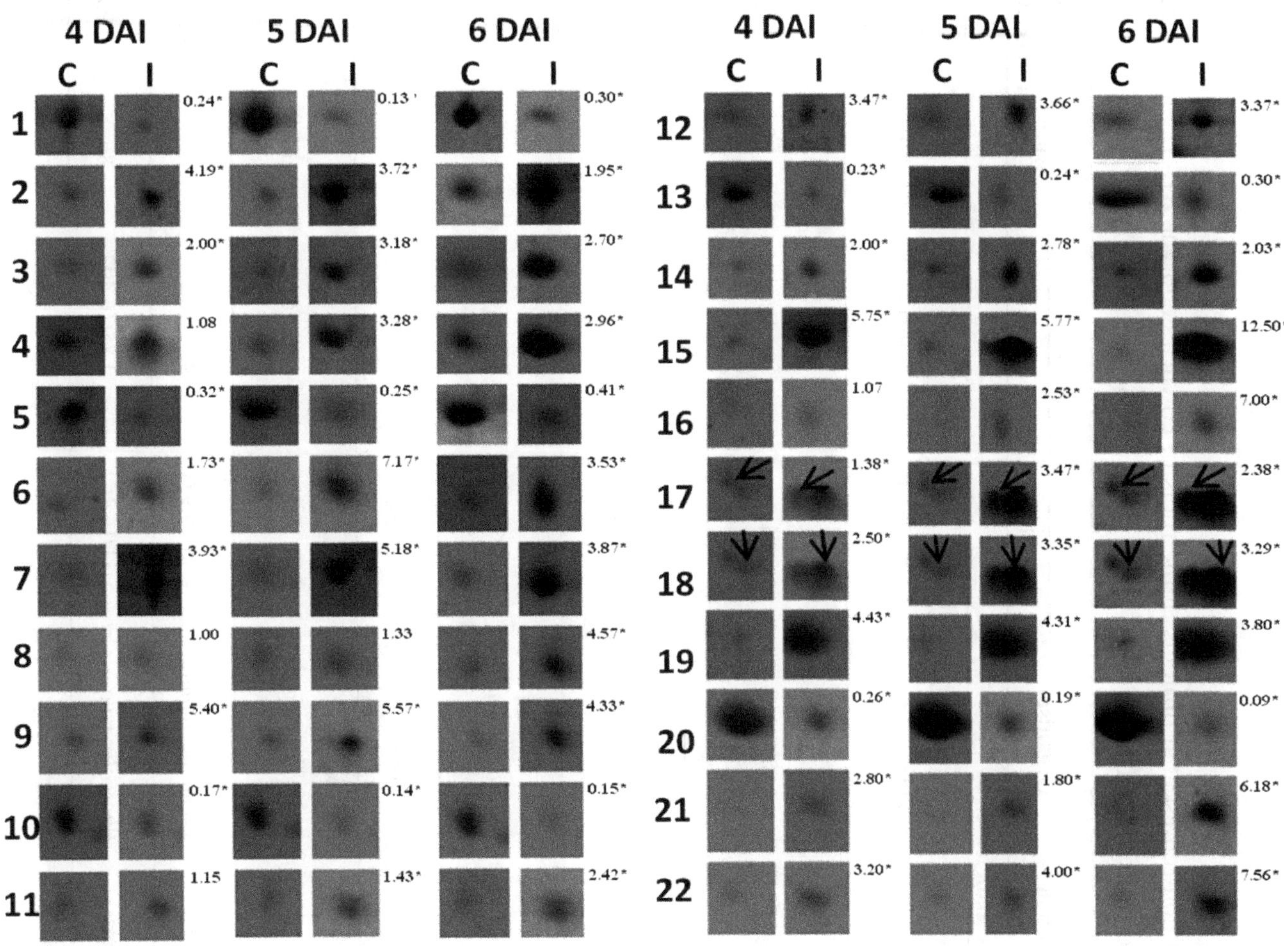

Table 3. Identification of up- and don-represented proteins of the roots of cowpea genotype CE-31 after being challenged with *M. incognita* (Race 3) in relation to non-infected control plants.

Spot * No.	Accession No. (NCBI)	Protein Identification	Organism	pI/MW (kDa) Experimental	pI/MW (kDa) Theoretical	Score	Sequence Covered (%)	Sequences
Functional category: Disease/defense								
2↑	gi/297493	ACC synthase	*Vigna radiata*	4.89/39.678	5.40/43.768	79	7.98	YFDGWK VHIVYSLSK VGTIYSYNDSVVTTAR
3↑	gi/86197901	ACC oxidase	*Vigna radiate*	5.05/38.056	5.86/40.236	111	12.30	GAAMEMIK EMVANK VSNYPPCPTPDLIK DDQWIDVPPMR
11↑	gi/42795352	Ascorbate peroxidase	*Vigna unguiculata*	6.62/28238	6.67/31.746	108	11.84	NCAPLMLR EIVALSGGHTLGR SGFDGPWTEDPLK
15↑	gi/13274148	CuZn-superoxide dismutase	*Populus tremula*	6.76/23.114	5.87/29.203	142	14.74	AVAVLK LTHGAPEDEIR GGHELSSTTGNAGGR
16↑	gi/123539	17.5 kDa HSP	*Glycine max*	5.87/22.268	6.12/17.412	177	18.18	DFHVPTSSVSAENSAFVSTR VLQISGER
17↑	gi/130829	PR-1	*Phaseolus vulgaris*	4.44/16.126	4.83/16.528	168	18.58	ALPDSFK ISFVEDGETK LSDGPNGGSLIK
18↑	gi/4850337	PR-3 (Chitinase)	*Vigna unguiculata*	4.55/15.971	4.75/16.265	122	14.28	ISFLEDGETK LSDGSNGGSVVK
19↑	gi/ 130835	PR-2 (β-1,3-glucanase)	*Phaseolus vulgaris*	5.11/16.352	4.85/16.402	86	10.96	ISIDSK GDAPPNEDELK

Table 3. *Cont.*

Spot * No.	Accession No. (NCBI)	Protein Identification	Organism	pI/MW (kDa)		Score	Sequence Covered (%)	Sequences
				Experimental	Theoretical			
Functional category: Secondary metabolism								
6↑	gi/27530706	Chalcone-flavone isomerase	*Lotus japonicus*	6.41/35666	5.94 /36.987	97	11.11	SYFLGGAGER STGTYGEAEAAAIGK
Functional category: Metabolism								
21↑	gi/ 20138591	Leghemoglobin	*Vigna unguiculata*	6.00/14.722	5.68/15.341	225	31.03	ADIPK NLFSFLANGVDATNPK ASGGVVADAALGAVHSQK EAVGDK
22↑	gi/ 26245403	Nucleoside diphosphate Kinase	*Glycine max*	6.89/15.964	6.30/16.254	183	21.08	PDGVQSGLIGEIISR IIGATNPAQSEPGTIR
Functional category: Protein destination and storage								
1↓	gi/ 4589396	Asparaginyl endopeptidase	*Vigna mungo*	4.38/49.316	5.14/52.982	77	6.83	FPIIFVVANLITLVSGGR NSLVPPSK APLGSSR
4↑	gi/ 1223922	Cysteinyl endopeptidase	*Vigna radiata*	5.10/35.289	5.44/37.332	64	6.35	LLWVVLSLSLVLGVANSFDFH EK
Functional category: Unknown/Predicted/Uncharacterized								
7↑	gi/297849580	Predicted protein	*Arabidopsis lyrata*	5.03/25.370	6.22/33.633	119	12.58	EVETLPEEAFEEEEDK EILENHGGEER IMDEAVNASR
9↑	gi/ 297824991	Predicted protein	*Arabidopsis lyrata*	5.26/25.967	7.97/39.819	58	5.54	QVVDETEPK VYGSIEEHYYR
10↓	gi/2970051	ARG 10	*Vigna radiata*	5.33/30.326	5.62/25.480	133	14.34	DEIFCLFEGALDNLGSLR VVCHLSGSFAFIVFDK
12↑	gi/416640	Auxin induced protein	*Vigna radiata*	4.81/19.354	4.65/21.345	104	11.34	EGLGLEITELR GYSDLAFALEK

* Arrows indicate up- (↑) and down-represented (↓) proteins. Functional categories are according to Bevan *et al.* [18].

3.2. RT-PCR Analyses of Cowpea Root Gene Expression

To validate the results obtained from the proteomic analysis, the levels of asparaginyl endopeptidase, ACC synthase, cysteinyl endopeptidase, chalcone-flavonone isomerase, ARG 10, ascorbate peroxidase (APX), superoxide dismutase (SOD) and leghemoglobin were analyzed by RT-PCR (Figure 3).

The transcript levels of asparaginyl endopeptidase (Spot 1) and ARG 10 (Spot 10) were downregulated between 4 and 6 days after *M. incognita* inoculation (Figure 3), as yet observed by 2D electrophoresis (Figures 2 and 3). On the other hand, increased accumulation patterns both at the protein (Figure 2: two-fold or more up-represented) and/or transcript levels (Figure 3) were observed for ACC synthase (Spot 2), cysteinyl endopeptidase (Spot 4), chalcone-flavone isomerase (Spot 6), Cu,Zn-superoxide dismutase (Spot 15), leghemoglobin (Spot 21) and chitinase type I, type IIIa and type IIIb. Amplification of *actin* cDNA, used as the endogenous control [23], showed homogeneity in RKN-inoculated and mock-inoculated cowpea root samples.

Figure 3. Kinetics of gene expression in the roots of the cowpea genotype CE-31 inoculated with *M. incognita* (Race 3) and the non-inoculated control. Roots were examined at the times indicated (4, 5 and 6 DAI) at the top of the figure. For experimental details, see the Experimental Section.

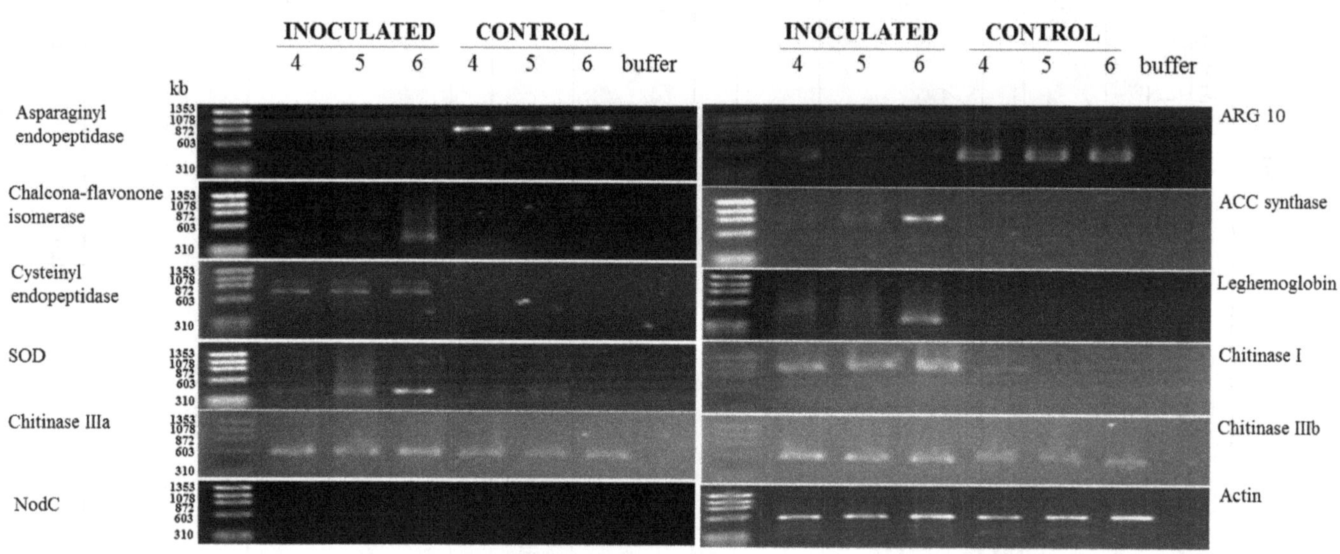

4. Discussion

In this study, 32 root proteins from the resistant cowpea cv. CE-31 were differentially regulated after challenge with the nematode *M. incognita* (Race 3). This genetic reprogramming was more noticeable between the fourth and sixth day after inoculation (DAI). Using a soybean genome array, Das *et al.* [7] showed that at 9 DAI, 141 genes were 1.5-fold or more upregulated, whereas 59 genes were downregulated in the *M. incognita* inoculated compared with the non-inoculated root of the resistant cowpea access CB46. These differences compared with our results might be due to the techniques used, the period examined, cowpea access-specific defense responses and, also, because we took into consideration protein spots that were two-fold or more up- or down-regulated after RKN inoculation compared with the corresponding control (non-inoculated). Certainly, using other quantitative mass spectrometry methods, such as multiplexed in-gel proteomics, label-free and selective or nonselective labeling of proteins, a much greater number of differential expressed proteins could have been identified in our study [24].

The asparaginyl endopeptidase level decreased within this time interval, compared with the non-inoculated controls (Spot 1: Figure 1; Figure 2; Table 3). It has been experimentally suggested that enzymes of this family also catalyze the transpeptidation by forming a peptide bond, leading to cyclization, as in the case of cyclotides [25]. Cyclotides belongs to a large family of macrocyclic plant proteins of 28–37 amino acids, with three intramolecular disulfide bonds. They have hemolytic, cytotoxic, antimicrobial, insecticidal and have molluscicidal and nematocidal activities, and in plants, their presumed role is to act as antibiotic agents to protect plants from pests or pathogens [26]. Taking into consideration the effects of cyclotides, the downregulation of asparaginyl endopeptidase observed in the cowpea roots challenged with *M. incognita* seems contradictory. Although some caution is needed while interpreting these results, the suppression of asparaginyl endopeptidase expression in the cowpea genotype CE-31 roots could be a nematode strategy to avoid damage by cyclotides. As a parasite, nematodes must protect themselves against plant defenses. Indeed, the potential ability of nematodes to mimic signals in natural plant pathways that manipulate various aspects of plant physiology, including plant defense responses, has been suggested [4]. A study of the *M. incognita* secretome by mass spectrometry identified 486 proteins, and several of these secreted proteins were homologous to plant proteins, which they may mimic, and contain domains that suggest effector functions toward regulating the plant cell cycle or growth, while others have regulatory domains that could reprogram host cells for its own purposes [27].

On the other hand, cysteinyl endopeptidase (Spot 4: Figure 1; Figure 2; Table 3), another proteinase, showed a strong accumulation in the roots of cowpea genotype CE-31 inoculated with *M. incognita*. In regard to the action of cysteine proteinases on nematodes, there is a patent for which it is proposed to use formulations based on at least one plant cysteine proteinase or active fragments to potentiate the anti-nematode effects of non-enzymatic nematicides [28]. Accordingly, cysteine proteinases from papaya latex, papain, stem bromelain and kiwi fruits could effectively reduce nematode infestation of host plants, as the cysteine proteinase attacks the protective cuticle of the nematode, causing blistering, and, eventually, total digestion. Therefore, the increased expression of cysteinyl endopeptidase demonstrated in this present work may have bearing on the resistance of the cowpea genotype CE-31 to *M. incognita* (Race 3).

ACC synthase (Spot 2: Figure 1; Figure 2; Figure 3; Table 3) and ACC oxidase (Spot 3: Figure 1; Figure 2; Table 3) were significantly upregulated, as observed by 2D gels of cowpea roots inoculated with root-knot nematodes compared to controls. The increased level of ACC synthase was also verified by RT-PCR, particularly at 6 DAI (Figure 3). ACC synthase is a key enzyme involved in the ethylene biosynthesis in plants. This sequential increase in ACC synthase and ACC oxidase suggests that the ethylene biosynthetic route was activated upon infection of the resistant cowpea CE-31 with *M. incognita* (Race 3). Thus, the increase in ethylene production after inoculation with *M. incognita* might contribute to the resistance of cowpea CE-31 to this nematode species. In contrast, in a compatible interaction, ACC oxidase was downregulated in the giant cells and surrounding cells seven days post-infection of *Medicago truncatula* cv. Jemalong A17, also a leguminous plant, with *M. incognita*, as observed by microarray hybridization using the Affymetrix GeneChip®Medicago genome [29].

According to these authors, this localized repression of the plant defense genes in cells of the host plant, *Medicago truncatula*, in direct contact with the nematode is in accordance with an effective suppression of defenses by secreted effectors of the pathogen, as previously commented. Nevertheless, as for other plant species, increased accumulation of defense transcripts of cowpea against *M. incognita* might result from gene regulation also by ethylene, although different plants may utilize different pathways for defense against a pathogen. Glazer *et al.* [30] have previously suggested that ethylene was closely associated with *M. javanica* infection, as infected tomato plants produced ethylene at a higher rate than uninfected plants and contained higher levels of the ethylene precursor, ACC.

There was a decrease in the abundance of the auxin downregulated ARG10 homologue (Spot 10: Figure 1; Figure 2; Table 3). Moreover, an auxin-induced protein (Spot 12: Figure 1; Figure 2; Table 3) was also upregulated upon *M. incognita* infection of cowpea CE-31. These findings suggest that the auxin level was augmented upon RKN-infection. The establishment and maintenance of nematode feeding sites are strongly influenced by the host plant ethylene and auxin signal transduction pathways [31].

In our previous studies with the pathosystem cowpea genotype CE-31 x *M. incognita*, the activities of the anti-oxidative enzymes, guaiacol peroxidase (POX) and superoxide dismutase (SOD), and those of the PR-proteins, β-1,3-glucanase (GLU), chitinase (CHI) and the cysteine protease inhibitor, were induced in the roots, within 4–8 DAI [10]. Using the proteomic approach and/or RT-PCR, it was confirmed here that CuZnSOD (Spot 15: Figure 1; Figure 2; Figure 3; Table 3), CHI (Spot 18: Figure 1; Figure 2; Figure 3; Table 3) and GLU (Spot 19: Figure 1; Figure 2; Table 3) were upregulated from 4 to 6 DAI. In addition, ascorbate peroxidase (Spot 11: Figure 1; Figure 2; Table 3) was also upregulated in RKN-inoculated cowpea cv. CE-31 in comparison with uninoculated controls. Copper/zinc superoxide dismutase (CuZnSOD) and APX are involved, together with other enzymes, such as CAT, glutathione peroxidase (GPX) and peroxiredoxin (PrxR), in the reactive oxygen species (ROS) network, more precisely with the fine control of hydrogen peroxide (H_2O_2) generation in plants, as SOD catalyzes the dismutation of superoxide anions to H_2O_2 and O_2, while APX converts H_2O_2 to water. H_2O_2 is a second messenger central in the activation of the mitogen-activated protein kinase (MAPK) cascade in plants. H_2O_2 is also involved in the cross-linking of cell wall proteins and plant cell wall bound-phenolics, lipid peroxidation, DNA and protein damage, HR, PCD and activation of defense genes and has microbicidal functions [32]. Accumulation of H_2O_2 in the leaves of the highly resistant (CE-31) cowpea genotype inoculated with *M. incognita* was previously noticed between 4 and 6 DAI and its decrease between 6 and 8 DAI [10]. *M. incognita* is a biotrophic organism, and therefore, tissue necrosis at the attempted site of nematode fixation caused by ROS during pathogen infection might increases host resistance. However, the persistence of high H_2O_2 levels could lead to excessive necrosis of the plant tissue. Thus, it is possible that at this stage (4–8 DAI), APX was enhanced in the studied cowpea to control excessive H_2O_2 generated by SOD activity and avoid excessive damage of the plant tissue. Indeed, in our previous enzyme kinetic studies, persistent high levels of SOD activity in the cowpea CE-31 roots between 2 and 10 DAI with *M. incognita* were also shown [10]. In soybean (*Glycine max*) roots infected with *M. incognita*, the increased SOD activity of the resistant centennial cultivar was also observed within 2–7 DAI over that of the respective uninoculated control [33].

The proteomic study of the cowpea CE-31 roots infected with *M. incognita* showed that chalcone-flavonone isomerase (CFI) significantly increased (Spot 6: Figure 1; Figure 2; Table 3) in relation to that of control plants. This finding at the protein level was in agreement with the gene induction observed by RT-PCR (Figure 3). CFI is directly related to the phenylpropanoid biosynthetic pathways, as it accelerates the spontaneous additional cyclization of chalcones to form the flavonoid core from which the antimicrobial compounds, phytoanticipins (constitutive) and phytoalexins (infection-induced), besides tannins and lignin, which also take part in the defense arsenal of plants, are derived [34]. Isoflavone reductase is a key enzyme involved in phytoalexin biosynthesis [34].

Increased accumulation of PR-1 (Spot 17: Figure 1; Figure 2; Table 3) and PR-2 (β-1,3-glucanase) (Spot 19: Figure 1; Figure 2; Table 3) in the roots of RKN-inoculated cowpea CE-31 was also noticed in comparison with control plants. Similarly, three PR-3 (chitinase) isoforms (class I, IIIa and IIIb) were upregulated, particularly chitinase I, as shown both by 2D electrophoresis (Spot 18: Figure 1; Figure 2; Table 3) and RT-PCR (Figure 3). In soybean challenged with *Meloidogyne incognita*, three chitinase isozymes with isoelectric points (pIs) of 4.8, 4.4 and 4.2 accumulated to a greater extent in the resistant (cv. Bryan) compared to the susceptible (cv. Brim) cultivar [35]. This increased accumulation of β-1,3-glucanase (PR-2) and chitinases (Figure 2) in cowpeas is in agreement with the time-course increase previously observed in the cowpea CE-31 roots by our research group [10]. PR-1, PR-2 and PR-3 belong to a protein group, designated pathogenesis-related proteins (PR-proteins), first discovered as being induced in tobacco mosaic virus (TMV)-infected tobacco plants and originally classified into five main groups (PR-1 to PR-5), based on decreasing electrophoretic mobility, but that today encompass seventeen classes numbered in the order of their discovery from PR-1 to PR-17 [36]. The PR-1 family is a dominant, highly conserved group of PRs in plants, induced by pathogens or salicylic acid (SA) and often associated with the establishment of systemic acquired resistance (SAR). Thus, it is plausible to speculate that the upregulation of PR-1 in the cowpea CE-31 root infected with *M. incognita* is associated with systemic acquried resistance (SAR).

Chitinases (PR-3) are enzymes that hydrolyze the beta-1,4-glycosidic linkage of chitin, present in filamentous fungi, insects and nematode eggshells. A great variety of studies have shown that chitinases play an important role in plant defense against biotic stresses. Of particular interest is that the development of eggs and hatching of *M. javanica* juveniles was blocked by proteases and chitinases secreted by *Paecilomyces lilacinus*, a parasite fungus that infects and assimilates eggs of the nematodes, *Meloidogyne* spp., *Globodera* spp. and *Heterodera* spp., as these enzymes drastically altered the eggshell structures when applied individually or in combination [37]. Therefore, overrepresentation of chitinase in the resistant cowpea cv. CE-31 might interfere with the morphofunctional state and hatching of nematode eggs.

β-1,3-Glucanases hydrolyzes β-1,3-glucans and represent the family of PR-2 proteins. Overaccumulation of β-1,3-glucanases together with upregulation of chitinases (PR-3) in response to various pathogen and insect attack has been reported to occur in several plants. PR-2 and PR-3 might contribute to plant defense by acting directly on the pathogen structure, leading to the release of elicitors, or eventually to pathogen death, or they can degrade endogenous plant substrates to generate signal molecules that may function as endogenous elicitors of active host defensive mechanisms [36].

In this present study, a nucleotide-diphosphate kinase (NDPK) was also overexpressed in the roots of the cowpea cv. CE-31 inoculated with *M. incognita*, when compared with mock-inoculated plants (Spot 22: Figure 1; Figure 2; Table 3). NDPKs catalyze the exchange of phosphate groups between different nucleoside diphosphates. Three groups of NDPKs (NDPK1, NDPK2, NDPK3) exist in plants. NDPK1 is localized in the cytosol, NDPK2 in the chloroplast stroma and NDPK3 in the chloroplasts (low abundance) and mitochondria (high abundance) [38]. As more than half of the NDPK transcript pool is represented by the cytosolic NDPK1 in the inflorescence, leaves and roots of *Arabidopsis thaliana* [37], it is supposed that the NADPK overexpressed in the cowpea CE-31 root challenge with *M. incognita* represents the NDPK1 group.

Nevertheless, plant NDPKs have been implicated in signal transduction events, UVB light signaling, hormone, heat shock response and interaction and seem to be involved in the mitogen-activated protein kinase (MAPK) pathway signaling [38]. TAB2, an NDPK of tomato, upregulated the expression of PR-1, PR-2 (β-1,3-glucanases) and PR-3 (chitinases) genes. Interestingly, a human NDPK isoform (Nm23) is a strong metastatic tumor suppressor [39]. In a compatible reaction of *M. incognita* with a host plant, one of the characteristic symptoms observed in the infected roots is the formation of the typical root gall (tumors) resulting from hyperplasia and hypertrophy of the cells surrounding the nematode feeding sites (giant cells) [1]. In the CE-31 resistant cowpea genotype, gall formation was a rare event [10].

A 17.5-kDa heat shock protein (HSP) class I (CI) was also upregulated in the cowpea CE-31 roots infected with *M. incognita* (Spot 16: Figure 1; Figure 2; Table 3). Based on its molecular mass, this HSP could be classed as small heat shock proteins (sHSPs). sHSPs are numerous and very diverse, both in sequence and where they function in the cell [40]. HSPs belong to a well-conserved class of molecules that function as molecular chaperones, playing key roles in protein folding and refolding, assembly and transport, stabilization of proteins and membranes under stress conditions and in the reestablishment of cellular homeostasis. Additionally, it has been reported that biotic stress can induce the gene expression of some, but not all, sHSPs [40].

Surprisingly, in the cowpea cv. CE-31 roots challenged with *M. incognita*, but not in control plants, there was the induction of the leghemoglobin (LegHb) biosynthesis (Spot 21: Figure 1; Figure 2; Table 3), which was linked to gene activation (Figure 3). To the best of our knowledge, this is the first time that induction of a LegHb by RKN-infection of non-rhizobium bacterized cowpea has been reported.

Since the cowpea plants were grown in autoclaved sand, the root system of the studied cowpea was not rhizobium bacterized, as proven by the absence of amplification (transcripts) of the NodC genes assessed by PCR using specific nucleotide primers from conserved regions of the *nodC* gene (Figure 3). Therefore, such upregulation of LegHb in the cowpea CE-31 was due to the nematode infection itself. LegHbs are essential for the symbiotic nitrogen fixation process in the legume root nodules induced by rhizobia, where the main function is to act as a carrier of oxygen from the atmosphere to the bacteroids for aerobic respiration [41].

It was previously shown [42] that cowpea seeds bacterized with a rhizobium strain and inoculated with *M. incognita* had a decrease in the LegHb content over that of the rhizobium bacterized cowpea not RKN-inoculated (control). Kinetic studies have shown that soybean LegHbs decompose H_2O_2 to H_2O with kinetics similar to that for the reactions of plant peroxidases [43]. Thus, it is possible that this enhanced legHb detected in our study must be also involved in the H_2O_2 homeostasis of cowpea plants infected with *M. incognita*. Nevertheless, further studies are needed to clarify both why LegHb was induced upon *M. incognita* challenge in non-rhizobium bacterized cowpea and what real physiological function LegHb plays within this scenario.

5. Conclusions

In conclusion, this work shows that the defense response of the resistant cowpea CE-31 to infection by root-knot nematodes, *M. incognita*, is complex and involves many different proteins and metabolic pathways (Figure 4). Nevertheless, the upregulated proteins, such as SOD, APX, PR-1, β-1,3-glucanase, chitinases, cysteine protease and secondary metabolism enzymes, key enzymes involved in the ethylene biosynthesis in plants, and proteins involved in the MAPK pathway signaling, amongst others, reinforce that they may contribute, directly or indirectly, to the resistance of cowpea to *M. incognita* attack.

Figure 4. Gene ontology (GO) for proteins in the roots of cowpea genotype CE-31 inoculated with *M. incognita* (Race 3) and non-inoculated control. Categories are according to [18] and represented as a percentage of total identified GO terms.

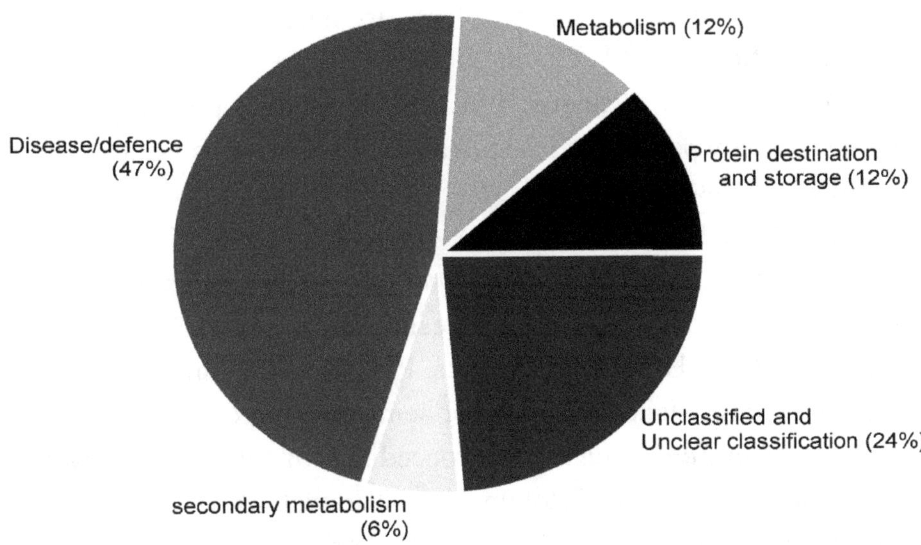

Acknowledgments

Araújo-Filho J. H. was supported by a doctoral grant from CNPq. The authors acknowledge the financial support by National Council for Scientific and Technological Development (CNPq), Ceara Foundation for Support of Scientific and Technological Development (FUNCAP), Coordination for the Improvement of Higher Education Personnel (CAPES).

Author Contributions

The conception of experimental design and cowpea/RKN treatment structure was performed by J.T.A.O., J.H.A.F. and I.M.V. Protein extractions were performed by J.H.A.F. and D.M.F.G. Protein preparation, mass spectrometry and sample analysis was done by J.H.A.F., D.M.F.G., J.S., P.M.P. and C.R.R.S.C. Sample preparations, primer design and RT-PCR analysis were performed by J.H.C., F.D.A.S., J.H.A.F., T.B.G. and M.D.P.L. J.T.A.O. and J.H.A.F. compiled the experimental results and had primary responsibility for manuscript preparation with the contribution of I.M.V., T.B.G. and J.H.C.

References

1. Castagnone-Sereno, P.; Danchin, E.G.J.; Perfus-Barbeoch, L.; Abad, P. Diversity and evolution of root-knot nematodes, genus *Meloidogyne*: New insights from the genomic era. *Annu. Rev. Phytopathol.* **2013**, *51*, 203–220.
2. Schulze-Lefert, P.; Ralph, P. A molecular evolutionary concept connecting nonhost resistance, pathogen host range, and pathogen speciation. *Trends Plant Sci.* **2011**, *16*, 117–125.
3. Quentin, M.; Abad, P.; Favery, B. Plant parasitic nematode effectors target host defense and nuclear functions to establish feeding cells. *Front. Plant Sci.* **2013**, *4*, e53.
4. Gheysen, G.; Mitchum, M.G. How nematodes manipulate plant development pathways for infection. *Curr. Opin. Plant Biol.* **2011**, *14*, 415–421.
5. Heil, M.; Bostock, R.M. Induced systemic resistance (ISR) against pathogens in the context of induced plant defences. *Ann. Bot.* **2002**, *89*, 503–512.
6. Barna, B.; Fodor, J.; Harrach, B.D.; Pogány, M.; Király, Z. The Janus face of reactive oxygen species in resistance and susceptibility of plants to necrotrophic and biotrophic pathogens. *Plant Physiol. Biochem.* **2012**, *59*, 37–43.
7. Das, S.; Ehlers, J.D.; Close, T.J.; Roberts, P.A. Transcriptional profiling of root-knot nematode induced feeding sites in cowpea (*Vigna unguiculata* L. Walp.) using a soybean genome array. *BMC Genomics* **2010**, *11*, e480.
8. Souza e Silva, S.M.; Maia, J.M.; Araújo, Z.B.; Freire-Filho, F.R. *Chemical Composition of 45 Cowpea [Vigna unguiculata (L.) Walp.] Genotypes*, technical report; The Brazilian Agricultural Research Corporation, Embrapa Meio-Norte: Teresina, Brazil, 2002; Volume 149, pp. 1–2.
9. Ehlers, J.D.; Matthews, W.C.; Hall, A.E.; Roberts, P.A. Breeding and evaluation of cowpeas with high levels of broad-based resistance to root-knot nematodes. In *Challenges and Opportunities for Enhancing Sustainable Cowpea Production*, Proceedings of the World Cowpea Conference III, Ibadan, Nigeria, 4–8 September 2000; Fatokun, C.A., Tarawali, S.A., Singh, B.B., Kormawa, P.M., Tamo, M., Eds.; International Institute of Tropical Agriculture (IITA): Ibadan, Nigeria, 2002; pp. 41–51.
10. Oliveira, J.T.A.; Andrade, N.C.; Martins-Miranda, A.S.; Soares, A.A.; Gondim, D.M.F.; Araújo-Filho, J.H.; Freire-Filho, F.R.; Vasconcelos, I.M. Differential expression of antioxidant enzymes and PR-proteins in compatible and incompatible interactions of cowpea (*Vigna unguiculata*) and the root-knot nematode *Meloidogyne incognita. Plant Physiol. Biochem.* **2012**, *51*, 145–152.
11. Silveira, J.A.G.; Costa, R.C.L.; Oliveira, J.T.A. Drought-induced effects and recovery of nitrate assimilation and nodule activity in cowpea plants inoculated with *Bradyrhizobium* spp. under moderate nitrate level. *Braz. J. Microbiol.* **2001**, *32*, 187–194.
12. Hussey, R.S.; Davis, E.L.; Baum, T.J. Secrets in secretions: Genes that control nematode parasitism of plants. *Braz. J. Plant Physiol.* **2002**, *14*, 183–194.
13. Atkinson, H.J. Nematodes. In *Molecular Plant Pathology*; Gurr, S.J., McPherson, M.J., Bowles, D.J., Eds.; Oxford University Press: Oxford, UK, 1992; Volume 1, pp. 99–107.
14. Bradford, M.M. A rapid and sensitive method for the quantification of microgram quantities of protein utilizing the principle of protein dye binding. *Anal. Biochem.* **1976**, *72*, 248–254.
15. Laemmli, U.K. Cleavage of structural proteins during the assembly of the head of bacteriophage T4. *Nature* **1970**, *227*, 680–685.

16. Candiano, G.; Bruschi, M.; Musante, L.; Santucci, L.; Ghiggeri, G.M.; Carnemolla, B.; Orecchia, P.; Zardi, L.; Righetti, P.G. Blue silver: A very sensitive colloidal Coomassie G-250 staining for proteome analysis. *Electrophoresis* **2004**, *25*, 1327–1333.
17. Shevchenko, A.; Wilm, M.; Vorm, O.; Mann, M. Mass spectrometric sequencing of proteins from silver-stained polyacrylamide gels. *Anal. Chem.* **1996**, *68*, 850–858.
18. Bevan, M.; Bancroft, I.; Bent, E.; Love, K.; Goodman, H.; Dean, C.; Bergkamp, R.; Dirkse, W.; van Staveren, M.; Stiekema, W.; *et al.* Analysis of 1.9 Mb contiguous sequence from chromosome 4 of *Arabidopsis thaliana*. *Nature* **1998**, *39*, 809–821.
19. Chang, S.; Puryear, J.; Cairney, J. A simple and efficient method for isolating RNA from pine trees. *Plant Mol. Biol. Rep.* **1993**, *11*, 113–116.
20. Sambrook, J.; Fritsch, E.F.; Maniatis, T. *Molecular Cloning: A Laboratory Manual*, 2nd ed.; Cold Spring Harbor Laboratory Press: Cold Spring Harbor, NY, USA, 1989; Volume 1, pp. 7.1–7.87.
21. Sarita, S.; Sharma, P.K.; Priefer, U.B.; Prell, J. Direct amplification of rhizobial *nodC* sequences from soil total DNA and comparison to *nodC* diversity of root nodule isolates. *FEMS Microbiol. Ecol.* **2005**, *54*, 1–11.
22. Warner, S. Genomic DNA isolation and lambda library construction. In *Plant Gene Isolation: Principles and Practice*; Foster, G.D., Twell, D., Eds.; John Wiley & Sons Ltd: Chichester, UK, 1996; pp. 51–53.
23. Costa, J.H.; Hasenfratz-Sauder, M.P.; Pham-Thi, A.T.; Silva Lima, M.G.; Dizengremel, P.; Jolivet, Y.; Fernandes de Melo, D. Identification in *Vigna unguiculata* (L.) Walp. of two cDNAs encoding mitochondrial alternative oxidase orthologous to soybean alternative oxidase genes 2a and 2b. *Plant Sci.* **2004**, *167*, 233–239.
24. Twyman, R.M. Strategies for protein quantitation. In *Principles of Proteomics*, 2nd ed.; Garland Science, Taylor & Francis Group, LLC: New York, NY, USA, 2014; pp. 69–86.
25. Xu, W.; Li, L.; Du, L.; Tan, N. Various mechanisms in cyclopeptide production from precursors synthesized independently of non-ribosomal peptide synthetases. *Acta Biochem. Biophys. Sin.* **2011**, *43*, 757–762.
26. Craik, D.J. Host-defense activities of cyclotides. *Toxins* **2012**, *4*, 139–156.
27. Bellafiore, S.; Shen, Z.; Rosso, M.-N.; Abad, P.; Shih, P.; Briggs, S.P. Direct identification of the meloidogyne incognita secretome reveals proteins with host cell reprogramming potential. *PLoS Pathog.* **2008**, *4*, e1000192.
28. Curtis, R.; Buttle, D.; Behnke, J.; Duce, I.; Shewry, P.; Kurup, S.; Kerry, B.; Kerry, M. Nematicidal effects of cysteine proteinases and methods of use thereof to treat nematode infestation. WO2008087555 A2, 24 July 2008.
29. Damiani, I.; Baldacci-Cresp, F.; Hopkins, J.; Andrio, E.; Balzergue, S.; Lecomte, P.; Puppo, A.; Abad, P.; Favery, B.; Hérouart, D. Plant genes involved in harbouring symbiotic rhizobia or pathogenic nematodes. *New Phytol.* **2012**, *194*, 511–522.
30. Glazer, I.; Epstein, E.; Orion, D.; Apelbaum, A. Interactions between auxin and ethylene in root-knot nematode (*Meloidogyne javanica*) infected tomato roots. *Physiol. Mol. Plant Pathol.* **1986**, *28*, 171–179.
31. Gutierrez, O.A.; Wubben, M.J.; Howard, M.; Roberts, B.; Hanlon, E.; Wilkinson, J.R. The role of phytohormones ethylene and auxin in plant-nematode interactions. *Russ. J. Plant Physiol.* **2009**, *56*, 1–5.

32. Neil, S.J.; Desikan, R.; Clarke, A.; Hurst, R.D.; Hancock, J.T. Hydrogen peroxide and nitric oxide as signaling molecules in plants. *J. Exp. Bot.* **2002**, *53*, 1237–1247.
33. Vanderspool, M.C.; Kaplan, D.T.; McCollum, T.G.; Wodzinski, R.J. Partial characterization of cytosolic superoxide dismutase activity in the interaction of *Meloidogyne incognita* with two cultivars of *Glycine max* *J. Nematol.* **1994** *26*, 422–429.
34. Ferrer, J.-L.; Austin, M.B.; Stewart, C., Jr.; Noel, J.P. Structure and function of enzymes involved in the biosynthesis of phenylpropanoids. *Plant Physiol. Biochem.* **2008**, *46*, 356–370.
35. Qiu, J.; Hallmann, J.; Kokalis-Burelle, N.; Waever, D.B.; Rodríguez-Kábana, R.; Tuzun, S. Activity and differential induction of chitinase isozymes in soybean cultivars resistant or susceptible to root-knot nematodes. *J. Nematol.* **1997**, *29*, 523–530.
36. Van Loon, L.C.; Rep, M.; Pieterse, C.M. Significance of inducible defense-related proteins in infected plants. *Annu. Rev. Phytopathol.* **2006**, *44*, 135–162.
37. Khan, A.; Williams, K.L.; Nevalainen, H.K.M. Effects of *Paecilomyces lilacinus* protease and chitinase on the eggshell structures and hatching of *Meloidogyne javanica* juveniles. *Biol. Control* **2004**, *31*, 346–352.
38. Hammargren, J.; Sundström, J.; Johansson, M.; Bergman, P.; Knorpp, C. On the phylogeny, expression and targeting of plant nucleoside diphosphate kinases. *Physiol. Plantarum* **2007**, *129*, 79–89.
39. Steeg, P.S.; Bevilacqua, G.; Kopper, L.; Thorgeirsson, U.P.; Talmadge, J.E.; Liotta, L.A.; Sobel, M.E. Evidence for a novel gene associated with low tumor metastatic potential. *J. Natl. Cancer Inst.* **1988**, *80*, 200–204.
40. Waters, E.R. The evolution, function, structure, and expression of the plant sHSPs. *J. Exp. Bot.* **2013**, *64*, 391–403.
41. Ott, T.; van Dongen, J.T.; Gunther, C.; Krusell, L.; Desbrosses, G.; Vigeolas, H.; Bock, V.; Czechowski, T.; Geigenberger, P.; Udvarvi, M.K. Symbiotic leghemoglobins are crucial for nitrogen fixation in legume root nodules but not for general plant growth and development. *Curr. Biol.* **2005**, *15*, 531–535.
42. Khan, T.A.; Husain, S.I. Proline and leghaemoglobin contents and water absorption capability of cowpea roots as influenced by infection with *Rotylenchus reniformis*, *Meloidogyne incognita* and *Rhizoctonia solani*. *Nematol. Medit.* **1989**, *17*, 135–137.
43. Job, D.; Zeba, B.; Puppo, A.; Rigaud, J. Kinetic studies of the reaction of ferric soybean leghemoglobins with hydrogen peroxide, cyanide and nicotinic acid. *Eur. J. Biochem.* **1980**, *107*, 491–500.

9

PAPE (Prefractionation-Assisted Phosphoprotein Enrichment): A Novel Approach for Phosphoproteomic Analysis of Green Tissues from Plants

Ines Lassowskat [1], Kai Naumann [2], Justin Lee [1,*] and Dierk Scheel [1]

[1] Leibniz Institute of Plant Biochemistry, Weinberg 3, Halle/Saale, D-06120, Germany; E-Mails: Ines.Lassowskat@ipb-halle.de (I.L.); dscheel@ipb-halle.de (D.S.)

[2] Riboxx GmbH, Meissner Str. 191, Radebeul, D-01445, Germany; E-Mail: Kai.Naumann@riboxx.com

* Author to whom correspondence should be addressed; E-Mail: jlee@ipb-halle.de

---

Abstract: Phosphorylation is an important post-translational protein modification with regulatory roles in diverse cellular signaling pathways. Despite recent advances in mass spectrometry, the detection of phosphoproteins involved in signaling is still challenging, as protein phosphorylation is typically transient and/or occurs at low levels. In green plant tissues, the presence of highly abundant proteins, such as the subunits of the RuBisCO complex, further complicates phosphoprotein analysis. Here, we describe a simple, but powerful, method, which we named prefractionation-assisted phosphoprotein enrichment (PAPE), to increase the yield of phosphoproteins from *Arabidopsis thaliana* leaf material. The first step, a prefractionation via ammonium sulfate precipitation, not only depleted RuBisCO almost completely, but, serendipitously, also served as an efficient phosphoprotein enrichment step. When coupled with a subsequent metal oxide affinity chromatography (MOAC) step, the phosphoprotein content was highly enriched. The reproducibility and efficiency of phosphoprotein enrichment was verified by phospho-specific staining and, further, by mass spectrometry, where it could be shown that the final PAPE fraction contained a significant number of known and additionally novel (potential) phosphoproteins. Hence, this facile two-step procedure is a good prerequisite to probe the phosphoproteome and gain deeper insight into plant phosphorylation-based signaling events.

Keywords: phosphoproteomics; LC-MS; *Arabidopsis thaliana*; phosphoprotein enrichment

---

1. Introduction

The completion of the genome sequencing in 2000 [1] has further propelled *Arabidopsis thaliana* into one of the most well-established model organisms to study plant molecular biology/biochemistry [2]. *Arabidopsis* is used for a wide range of "OMICS" analysis concerning genes (genomics; [3,4]), proteins (proteomics; [5–7]) and metabolites (metabolomics, [8]). One sub-topic of proteomics, rising in the last few years, is the field of phosphoproteomics [9]. The strong interest originates from the importance of protein phosphorylation for the biochemistry of all organisms, especially in regulating cellular processes, ranging from cell differentiation, development, cell cycle control, metabolism and signal transduction [10–12]. Probably 30% of all proteins are phosphorylated at any given time and state [13], indicating the immense dimension of the phosphoproteome. Beside its different roles in the regulation of protein synthesis, gene expression and apoptosis, phosphorylation events exhibit a pivotal role in defense responses [14]. An example is the activation of mitogen-activated protein kinase (MAPK)-mediated phosphorylation signaling cascades upon stress or other environmental signals [15–17]. The corresponding downstream targets of such a cascade are, to a great extent, unknown. For further understanding of defense mechanisms in plants, more knowledge about signaling cascades is of high significance. Therefore, a fully developed strategy for phosphoprotein/peptide enrichment is necessary.

Unfortunately, plant phosphoproteomics using leaf material can be a challenging task. Not only the presence of highly abundant proteins, like RuBisCO, but also the low levels of phosphorylated signaling proteins limit their visualization and detection on PAGE-gels. Even highly advanced mass spectrometry is often unable to recover large numbers of phosphopeptides in complex samples. Common methods frequently describe the enrichment of phosphopeptides prior to measurement to overcome this challenge. Most methods use metal ions for the binding of phosphopeptides, for instance, chelated metal ions (immobilized metal affinity chromatography IMAC); [18,19]) or metal oxides (metal oxide affinity chromatography (MOAC); [20]). Other methods describe the use of multi-step procedures, in which a first enrichment of phosphoproteins should assist the subsequent phosphopeptide enrichment [21]. Nevertheless, one disadvantage of such an approach is that not all phosphopeptides are efficiently captured, and also, information concerning the non-phosphorylated peptides is lost, which may impede target identification, for instance, in the cases of highly similar proteins of multigene families [22]. Other approaches first remove highly abundant proteins that might interfere with the applied phospho-enrichment matrix. In plants, this means the reduction or depletion of RuBisCO prior to phosphoprotein enrichment [23,24]. A popular way to accomplish the fractionation of proteins is salting out with chemicals. Polyethylene-glycol (PEG)-based fractionation, for instance, has been successfully employed for improved proteome coverage, leading to the detection of differentially-expressed proteins of low abundance [25,26]. However, since the remaining PEG can interfere in MS analysis, we tested here another commonly used fractionation, namely, ammonium sulfate (AS) precipitation. In previous work done in our laboratory, it could be shown that a reduction of the RuBisCO content via AS precipitation had a positive effect on the preparation of 2D-PAGE, as well as the enrichment of phosphoproteins [27]. As a further improvement for phosphoprotein analysis, we now incorporated the metal oxide affinity chromatography (MOAC) method [20] to the AS-based RuBisCO removal step, which, by itself, already acts as prefractionation/enrichment of phosphoproteins. This led to a facile, but efficient, phosphoproteome analysis procedure, which we termed prefractionation-assisted phosphoprotein enrichment (PAPE).

2. Experimental

2.1. Plant Growth

Arabidopsis thaliana (Col-0) seeds were grown in soil. After two days of stratification at 4 °C, the plants were maintained under short-day conditions (8 h, 200 µE, 23 °C) for six weeks prior to protein extraction.

2.2. Protein Extraction

Leaf material was ground to a fine powder in liquid nitrogen and mixed vigorously with 3 volume of extraction buffer (100 mM HEPES-KOH, pH 7.5; 5% glycerol; 5 mM EDTA; with freshly added 0.1% mercaptoethanol, 1% proteinase and phosphatase inhibitors 2 + 3 from Sigma Aldrich, Taufkirchen, Germany) for 20 min (4 °C). The suspension was centrifuged at 3,220 ×*g* for 15 min. The supernatant was filtered [0.45 µm cellulose mixed ester (CME) filter, Roth, Karlsruhe, Germany].

2.3. Precipitation of Protein Extract

(a) *Fractionation of proteins.* Ammonium sulfate was added to a final concentration of 40% saturation and incubated for half an hour (4 °C). After centrifugation (3,220 × *g*, 15 min, 4 °C), the pellet was washed twice with wash solution (80% acetone, 20% Tris-HCl (50 mM, pH 7.5); −20 °C) and once with ice-cold acetone.

(b) *Precipitation of total proteins.* An equal volume of Tris-EDTA-buffered phenol (Roth) was added, mixed vigorously for 1 min and incubated for 5 min at 4 °C. After centrifugation (3,220 × *g*, 15 min, 4 °C), the phenolic phase was transferred and re-extracted twice with 1 volume of re-extraction buffer (100 mM Tris-HCl, pH 8.4, 20 mM KCl, 10 mM EDTA and freshly added 0.4% (v/v) β-mercaptoethanol). The final phenolic phase was mixed with 5 volume of precipitation solution (100 mM ammonium acetate in methanol; −20 °C), incubated over night at −20 °C, and the proteins pelleted by centrifugation (3,220 ×*g*, 15 min, 4 °C). The pellet was washed once with precipitation solution and twice with wash solution. The pellets (from a and b) were air dried and solubilized in LysShot buffer (8 M urea, 50 mM Tris, pH 8.5) or in MOAC incubation buffer (30 mM MES, 20 mM imidazole, 200 mM aspartate, 200 mM glutamate, 0.25% 3-[(3-cholamidopropyl)dimethylammonio]-1-propanesulfonate (CHAPS), 8 M urea, pH 6.1) for samples to be processed by the MOAC step.

2.4. Phosphoprotein Enrichment (MOAC)

Forty milligrams of $Al(OH)_3$ matrix (Sigma-Aldrich) were equilibrated with 1.8 mL of incubation buffer (see Section 2.3). A 1.5-mL sample with a protein concentration of 0.5 µg/µL was loaded and incubated by rotating for 30 min (4 °C). After incubation, the matrix was washed four times with incubation buffer. The proteins were eluted twice (800 µL and 400 µL) with tetrapotassium pyrophosphate (TKPP) buffer (8 M urea, 100 mM TKPP, pH 9.0) for 45 min at room temperature [20], and centrifuged (18,514 ×*g*, 2 min, 15 °C) to pellet the $Al(OH)_3$ matrix. The pooled eluates were centrifuged twice (18,514 ×*g*, 2 min, 15 °C), to pellet any remaining matrix, and, subsequently, concentrated with centricon filter devices (3 kDa cut-off; Millipore, Bilterica, MA, USA). Proteins were precipitated with a 2D-CleanUp kit (GE Healthcare, Hercules, CA, USA), according to the manufacturer's instructions, and solubilized in LysShot (see Section 2.3).

2.5. SDS-PAGE and Phosphoprotein Staining

Protein concentration was determined by a 2-D Quant Kit (GE Healthcare). SDS-PAGE was carried out according to [28] by using Precast Gels (Criterion Tris-HCl 12.5%; Biorad, Munich, Germany). Ten micrograms of each sample in loading buffer (0.313 M Tris-HCl, pH 6.8, 50% glycerol, 10% SDS, 0.05% (w/v) bromophenol blue, 0.5 M dithiothreitol (DTT) were heated for 5 min at 95 °C and cooled to room temperature prior to loading. Peppermint Stick™ Phosphoprotein Molecular Weight Standard (Life technologies, Darmstadt, Germany) was used as the molecular weight marker. Pro-Q Diamond (Life technologies) staining was carried out according to a modified protocol [29]. Fluorescent images were obtained using the Typhoon scanner (GE Healthcare) with the settings: 532 nm excitation, 580 nm band pass emission filter and the photo multiplier tube at 550. ImageJ software (National Institute of Health, Bethesda, MD, USA) was used for false color representation. Total protein was visualized with Novex® Colloidal Blue Staining Kit (Life Technologies).

2.6. In-Solution Digestion

Protein concentration was determined by a 2-D Quant Kit (GE Healthcare), and the proteins (in LysShot) were reduced with 200 mM DTT (in 100 mM Tris, pH 7.8) for 1 h and, subsequently, alkylated with 200 mM iodoacetamide (in 100 mM Tris, pH 7.8) for 1 h at room temperature. The solution was diluted to 0.5 M urea with 50 mM NH_4HCO_3 (pH 8) and digested overnight with sequencing grade trypsin (Promega, Mannheim, Germany) at a ratio of 1:50 at 37 °C. Peptides were desalted on C18 tips or columns (Protea, Morgantown, WV, USA; Thermo, Bonn, Germany) and reconstituted in solution containing 5% acetonitrile (ACN) and0.1% trifluoroacetic acid (TFA).

2.7. Mass Spectrometry

Tryptic digests were analyzed with an LC-MS system consisting of a nano-LC (Easy-nLC II, Thermo Fisher Scientific, Bremen, Germany) coupled to a hybrid-Fourier Transform (FT)-mass spectrometer [Linear Trap Quadrupole (LTQ) Orbitrap Velos, Thermo Fisher Scientific]. Peptide separations were performed on a C18 column (EASY column; 10 cm, ID 75 μm, particle diameter: 3 μm) at a flow rate of 300 nL/min and a linear gradient of 5% to 40% B in 150 min (A: 0.1% formic acid in water, B: 0.1% formic acid in ACN). A voltage of +1.9 kV was applied to electrospray peptide ions. A capillary temperature of 275 °C for peptide transfer and a lock mass of 445.120024 *m/z* were used. Precursor mass scanning was performed from 400 to 1,850 *m/z* in the Orbitrap with a resolution of 30,000, and the 20 most intense precursor ions were selected for subsequent collision-induced dissociation (CID) fragmentation in the linear quadrupole mass analyzer (LTQ). Singly-charged ions were rejected from fragmentation. Dynamic exclusion was enabled (repeat count: 1; repeat duration: 20 s; exclusion list size: 500; exclusion duration: 30 s).

2.8. Spectral Data Analysis

MS raw data were searched against an *A. thaliana* protein database based on The Arabidopsis Information Resource (TAIR) 10 with the Proteome Discoverer 1.3 using an in-house Mascot server (precursor mass tolerance: 7 ppm; fragment mass tolerance: 0.8 Da; missed cleavages: 2). Carbamidomethylation of cysteine was set as a static modification. Variable modifications were oxidation (Methionine), acetylation (protein *N*-terminus), deamidation (Asparagine/Glutamine) and phosphorylation (Serine/Threonine). Further data evaluation was carried out with the Scaffold software (Version 3.3, Proteome Software Inc., Portland, OR, USA), Proteome Discoverer 1.3 with phosphoRS 1.0 (Thermo Fisher Scientific) and DanteR [30] for total protein content. Phosphopeptides were identified with the Proteome Discoverer 1.3 software, which includes the phosphoRS 1.0 algorithm (Thermo Fisher Scientific) for phospho-site mapping. A false discovery rate (FDR) was calculated by searching a "decoy" database containing all the target database sequences in reverse order. Peptide-spectrum match (PSM) was set at a q-value <0.05 (*i.e.*, a corrected significance threshold employing the Benjamini-Hochberg FDR procedure to control for a family-wise error rate). Protein grouping was enabled. Gene ontology (GO) annotation was achieved with the tool on TAIR [31]. The mass spectrometry proteomics data have been deposited to the ProteomeXchange Consortium (http://proteomecentral.proteomexchange.org) via the PRIDE (PRoteomics IDEntifications) partner repository [32] with the dataset identifier PXD000421.

3. Results and Discussion

3.1. Prefractionation of Arabidopsis Leaf Proteins

A stepwise fractionation with ammonium sulfate (AS) was used to salt out proteins in solution. This was done with 20% AS increment steps, while pelleting precipitated proteins by centrifugation after every step. The molecular weight distribution of proteins in the AS steps was determined with 1D-PAGE (Figure 1B). The large subunit of RuBisCO (boxed, Figure 1B, lower panel), which is one of the most abundant proteins in the non-fractionated sample (crude extract), is predominantly located in the fractionation steps using more than 40% AS. Serendipitously, the fractions produced with 20% and 40% AS (with little or no apparent RuBisCO content) also contained the most phosphoproteins, as evidenced by phospho-specific Pro-Q Diamond staining (Figure 1B, upper panel). In contrast, the samples from the 60%–100% AS fractionation steps showed only very low levels of phosphoproteins. Therefore, the sample precipitated with 40% AS is an excellent source for subsequent phosphoprotein enrichment and represents the first step of the method described below, which we called prefractionation-assisted phosphoprotein enrichment (PAPE).

3.2. PAPE: Prefractionation-Assisted Phosphoprotein Enrichment

Crude extract and the 40% AS fraction (40% AS) were subjected to phosphoprotein enrichment with metal oxide affinity chromatography (MOAC) [20] (Figure 1A) with minor modifications, as described in the Experimental section. To evaluate the reproducibility and efficiency of the PAPE procedure (a combination of AS precipitation followed by MOAC), the total extract, the 40% AS fraction and the corresponding MOAC-enriched fractions were each prepared three times, separated on a 1D-PAGE and visualized by coomassie brilliant blue and Pro-Q Diamond phosphospecific staining (Figure 1C). As observed in the stepwise fractionation, the non-fractionated samples had the least visible phosphoprotein content (crude extract). While a faint enrichment effect could be achieved via MOAC (crude extract + MOAC), the prefractionation (40% AS) already had a high phosphoprotein content, which was dramatically increased in combination with the additional MOAC phosphoprotein enrichment step (40% AS + MOAC, Figure 1C). We will hereafter refer to this "40% AS + MOAC" fraction as the PAPE fraction.

Figure 1. Prefractionation-assisted phosphoprotein enrichment (PAPE). (**A**) Experimental setup. *Arabidopsis* soluble leaf proteins were either extracted with phenol to obtain the crude extract (total protein) or pre-fractionated by 40% ammonium sulfate precipitation. Subsequently, phosphoproteins were enriched using metal oxide affinity chromatography (MOAC). Quality control was assessed with SDS-PAGE and phospho-specific Pro-Q Diamond staining. Mass spectrometry analysis was performed with on-line nano-LC (Easy-nLC II) FT-mass spectrometry (LTQ Orbitrap Velos); (**B**) SDS-PAGE showing the crude extract and stepwise ammonium sulfate (AS) fractionations. Each step was performed three times (lanes labeled 1, 2 and 3). Visualization of proteins was achieved with (I) Pro-Q Diamond phosphoprotein staining in false-color representation and (II) colloidal coomassie staining. Protein molecular weights are indicated on the left-hand margin. (**C**) SDS-PAGE of different extraction and phosphoprotein enrichment steps, with visualization of the proteins as described above. Boxed areas mark the position of the large subunit of RuBisCO. (**D**) Boxplot depiction of protein yield (log10 microgram of proteins) from 25 g of leaf material (n = 6).

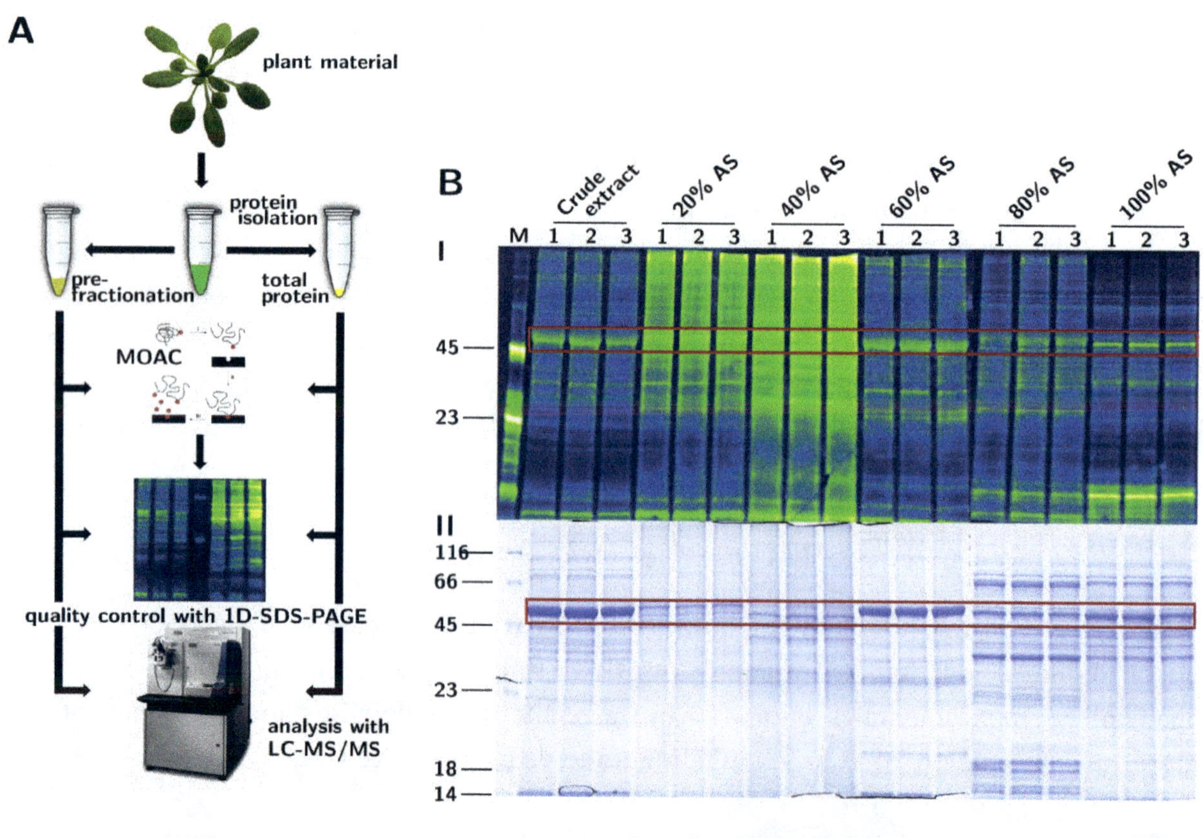

Figure 1. *Cont.*

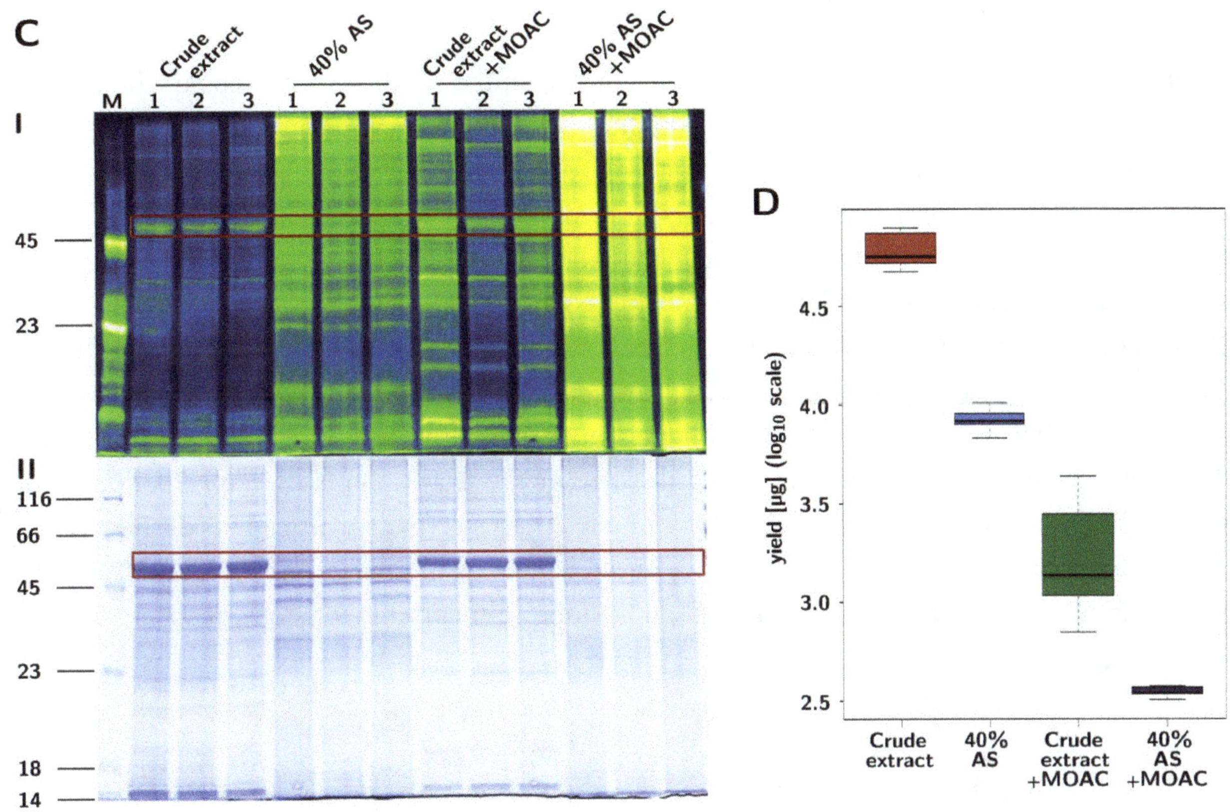

Notably, on the basis of the prepared replicates shown here, the "MOAC-only" method (crude extract + MOAC) had a larger variability in phosphoprotein enrichment compared to the other procedures (Figure 1C, upper panel; see, also, the standard deviation of the box plot in Figure 1D). Moreover, the "MOAC-only" fractions contained substantial amounts of RuBisCO (Figure 1C, lower panel), which may be a hindrance in subsequent mass spectrometry-based detection of less abundant proteins [33]. The PAPE fraction showed no distinct bands, but a uniform distribution across all molecular masses in the coomassie, as well as in the phosphospecific stain. Hence, the combination of 40% AS fractionation served both to remove RuBisCO and to enrich for phosphoproteins. The final protein yield by the PAPE procedure is about 0.6% of the total crude extract (Figure 1D); assuming all these are phosphoproteins, this is in agreement with the total phosphoprotein amount expected.

3.3. Reproducibility and Robustness of PAPE on the Basis of Mass Spectrometry Analysis

In addition to the coomassie and phosphostain gel-based analysis, mass spectrometry may provide a more qualitative estimation of the PAPE efficiency. Tryptic peptides derived from two micrograms of proteins from each of the three replicates of the four fractionation steps (*i.e.*, crude extract, 40% AS fraction and the corresponding MOAC-treated samples of these two fractions) were measured with shotgun LC-MS. Each sample was measured in two LC-MS runs and the proteins identified for each fractionation step pooled from both runs. This led to the identification of 850, 1,024, 1,151 and 803 proteins from the crude extract, the 40% AS fraction and their corresponding MOAC-treated samples, respectively (SCAFFOLD Software parameters: minimum protein probability 99.0%/minimum number of peptides 2/minimum peptide probability 90%). These represent a total of 1,928 unique proteins, and the distribution in the four fractionation steps is illustrated in Figure 2A. The identities of these 1,928 proteins are listed in Supplemental Table S1.

Figure 2. Mass spectrometry analysis of proteins from the fractionation steps of the PAPE procedure. **(A)** Flower plot showing the qualitative differences in the protein composition of the various PAPE fractions. The numbers are the total number of proteins identified from three experiments, with each sample being measured twice; **(B)** Variability and reproducibility of the PAPE procedure. Each small square represents a scatter plot of protein abundance (quantitative values based on spectral counting, SCAFFOLD; DanteR [30]) of the intersecting samples from the various fractionation steps. The letters A–D denote crude extract, crude extract + MOAC, 40% AS and 40% AS + MOAC, respectively, and the numbers 1–3 correspond to the three replicate experiments for each fractionation step. Note the strong positive correlation within the three replicate experiments of each fractionation step (colored boxes); **(C)** Principal Component Analysis (PCA) plot. The dashed lines divide the plot into sectors along the weight of the principal components separating with/without prefractionation and MOAC phosphoprotein enrichment steps, respectively.

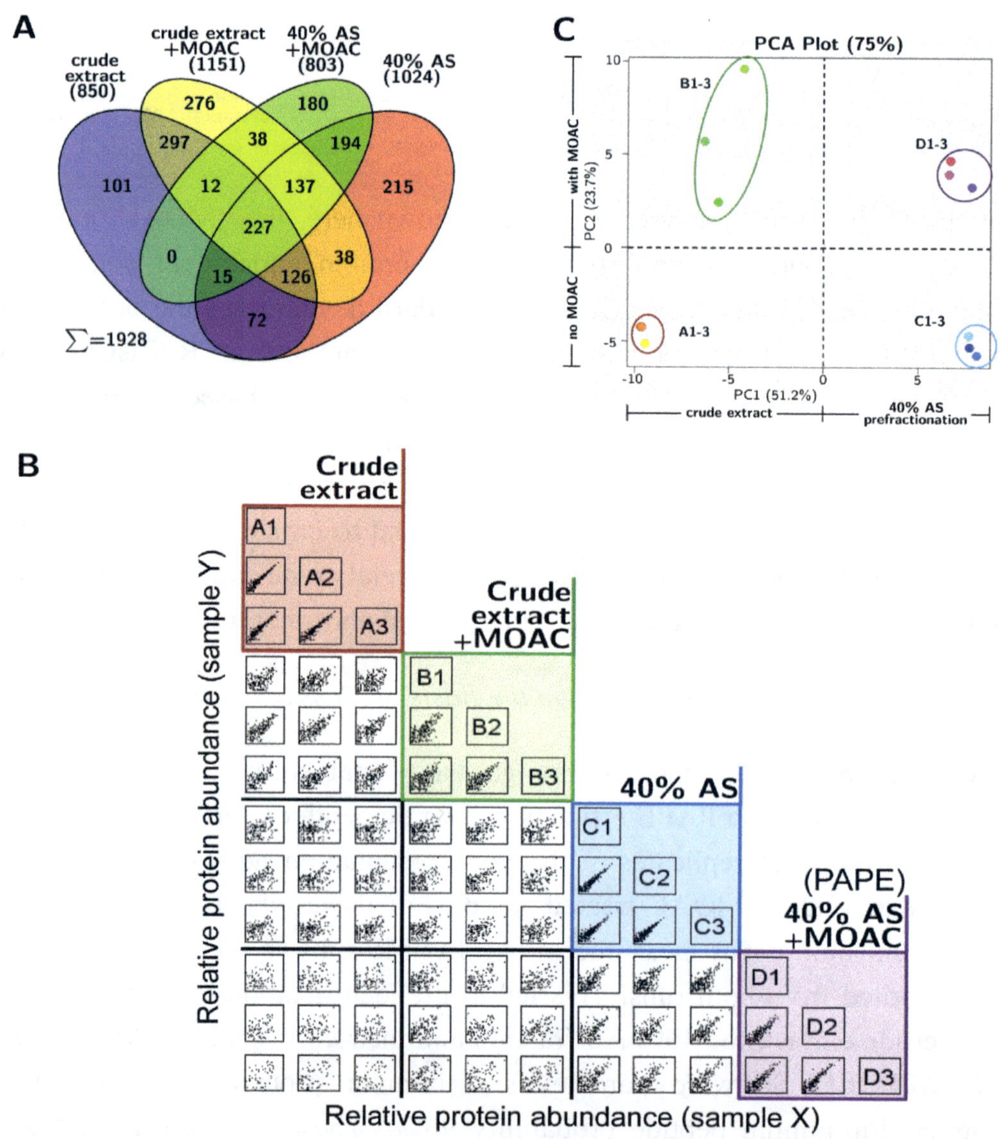

The qualitative protein composition varied greatly between fractions. In fact, of the total 1,928 proteins identified, only 227 proteins were common to all fractions, thus suggesting that the fractions contain different subsets of proteins (Figure 2A). The overlap between the crude extract and the 40% AS fraction was 440 proteins (~50% of the crude extract), indicating that 40% AS precipitated a subset of the total proteins, as is expected when considering the wide range of protein solubility in aqueous solvents [34]. Surprisingly, the overlap between the crude extract and the MOAC-enriched fraction revealed 662 proteins, which represents ~78% of the crude extract. Since it is unlikely that 78% of the identified proteins in the crude extract are phosphoproteins, it hints at substantial unspecific binding to the metal oxide. For instance, this might be due to binding to the negative charges provided by carboxylate moieties within proteins [35], which can exacerbate the binding of phosphoproteins in complex protein mixtures. These problems of the MOAC step in capturing non-phosphorylated targets is partially alleviated by the PAPE procedure described here, since the AS-prefractionation is already enriched for phosphoproteins (see Figure 1C). Therefore, the PAPE procedure is clearly advantageous compared to using only MOAC in phosphoproteomics.

The high technical reproducibility of each fractionation step can be seen in the positive linear relationship in the scatter plot of the quantitative value (based on spectral counting, SCAFFOLD; DanteR [30]) of each identified protein between the replicate experiments (see the colored boxes in Figure 2B). Notably, the tighter clustering of the replicates from the PAPE procedure when compared to the MOAC samples (purple box *versus* green box, respectively; Figure 2B), as well as the grouping within a Principal Component Analysis (Figure 2C) supports the robustness of the PAPE method over the MOAC method.

3.4. Validation of Phosphoprotein Enrichment by the PAPE Procedure

Figure 1C demonstrated that the PAPE procedure precipitated and enriched phosphoproteins. To further support this Pro-Q Diamond phosphostain evidence (Figure 1C), we determined if there was indeed an increase in the identification of known phosphoproteins from the fractionation steps. Using P3DB, a curated plant phosphoprotein database that contains only experimentally verified high quality entries, we found that the 40% AS, MOAC and PAPE fractions contain significantly higher numbers of known phosphoproteins than the crude extract (Figure 3B). Since the number of identified proteins varied between fractions, we also calculated the identified known phosphoproteins as a percentage of all identified proteins within each fraction (grey line in Figure 3B) in order to circumvent any misrepresentation.

This demonstrated that the PAPE fraction had proportionally more known phosphoproteins than the MOAC fraction (36% and 23%, respectively), thus suggesting the improvement of the PAPE procedure over MOAC alone to enrich phosphoproteins. Gene ontology (GO) annotation of the proteins identified in the PAPE fraction showed an enrichment of proteins involved in response to abiotic and biotic stimuli and to stress (Figure 3A). Since protein phosphorylation regulates many of these processes, it supports the effectiveness of PAPE to enrich lowly abundant phosphorylated proteins that are also involved in cellular signaling.

Figure 3. Changes of the protein/phosphoprotein composition. (**A**) Gene ontology functional categorization (based on the The Arabidopsis Information Resource (TAIR) gene ontology (GO) web-tool) of the proteins detected with the PAPE procedure; (**B**) The number of identified proteins in the various fractionation steps that are annotated as known phosphoproteins in the P3DB database. The grey line represents the percentage of identified known phosphoproteins to the total number of identified proteins in each fraction (see Figure 2A); (**C**) The number of phosphopeptides identified in the various fractionation steps. (Only high-confidence phosphopeptides with a phosphorylation site probability (pRS) score >30 are considered; for a full list, see Table S2). Each experiment was performed three times and measured twice. Black bars are the average number of phosphopeptides (+/−standard deviation) detected in each fraction, while grey bars depict the total number of non-identical phosphopeptides identified from all replicates. The grey line depicts the percentage of identified phosphopeptides to the total number of identified proteins in each fraction.

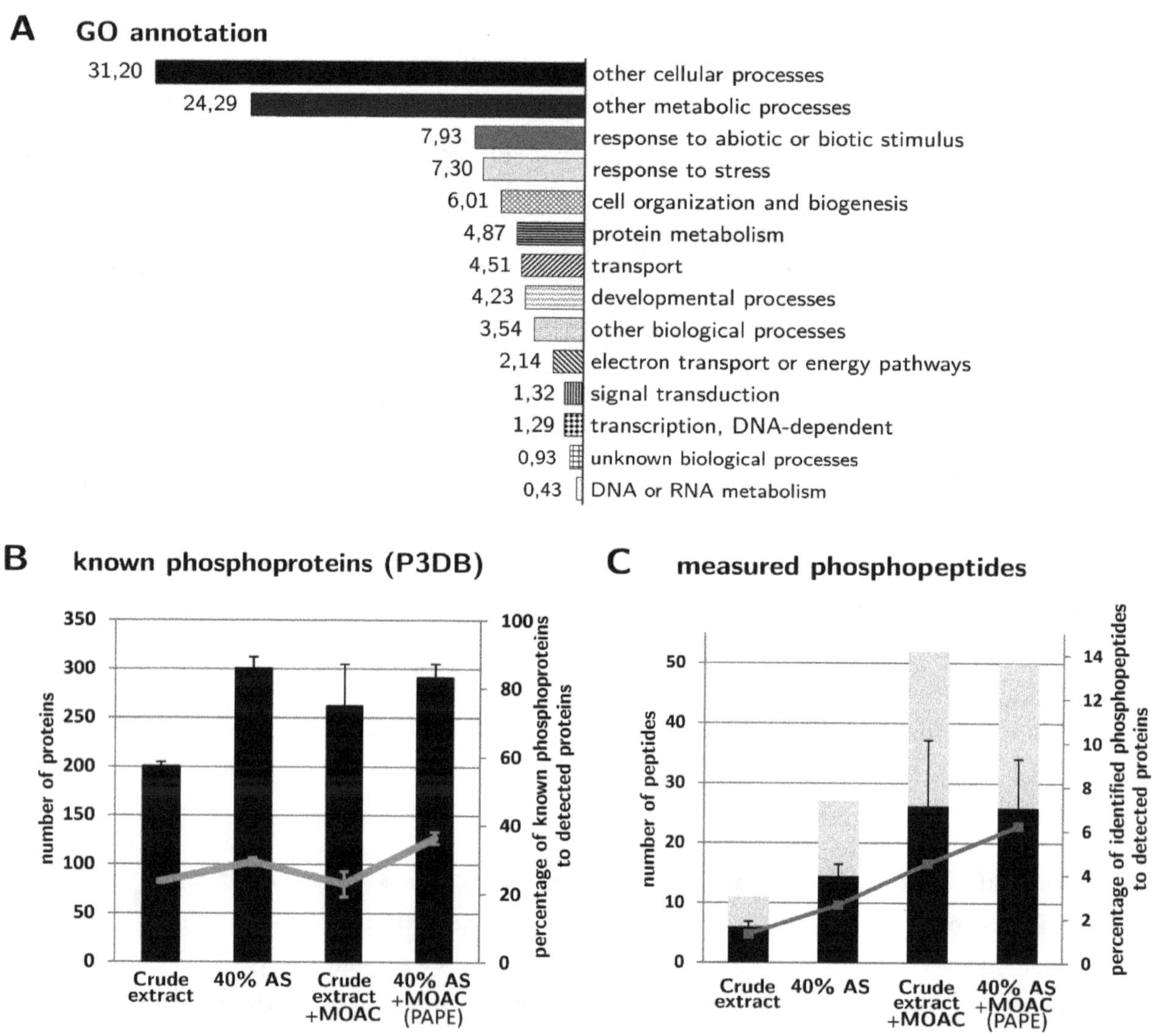

Correspondingly, we identified more phosphopeptides in the 40% AS, MOAC and PAPE fractions than in the crude extract (Figure 3C). In particular, when represented as the percentage of phosphopeptides relative to the total number of proteins identified in each fraction, more phosphopeptides were recorded in the PAPE than the MOAC fraction (6.2% and 4.5%, respectively). Interestingly, many of the phosphopeptides detected in the PAPE fractions were not listed in the P3DB [36,37] and PhosPhAt 3.0 [38,39] databases, which includes both novel phosphopeptides in proteins that are, so far, not annotated as phosphoproteins, as well as novel phosphopeptides in other regions of known phosphoproteins (see Tables 1 and S2). Note that Table 1 lists only the novel phosphopeptide with a high-confidence pRS score cutoff (>30); a longer list of all potential phosphopeptides is shown in Table S2. Additionally, Table S3 (a modified version of Table S2) links the identified phosphopeptides and the associated phosphoproteins. Inspection of these tables also reveals a progressive increase in the number of phosphopeptides associated with a particular (phospho)protein from the crude extract to the final PAPE fraction. Examples include RD29A (desiccation-responsive protein 29; also known as low-temperature-responsive protein 78, At5g52310), NR2 (nitrate reductase 2, At1g37130) and two proteins with tetratricopeptide repeat (TPR) domains (At1g01320 and At4g28080) (Figure 4). Taken together, these phosphopeptide detection data demonstrate the efficacy of the PAPE procedure to identify (novel) phosphoproteins.

Figure 4. Examples of the increased detection of phosphopeptides associated with a particular protein in the PAPE fraction. A progressively increasing number of phosphopeptide detections is seen for the listed proteins from the crude to the PAPE fraction. (Abbreviations: NR2, nitrate reductase 2; RD29, desiccation-responsive protein 29, which is also known as low-temperature-responsive protein 78; TPR-like, proteins from the tetratricopeptide repeat superfamily).

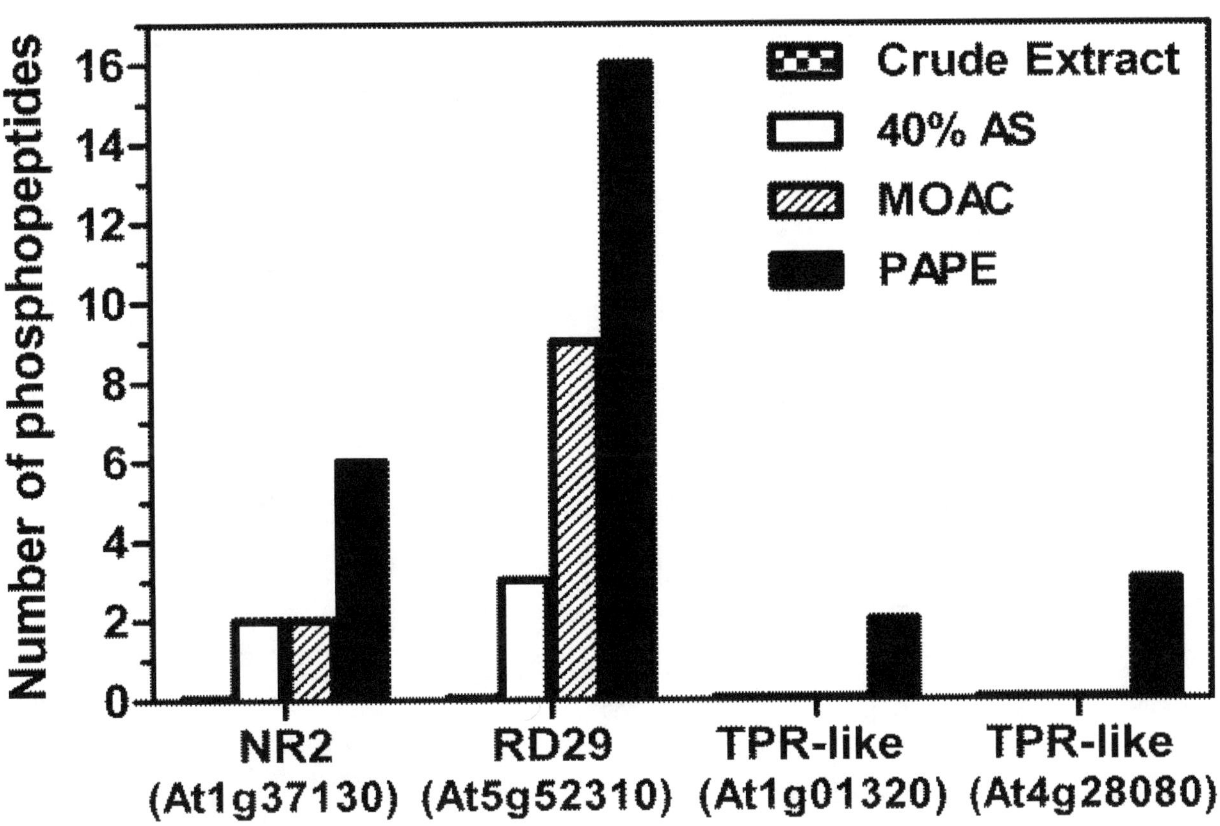

Table 1. List of novel phosphopeptides identified in this study (q-value < 0.05; pRS score > 30), which are not found in the P3DB or PhosPhAt 3.0 phosphoprotein databases.

	No.	Protein code	Description	Sequence	MH+[Da]	q-Value	PEP	pRS Score	# PSMs	pRS Site Probabilities
Crude extract	1	AT1G14010.1	emp24/gp25L/p24 family/GOLD family protein	SSIVLLILSILSPVTLSIR###	2,184.20708	0.016281	0.3789499	58	11	S(1): 15.3; S(2): 15.3; S(9): 84.3; S(12): 84.6; T(15): 0.5; S(17): 0.0
	2	AT2G40840.1	disproportionating enzyme 2	VEKPLGVFMNKSDQDDSVVVQFK	2,689.27768	0.021343	0.6031149	33	1	S(12): 0.4; S(17): 99.6
	3	AT2G38280.1	AMP deaminase, putative/myoadenylate deaminase, putative	SNGHVYVDEIPPGLPRLHTPSEGRASVHGASSIR	3,832.73672	0.022388	0.6381906	32	2	S(1): 33.1; Y(6): 33.1; T(19): 33.1; S(21): 4.1; S(26): 95.3; S(31): 50.7; S(32): 50.7
	4	AT4G38740.1; AT2G21130.1	rotamase CYP 1/ Cyclophilin-like peptidyl-prolyl cis-trans isomerase family protein	HTGPGILSMANAGANTNGSQFFICTVK	2,873.30117	0.027543	0.4591594	37	3	T(2): 0.7; S(8): 95.1; T(16): 2.0; S(19): 2.0; T(25): 0.1
	5	AT4G23670.1	polyketide cyclase/dehydrase and lipid transport superfamily protein	ATSGTYVTEVPLKGSAEK###	1,917.91213	0.032722	0.4923883	48	4	T(2): 24.5; S(3): 24.5; T(5): 24.5; Y(6): 24.5; T(8): 1.9; S(15): 0.0
	6	AT1G70200.1	RNA-binding (RRM/RBD/RNP motifs) family protein	QFTGQSLAFGKVIKQIK	1,973.05167	0.047422	0.6522376	35	12	T(3): 86.5; S(6): 13.5

Table 1. *Cont.*

	No.	Protein code	Description	Sequence	MH+[Da]	q-Value	PEP	pRS Score	# PSMs	pRS Site Probabilities
40% AS	1	AT5G56740.1	histone acetyltransferase of the GNAT family 2	LSQILVLPSFQGK	1,509.80133	0.011279	0.2471017	30	4	S(2): 0.4; S(9): 99.6
	2	AT1G23740.1	oxidoreductase, zinc-binding dehydrogenase family protein	NAALATTTATTPVLRR	1,736.90842	0.015443	0.4098554	40	1	T(6): 12.5; T(7): 12.5; T(8): 59.5; T(10): 12.5; T(11): 3.0
	3	AT5G52790.1	CBS domain-containing protein with a domain of unknown function (DUF21)	LLDLLLGKRHSTLLGR###	1,885.07854	0.023892	0.3993647	51	11	S(11): 1.7; T(12): 98.3
	4	AT4G28000.1	P-loop containing nucleoside triphosphate hydrolases superfamily protein	HTRNLAPGSK	1,160.55168	0.03559	0.754117	50	3	T(2): 0.0; S(9): 100.0
	5	AT1G72150.1	PATELLIN 1	SSFVFVSDFRNAPGLGKR	2,064.01040	0.039425	0.6264254	39	1	S(1): 1.0; S(2): 1.0; S(7): 98.0
	6	AT2G04842.1	threonyl-tRNA synthetase, putative/threonine-tRNA ligase, putative	SRFGGELGTIPVDDLINKINIAVETR###	3,067.42545	0.041246	0.6307821	38	1	S(1): 100.0; T(9): 100.0; T(25): 100.0
	7	AT3G22760.1	tesmin/TSO1-like CXC domain-containing protein	VIRNSDSIIEVGEDASK###	1,911.89481	0.042639	0.8216446	52	1	S(5): 0.0; S(7): 0.1; S(16): 99.9
	8	AT3G16950.1; AT4G16155.1	lipoamide dehydrogenase 1/ dihydrolipoyl dehydrogenases	DIIIATGSVPFVPK	1,536.80143	0.043742	0.5158506	38	2	T(6): 12.3; S(8): 87.7

Table 1. *Cont.*

	No.	Protein code	Description	Sequence	MH+[Da]	q-Value	PEP	pRS Score	# PSMs	pRS Site Probabilities
Crude extract + MOAC	1	AT1G56220.4	dormancy/auxin associated family protein	HHTFSFRPSSGNDQSEAGSAR###	2,354.98525	0	7.2906E-05	36	16	T(3): 13.7; S(5): 57.9; S(9): 13.7; S(10): 13.7; S(15): 1.1; S(19): 0.1
	2	AT2G17410.2	ARID/BRIGHT DNA-binding domain-containing protein	HSEENQSPHHHANNVMEQDQAAEER	3,004.19179	0	0.00011139	60	9	S(2): 97.1; S(7): 2.9
	3	AT5G52310.1	low-temperature-responsive protein 78 (LTI78)	MDQTEEPPLNTHQQHPEEVEHHENGATK	3,342.38857	0	2.1345E-05	36	16	T(4): 96.1; T(11): 3.8; T(27): 0.0
	4	AT5G55160.1	small ubiquitin-like modifier 2	SATPEEDKKPDQGAHINLK###	2,237.97500	0.000487	0.00798287	31	2	S(1): 100.0; T(3): 100.0
	5	AT2G24270.2	aldehyde dehydrogenase 11A3	AGTGLFAEILDGEVYK###	1,762.82077	0.000503	0.0115286	38	3	T(3): 100.0; Y(15): 0.0
	6	AT1G45207.2	remorin family protein	GWSSERVPLR	1,266.59882	0.00075	0.01489954	38	24	S(3): 50.0; S(4): 50.0
	7	AT1G01100.1; AT5G47700.1	60S acidic ribosomal protein family	STVGELACSYAVMILEDEGIAITADK	2,836.31522	0.000976	0.01672018	48	4	S(1): 25.0; T(2): 25.0; S(9): 25.0; Y(10): 25.0; T(23): 0.0
	8	AT4G12420.1	cupredoxin superfamily protein	RPLTGPAKVATSIINGTYR	2,175.06754	0.002939	0.08369295	36	3	T(4): 99.1; T(11): 7.4; S(12): 91.8; T(17): 0.9; Y(18): 0.9

Table 1. *Cont.*

No.	Protein code	Description	Sequence	MH+[Da]	q-Value	PEP	pRS Score	# PSMs	pRS Site Probabilities
9	AT1G74920.1; AT3G48170.1	aldehyde dehydrogenase 10A8/9	SPLIVFDDVDLDK	1,555.73259	0.009502	0.2584146	70	2	S(1): 100.0
10	AT3G28710.1	ATPase, V0/A0 complex, subunit C/D	AVNITINSIGTELTR###	1,681.86216	0.015828	0.3155473	37	30	T(5): 0.0; S(8): 77.3; T(11): 11.3; T(14): 11.3
11	AT1G73610.1	GDSL-like Lipase/Acylhydrolase superfamily protein	SYETIAPQIIENIKAK###	1,977.93410	0.017056	0.3870838	30	18	S(1): 50.3; Y(2): 50.3; T(4): 99.3
12	AT2G41110.1	calmodulin 2	ADQLTDDQISEFK	1,589.66288	0.019757	0.513722	50	5	T(5): 100.0; S(10): 0.0
13	AT1G70200.1	RNA-binding (RRM/RBD/RNP motifs) family protein	QFTGQSLAFGKVIKQIK	1,973.05405	0.02041	0.4607051	40	7	T(3): 50.0; S(6): 50.0
14	AT2G22400.1	S-adenosyl-L-methionine-dependent methyltransferases superfamily protein	EIRKNQTLER	1,366.68921	0.02499	0.5231273	38	1	T(7): 100.0
15	AT4G30630.1	unknown protein	LSESGGLEVPRKPSGERK###	2,006.01230	0.031388	0.6183366	32	1	S(2): 0.1; S(4): 0.2; S(14): 99.7
16	AT5G64090.1	unknown protein	ASHDLNPQAILATR	1,586.76555	0.035	0.6582299	57	1	S(2): 0.0; T(13): 100.0
17	AT1G80380.3	P-loop containing nucleoside triphosphate hydrolases superfamily protein	GNAGSHDLKLSVETLEALSKLTK###	2,491.28416	0.043032	0.7887968	36	1	S(5): 0.2; S(11): 0.1; T(14): 0.2; S(19): 9.4; T(22): 90.1

Table 1. *Cont.*

	No.	Protein code	Description	Sequence	MH+[Da]	q-Value	PEP	pRS Score	# PSMs	pRS Site Probabilities
40% AS + MOAC (PAPE)	1	AT5G52310.1	low-temperature-responsive protein 78 (LTI78)	SHELDLKNESDIDK DVPTGFDGEPDFLA K	3,311.49355	0	0.0027675	58	8	S(1): 0.6; S(10): 98.8; T(18): 0.6
	2	AT1G01320.2	tetratricopeptide repeat (TPR)-like superfamily protein	STQPSSGNAKTAGE TSEEDGLKTDASSV EPPTLSSTVQSEAY HTK###	4,690.11245	0.000811	0.04504684	41	10	S(1): 2.7; T(2): 2.7; S(5): 2.7; S(6): 2.7; T(11): 2.7; T(15): 17.1; S(16): 17.1; T(23): 17.1; S(26): 17.1; S(27): 17.1; T(32): 0.5; S(34): 0.1; S(35): 0.1; T(36): 0.1; S(39): 0.0; Y(42): 0.0; T(44): 0.0
	3	AT3G11130.1; AT3G08530.1	clathrin, heavy chain	EYSGKVDELIK###	1,360.63437	0.000811	0.04561926	49	6	Y(2): 99.7; S(3): 0.3
	4	ATMG00285.1	NADH dehydrogenase 2A	KSEFSTEAGSK###	1,250.52198	0.001809	0.09635145	66	3	S(2): 0.0; S(5): 9.0; T(6): 91.0; S(10): 0.0
	5	AT1G20620.1	catalase 3	MDPYKYRPSSAYN APFYTTNGGAPVSN NISSLTIGER	4,118.89516	0.002241	0.04873965	47	14	Y(4): 2.0; Y(6): 2.0; S(9): 15.6; S(10): 15.6; Y(12): 15.6; Y(17): 15.6; T(18): 15.6; T(19): 15.6; S(26): 2.0; S(30): 0.3; S(31): 0.1; T(33): 0.0

Table 1. *Cont.*

No.	Protein code	Description	Sequence	MH+[Da]	q-Value	PEP	pRS Score	# PSMs	pRS Site Probabilities
6	AT3G18780.2	actin 2	AEADDIQPIVCDNGTGMVKAG FAGDDAPR###	3,070.31753	0.00444	0.09388046	45	2	T(15): 100.0
7	AT5G09810.1; AT2G37620.1	actin 1/7	ADGEDIQPLVCDNGTGMVKA GFAGDDAPR###	3,056.32534	0.005154	0.1148702	36	5	T(15): 100.0
8	AT5G56180.1	actin-related protein 8	TVVLTGGSACLPGLSER###	1,796.85564	0.005762	0.1187534	68	2	T(1): 0.0; T(5): 0.0; S(8): 0.1; S(15): 99.8
9	AT1G49240.1	actin 8	ADADDIQPIVCDNGTGMVKAG FAGDDAPR###	3,056.32534	0.008319	0.1784335	36	3	T(15): 100.0
10	AT3G02830.1	zinc finger protein 1	NKAGIAGRVSLNMLGYPLR	2,110.10227	0.016331	0.3097203	47	1	S(10): 100.0; Y(16): 0.0
11	AT1G64790.1	ILITYHIA	SPIVSAAAFENLVK	1,525.75934	0.017784	0.3293382	48	5	S(1): 10.3; S(5): 89.7
12	AT4G38770.1	proline-rich protein 4	KEVPPPVPVYKPPPK###	1,751.95328	0.026158	0.4063287	34	1	Y(10): 100.0
13	AT4G31120.2	SHK1 binding protein 1	DVHLGIEPTTATPNMFSW###	2,095.92093	0.031227	0.430492	31	1	T(9): 0.5; T(10): 0.5; T(12): 1.4; S(17): 97.6
14	AT5G16330.1	NC domain-containing protein-related	RGTCTIAPSDPCDEVISR###	2,193.86269	0.039343	0.684857	64	3	T(3): 5.3; T(5): 0.4; S(9): 94.3; S(17): 100.0

###: Peptides marked with "###" are from proteins that are, so far, not annotated as phosphoproteins. All other peptides belong to known phosphoproteins, but are themselves not reported as being phosphorylated. Abbreviations: MH+ = Positive ion mode pseudo-molecular ion; Da = Dalton; PEP = Posterior error probability; PSM = peptide spectrum match.

However, there are also cases where no phosphopeptides could be identified for the putative phosphoprotein enriched in the PAPE fraction (e.g., MPK4 or MPK6). This is possibly one of the caveats of the present study, which is that when compared to the reproducibility in identification of the (putative) phosphoprotein (see Table S1, Figure 2B), there is often difficulties or variation in the phosphopeptide identification between replicate measurements. Contrary to expectation, the absolute number of phosphopeptide identified is not particularly high, despite the increased phosphoprotein detection (Figure 3C). However, such limitations can be attributed to the fact that the subsequent tryptic digestion reintroduced a complex peptide mixture, thereby hindering the phosphopeptide identification by MS as a consequence of the over-representation of non-phosphorylated peptides over phosphopeptides [40]. It is known that phosphoprotein enrichment procedures will increase the number of phosphorylated proteins, but this does not necessarily translate to larger numbers of identified phosphorylated peptides [9]. For this purpose, an additional phosphopeptide enrichment step to the current PAPE procedure may be included to enhance phosphopeptide identification. However, due to the different efficiencies in capturing mono-phosphorylated and multiple phosphorylated peptides from complex peptide mixtures [41], this was not done in the current study to avoid losing the identification of certain phosphoproteins. The current PAPE procedure is mainly designed to detect phosphoproteins from green plant tissues.

4. Conclusions

We report here that a simple ammonium sulfate fractionation step can be used to eliminate abundant RuBisCO proteins and simultaneously enrich phosphoproteins from *Arabidopsis* leaves. A combination of this step with MOAC phosphoprotein enrichment, which we termed PAPE, enabled the identification of low abundance phosphoproteins, including several that are not annotated in the P3DB and PhosPhAt 3.0 databases. Overall, the PAPE procedure performed better than MOAC alone to enrich phosphoproteins. While some proteins will be missed by the PAPE procedure, because of removal during the ammonium sulfate precipitation step, the Pro-Q Diamond phosphostain indicated that the bulk of phosphoproteins are actually within the fraction used for analysis (see Figure 1B). Thus, by eliminating RuBisCO and enriching phosphoproteins, the PAPE procedure reduces the effective dynamic range of protein abundance in the plant proteome and ameliorates the detection of phosphoproteins. Its facile handling allows it to be implemented in any laboratory. We also envisage that the inclusion of a phosphopeptide enrichment step to the current PAPE fraction would further improve the mapping of the plant phosphoproteome.

Author Contributions

I.L. performed the experiments. The conception of the PAPE procedure was by I.L. and K.N. D.S. and J.L. supervised the project. I.L., J.L. and D.S. wrote the paper.

Acknowledgments

This work was financed through a *Deutsche Forschungsgemeinschaft* (DFG) Grant to the European Research Area Network – Plant Genomics (ERA-PG) “PathoNET” project (SCHE 235/15-1) to D.S. and

J.L. We thank Petra Majovsky and Sylvia Krüger for technical assistance, Wolfgang Hoehenwarter for proof reading and suggestions and members from our laboratory for helpful discussion.

References

1. The Arabidopsis Genome Initiative. Analysis of the genome sequence of the flowering plant *Arabidopsis thaliana*. *Nature* **2000**, *408*, 796–815.
2. Lamesch, P.; Berardini, T.Z.; Li, D.; Swarbreck, D.; Wilks, C.; Sasidharan, R.; Muller, R.; Dreher, K.; Alexander, D.L.; Garcia-Hernandez, M.; *et al.* The *Arabidopsis* information resource (TAIR): Improved gene annotation and new tools. *Nucleic Acids Res.* **2011**, *40*, D1202–D1210.
3. Richards, C.L.; Rosas, U.; Banta, J.; Bhambhra, N.; Purugganan, M.D. Genome-wide patterns of *Arabidopsis* gene expression in nature. *PLoS Genet.* **2012**, *8*, e1002662.
4. Snoek, L.B.; Terpstra, I.R.; Dekter, R.; van den Ackerveken, G.; Peeters, A.J. Genetical genomics reveals large scale genotype-by-environment interactions in *Arabidopsis thaliana*. *Front. Genet.* **2013**, *3*, e317.
5. Heazlewood, J.L. The green proteome: Challenges in plant proteomics. *Front. Plant Sci.* **2011**, *2*, e6.
6. Joshi, H.J.; Hirsch-Hoffmann, M.; Baerenfaller, K.; Gruissem, W.; Baginsky, S.; Schmidt, R.; Schulze, W.X.; Sun, Q.; van Wijk, K.J.; Egelhofer, V.; *et al.* MASCP Gator: An aggregation portal for the visualization of *Arabidopsis* proteomics data. *Plant Physiol.* **2011**, *155*, 259–270.
7. Petricka, J.J.; Schauer, M.A.; Megraw, M.; Breakfield, N.W.; Thompson, J.W.; Georgiev, S.; Soderblom, E.J.; Ohler, U.; Moseley, M.A.; Grossniklaus, U.; *et al.* The protein expression landscape of the Arabidopsis root. *Proc. Natl. Acad. Sci. USA* **2012**, *109*, 6811–6818.
8. Sakurai, T.; Yamada, Y.; Sawada, Y.; Matsuda, F.; Akiyama, K.; Shinozaki, K.; Hirai, M.Y.; Saito, K. PRIMe Update: Innovative content for plant metabolomics and integration of gene expression and metabolite accumulation. *Plant Cell Physiol.* **2013**, *54*, e5.
9. Nakagami, H.; Sugiyama, N.; Ishihama, Y.; Shirasu, K. Shotguns in the front line: Phosphoproteomics in plants. *Plant Cell Physiol.* **2012**, *53*, 118–124.
10. Guo, M.; Huang, B.X. Integration of phosphoproteomic, chemical, and biological strategies for the functional analysis of targeted protein phosphorylation. *Proteomics* **2013**, *13*, 424–437.
11. Manning, G.; Plowman, G.D.; Hunter, T.; Sudarsanam, S. Evolution of protein kinase signaling from yeast to man. *Trends Biochem. Sci.* **2002**, *27*, 514–520.
12. Wang, D.; Harper, J.F.; Gribskov, M. Systematic trans-genomic comparison of protein kinases between *Arabidopsis* and *Saccharomyces cerevisiae*. *Plant Physiol.* **2003**, *132*, 2152–2165.
13. De la Fuente van Bentem, S.; Roitinger, E.; Anrather, D.; Csaszar, E.; Hirt, H. Phosphoproteomics as a tool to unravel plant regulatory mechanisms. *Physiol. Plant.* **2006**, *126*, 110–119.
14. Hunter, T. Signaling—2000 and beyond. *Cell* **2000**, *100*, 113–127.
15. Herskowitz, I. MAP kinase pathways in yeast: For mating and more. *Cell* **1995**, *80*, 187–197.
16. Hamel, L.P.; Nicole, M.C.; Sritubtim, S.; Morency, M.J.; Ellis, M.; Ehlting, J.; Beaudoin, N.; Barbazuk, B.; Klessig, D.; Lee, J.; *et al.* Ancient signals: Comparative genomics of plant MAPK and MAPKK gene families. *Trends Plant Sci.* **2006**, *11*, 192–198.
17. Meng, X.; Zhang, S. MAPK cascades in plant disease resistance signaling. *Annu. Rev. Phytopathol.* **2013**, *51*, 245–266.

18. Di Palma, S.; Zoumaro-Djayoon, A.; Peng, M.; Post, H.; Preisinger, C.; Munoz, J.; Heck, A.J. Finding the same needles in the haystack? A comparison of phosphotyrosine peptides enriched by immuno-affinity precipitation and metal-based affinity chromatography. *J. Proteomics* **2013**, *91*, 331-337.
19. Yue, X.S.; Hummon, A.B. Combination of multistep IMAC enrichment with high-pH reverse phase separation for in-depth phosphoproteomic profiling. *J. Proteome Res.* **2013**, *12*, 4176-4186.
20. Wolschin, F.; Wienkoop, S.; Weckwerth, W. Enrichment of phosphorylated proteins and peptides from complex mixtures using metal oxide/hydroxide affinity chromatography (MOAC). *Proteomics* **2005**, *5*, 4389–4397.
21. Güzel, Y.; Rainer, M.; Mirza, M.R.; Messner, C.B.; Bonn, G.K. Highly selective recovery of phosphopeptides using trypsin-assisted digestion of precipitated lanthanide-phosphoprotein complexes. *Analyst* **2013**, *138*, 2897–2905.
22. Fila, J.; Honys, D. Enrichment techniques employed in phosphoproteomics. *Amino Acids* **2011**, *43*, 1025–1047.
23. Cellar, N.A.; Kuppannan, K.; Langhorst, M.L.; Ni, W.; Xu, P.; Young, S.A. Cross species applicability of abundant protein depletion columns for ribulose-1,5-bisphosphate carboxylase/oxygenase. *J. Chromatogr. B Analyt. Technol. Biomed. Life Sci.* **2008**, *861*, 29–39.
24. Xi, J.; Wang, X.; Li, S.; Zhou, X.; Yue, L.; Fan, J.; Hao, D. Polyethylene glycol fractionation improved detection of low-abundant proteins by two-dimensional electrophoresis analysis of plant proteome. *Phytochemistry* **2006**, *67*, 2341–2348.
25. Aryal, U.K.; Krochko, J.E.; Ross, A.R. Identification of phosphoproteins in *Arabidopsis thaliana* leaves using polyethylene glycol fractionation, immobilized metal-ion affinity chromatography, two-dimensional gel electrophoresis and mass spectrometry. *J. Proteome Res.* **2011**, *11*, 425–437.
26. Widjaja, I.; Naumann, K.; Roth, U.; Wolf, N.; Mackey, D.; Dangl, J.L.; Scheel, D.; Lee, J. Combining subproteome enrichment and Rubisco depletion enables identification of low abundance proteins differentially regulated during plant defense. *Proteomics* **2009**, *9*, 138–147.
27. Lassowskat, I. Optimisation, Extension and application of metal-oxide affinity chromatography for Proteome analysis of *Arabidopsis thaliana. Diplom* (Master) Thesis (in German), Martin-Luther-Universität Halle-Wittenberg, Halle/Saale, Germany, 2008.
28. Laemmli, U.K. Cleavage of structural proteins during the assembly of the head of bacteriophage T4. *Nature* **1970**, *227*, 680–685.
29. Agrawal, G.K.; Thelen, J.J. Development of a simplified, economical polyacrylamide gel staining protocol for phosphoproteins. *Proteomics* **2005**, *5*, 4684–4688.
30. Taverner, T.; Karpievitch, Y.V.; Polpitiya, A.D.; Brown, J.N.; Dabney, A.R.; Anderson, G.A.; Smith, R.D. DanteR: An extensible R-based tool for quantitative analysis of -omics data. *Bioinformatics* **2012**, *28*, 2404–2406.
31. The Arabidopsis information resource. Available online: http://arabidopsis.org/tools/bulk/go/index.jsp (accessed on 1 February 2013).
32. Vizcaino, J.A.; Cote, R.G.; Csordas, A.; Dianes, J.A.; Fabregat, A.; Foster, J.M.; Griss, J.; Alpi, E.; Birim, M.; Contell, J.; *et al.* The PRoteomics IDEntifications (PRIDE) database and associated tools: Status in 2013. *Nucleic Acids Res.* **2013**, 41, D1063–D1069.
33. Görg, A.; Weiss, W.; Dunn, M.J. Current two-dimensional electrophoresis technology for

proteomics. *Proteomics* **2004**, *4*, 3665–3685.

34. Pace, C.N.; Trevino, S.; Prabhakaran, E.; Scholtz, J.M. Protein structure, stability and solubility in water and other solvents. *Philos. Trans. R Soc. Lond. B Biol. Sci.* **2004**, *359*, 1225–1235.
35. Stensballe, A.; Andersen, S.; Jensen, O.N. Characterization of phosphoproteins from electrophoretic gels by nanoscale Fe(III) affinity chromatography with off-line mass spectrometry analysis. *Proteomics* **2001**, *1*, 207–222.
36. Gao, J.; Agrawal, G.K.; Thelen, J.J.; Xu, D. P3DB: A plant protein phosphorylation database. *Nucleic Acids Res.* **2009**, *37*, D960–D962.
37. Yao, Q.; Bollinger, C.; Gao, J.; Xu, D.; Thelen, J.J. P(3)DB: An integrated database for plant protein phosphorylation. *Front. Plant Sci.* **2012**, *3*, e206.
38. Durek, P.; Schmidt, R.; Heazlewood, J.L.; Jones, A.; MacLean, D.; Nagel, A.; Kersten, B.; Schulze, W.X. PhosPhAt: The *Arabidopsis thaliana* phosphorylation site database. An update. *Nucleic Acids Res.* **2009**, *38*, D828–D834.
39. Heazlewood, J.L.; Durek, P.; Hummel, J.; Selbig, J.; Weckwerth, W.; Walther, D.; Schulze, W.X. PhosPhAt: A database of phosphorylation sites in *Arabidopsis thaliana* and a plant-specific phosphorylation site predictor. *Nucleic Acids Res.* **2008**, *36*, D1015–D1021.
40. Boersema, P.J.; Mohammed, S.; Heck, A.J. Phosphopeptide fragmentation and analysis by mass spectrometry. *J. Mass Spectrom.* **2009**, *44*, 861–878.
41. Thingholm, T.E.; Jensen, O.N.; Larsen, M.R. Enrichment and separation of mono- and multiply phosphorylated peptides using sequential elution from IMAC prior to mass spectrometric analysis. *Methods Mol. Biol.* **2009**, *527*, 67–78.

10

Proteomics Advances in the Understanding of Pollen–Pistil Interactions

Ziyang Fu and Pingfang Yang *

Key Laboratory of Plant Germplasm Enhancement and Speciality Agriculture, Wuhan Botanical Garden, Chinese Academy of Sciences, Wuhan 430074, China; E-Mail: ziyangrd@163.com

* Author to whom correspondence should be addressed; E-Mail: yangpf@wbgcas.cn

---

Abstract: The first key point to the successful pollination and fertilization in plants is the pollen-pistil interaction, referring to the cellular and molecular levels, which mainly involve the haploid pollen and the diploid pistil. The process is defined as "siphonogamy", which starts from the capture of pollen by the epidermis of stigma and ends up with the fusion of sperm with egg. So far, the studies of the pollen-pistil interaction have been explicated around the self-compatibility and self-incompatibility (SI) process in different species from the molecular genetics and biochemistry to cellular and signal levels, especially the mechanism of SI system. Among them, numerous proteomics studies based on the advanced technologies from gel-system to gel-free system were conducted, focusing on the interaction, in order to uncover the mechanism of the process. The current review mainly focuses on the recent developments in proteomics of pollen-pistil interaction from two aspects: self-incompatible and compatible pollination. It might provide a comprehensive insight on the proteins that were involved in the regulation of pollen-pistil interaction.

Keywords: proteomics; interaction; pollen; pistil; self-incompatibility

---

1. Introduction

Deep in the evolutionary history, the plant kingdom goes through trends from low to high, simple to complex, aquatic to terrestrial. Corresponding to the life cycles, it can be of three-types of propagation:

vegetative propagation, asexual reproduction, and sexual reproduction. The sexual reproduction is independent of water during fertilization. The transporting of the sperm to the egg is via the pollen tube. Angiosperms are the largest and the most diverse groups in the plant kingdom. They are placed on top of the evolution table, showing alternation of generation in their life cycle. Their sporophyte produce spores meiotically in their asexual reproduction [1]. Angiosperms sexual reproduction occurs when the female and male gametes fuse during the process of fertilization to produce viable offspring. The formation of a viable zygote in angiosperms is dependent on successful pollination and fertilization, which begins with pollen grain landing on the stigma of the pistil. The first key point of this complex process of fertilization is the pollen-pistil interaction, referring to a cellular and molecular interaction where the haploid pollen and the diploid pistil fuse. The specific steps of the interaction are as follows: before the substantive contact, the immobile pollen is transferred by external environmental forces, such as wind or insects. Upon pollen grain landing on the surface of stigma, the recognition starts at molecular and cellular levels. The "compatible" pollen germinates if the surrounding conditions are suitable. Then, the pollen tube grows and extends through the style into the ovule; the sperm cells are discharged by female gametes. The series of process is also defined as "siphonogamy" [2]. It has significant role in reproductive biology of flowering plants. This kind of interaction is vital in sexual reproduction in order to obtain genetic diversity in a population. In addition, the interaction between the pollen and pistil is regarded as a series of stages while studying the molecular mechanisms underlying cell recognition, cell germination and cell-to-cell communication. Scrutinizing the mechanisms of interaction during fertilization will be beneficial to crop breeding and reproduction. So far, the studies of the pollen-pistil interaction have been explicated around the self-compatibility and self-incompatibility process in different species from the molecular genetics and biochemistry to cellular and signal level, especially the generation mechanism of SI system [3–5].

Thanks to the completed genome sequences of some model plants, such as Arabidopsis, the genus is Populus and rice, and the field has been expanded into genomics and proteomics [6]. Among them, proteomics, as the study of how proteins work, interact, diversify and specialize on a global scale, has been widely applied in analyzing the biological processes drawing support from the rapid development of mass spectrometry (MS). By these new and powerful proteomic techniques, the interaction between pollen and pistil can be improved and studied with new insight [7]. Looking back into the history of its development, the techniques in proteomic research can be divided into two categories: gel-based strategy, developed from 2-D gel electrophoresis (2-DE) and isoelectric focusing PAGT (IEF-PAGE) to differential in-gel electrophoresis (DIGE), and gel-free system based on the liquid chromatography-mass spectrometry (LC-MS), which analyzes proteins on a large-scale [8,9]. Among them, the differential in-gel electrophoresis (DIGE) uses different fluorescent dyes on the 2D-GE to enhance the accuracy of quantifying the expression level of different proteins. Additionally, LC-MS/MS-based quantitative methods can be improved with the different labeling strategies, such as isobaric tag for relative and absolute quantitation (iTRAQ) [10]. Recently, a new technique label-free quantification has been improved, and it is termed "label-free LC MS/MS" [11].

This review focuses on the advance of proteomics studies in the interaction between pollen and pistil to reveal the complex biological process from a pivotal gene or enzyme to an integral pathway of regulatory or signal transduction.

2. Proteomic Analysis of Pollen-Pistil Interaction with Successful Fertilization

2.1. Proteome Dynamics in Pollen at Different Points from Development to Germination

Before cell-to-cell interaction, pollen has to undergo a complex and necessary development to form a suited grain with high quality. Anther with four locules is the progenitor cells to release microspores, which is generated from microsporocyte via two continuous meiotic divisions [12]. The microspores released in the locule undergo cytoplasmic reorganization mediated by the cytoskeleton, and it also termed as polarizability. Subsequently, the polarized cell forms a male germ unit not only with two sperm cells that fuse with egg cell and central cell, respectively, but also with a vegetative nucleus via two mitotic divisions. The pollen with the three cells is called mature pollen, which can interact with gynoecia and be selected into the style for tube growth to access the embryo sac for double fertilization [13,14]. Mature pollen germinates fast after pollination from quiescent to active state, and the vegetative cell grows a pollen tube transporting the sperm cells into the embryo sac to complete the double fertilization process.

According to previous study on the pollen transcriptome in the model species *Arabidopsis*, 992 pollen-expressed mRNAs were identified by comparing with that of the sporophyte. Among them, nearly 40% were detected specifically in mature pollen, encoding proteins involved in carbohydrate metabolism and cytoskeleton dynamics, which would be a key to ensure the quality of pollen through regulatory and functional specialization [15]. Furthermore, one-third of the genes constitutively expressed in the vegetative tissues were not expressed in pollen. These results revealed that the transcriptome of gamete showed more simplicity and higher proportion of selectively expressed genes than sporophytic tissues [16]. Soon afterwards, a proteome map for mature pollen of *Arabidopsis thaliana* was drawn using 2-DE followed by ESI-MS/MS to expound the biochemistry of pollen [17]. Among the 135 identified proteins, there were almost 20% involved in metabolism, 17% in energy generation, or 12% in cell structure, which is similar to the study at transcriptional level. Moreover, seven proteins whose RNAs were not shown in the transcriptome have functions in metabolism, energy generation or cell structure. To resolve the omission of short and low abundant proteins, a new generic deterministic peptide classification scheme was set up to identify the proteins with minimized error rate on the Arabidopsis pollen [18]. In addition to the large scale on the pollen grains, the analysis on the pollen coat at gene level and protein levels had been taken [19]. There were 322 special proteins in mature pollen of rice using MS technologies, which had been classified into at least 14 functional categories [20]. Among them, 38 unique proteins were beta-1,4-xylanase and beta-glucanase in pollen coats, which were major proteins in the pollen coat of maize [21]. The released proteins from the pollen might contribute to the germination and growth of the grains and pollen tube.

In a way, the revelation of the gene level cannot directly link to the discipline of the protein level and activity. The proteomics work on developmental pollen used two-dimensional gel to find disparate points and then identified them by MALDI-TOF MS based on the sequence dates. For example, in *Oryza sativa*, the comparative proteins of the different stages of pollen (pollen mother cell, tetrad, early young microspore, middle young microspore, early binucleate, late binucleate, heading stage) were taken into analysis by similar proteomics technologies. The 33 unique proteins with the same changing trend participated in sugar metabolism, cell elongation and expansion, which were essential to the pollen germination [22]. Analogously, rice mature (MPG) and germinated (GPG) pollen grains were selected artificially and then flowed to 2D-gels to obtain protein spots. Comparing proteins of the two different growth stages, 186 proteins from almost 2300 proteins were differentially expressed. These proteins are involved in regulatory and metabolic processes, such as the dynamics of protein and cytoskeleton [23]. Comparing the protein expression of the germinating pollen (GP) and the mature pollen (MP) of canola (*Brassica napus*) via DIGE associated to MALDI-TOF/TOF, the up-regulated proteins play roles in carbohydrate, nucleotide and protein metabolism, signal transduction and stress responses. On the contrary, some catalases and LEA proteins were showed to be down-regulated. These showed that the proteins associated in macromolecules' metabolism and enzymes involved in the signal pathways were essential to the pollen germination [24]. The differentially expressed proteins in pollen tubes compared to the un-germinated pollens could demonstrate the special physiological processes that occur during the development of the pollen tubes [25]. Recently, a proteomics analysis was also conducted in pollen (before-pollinated and/or pollinated) and pollen tube of the *Picea meyer* [26], *Triticosecale wittmack* [27] and *Lilium davidii* [28]. These researches are better conducted in two aspects: pollen collection, including the condition of pollen grains germination *in vitro*, the nutrition and temperature for plants growth, pollen review by light microscopy, and the identification and quantification of proteomics, covering the ITRAQ labeling method and MALDI analysis by TOF/TOF instrument. The proteomics for male gametophyte contribute a new slight on examining the molecular mechanism and lay a great foundation for the study of fertilization.

2.2. Protein Analysis on Pistils by Comparative Proteomics

Pistil, another participant, is composed of stigma, style, and ovary [29]. The apical stigma is in the position of capturing and ingesting pollen grains. The style can be regarded as a subtle bridge linking the stigma and the ovary, where the transmitting tissue plays role accompanying the extension of the pollen tube. The basal ovary is an organ containing the ovules, which gestate the embryo sac to grow the eggs [30,31]. As the name suggests, pollination depends on transferring male gametophyte (pollen grain) from the anther to the sigma of pistil. That is the space where interaction occurs. Judging by whether or not there is secretion on the surface, the stigmas can be classified into two broad categories, wet and dry. Comparing the two biochemical reactions based on the different structure, up to date, pollen capture by the wet stigmatic secretion is nonspecific while the recognition process shows a degree of species specificity in the dry-type stigma. Besides, pollen hydration within the secretion is passive and unregulated. On the contrary, it is a regulated process in dry stigma [2].

Crucifer is the typical family with dry stigma, which is selected as model plants for study of pollen-pistil interaction. Through comparing the whole-genome transcriptional profiles of stigmas and ovaries isolated separately from wild-type *Arabidopsis* and transgenic plants, in which cells of the stigma epidermis and transmitting tract were ablated by expression of a cellular toxin on the microarray platform, 115 and 34 genes were identified from 23,000 genes on the array to be expressed specifically in the stigma epidermis and transmitting tract. The proteins encoded by these genes were functional classified in signal transduction pathway, regulating the components of the extracellular matrix during pollination. Among them, S-locus receptor kinase (SRK), M-locus protein kinase (MLPK) and arm repeat containing (ARC1) play roles in self-incompatibility [32].

The affymetrix ATH1 whole genome array were used in comparing the different gene levels in un-pollinated pistils and un-fertilized ovules of *Arabidopsis thaliana*, as well as the pollinated pistils in special timing points that represented the most significant development from pollination to fertilization. The result showed 1373 genes were differentially expressed during pollen–pistil interaction, whose function were explained and projected to the extent necessary for successful fertilization [33]. The temporal and spatial gene expression profile of *in vivo* pollen-pistil interaction provided a detailed evidence of changing in gene expression pattern, further supporting the molecular mechanisms operating during pollination.

Recent researches on proteomic analysis of the pistil in crucifer were mainly taken in SI response to find the candidate proteins. This implies that the mechanism of compatible interaction between pollen and pistil is not still directly addressed, especially the proteome in pistil. After searching—a comparative proteomic analysis of pistil before and after pollination in Soybean cooperated with the proteome database [34] and transcriptomics analyses [35]—a strict self-pollination plant was found. According to the MALDI-TOF-MS results based on the 2D-gel, 58 differently expressed proteins were identified, of which there were 22 up-regulated proteins and 36 down-regulated proteins after pollination. After functional classification, the largest group was metabolism-related proteins. Among them, the sucrose-phosphate synthase (U18) was increased in expression, while some isoforms of glutamine synthetase were decreased. These indicated that the primary metabolisms were enhanced to facilitate the pollination and the following pollen tube growth [36]. This study enhances our understanding of the level of proteins expression, as well as the participated biological processes.

The representative plants with the wet stigma are mainly Solanaceae, Rosaceae and Liliaceae [2]. The stigmatic secretion (SE) on the surface of the stigma plays an important role in ingesting the compatible pollen grains. Except the lipid and carbohydrate, SE also contains a wide range of proteins with profound functions, such as the sigma-specific protein 1 (STIG1), regulating the timing of the accumulation of SE in tobacco and petunia [37]. The first analysis of SE proteins on a large-scale level was taken in the *Lilium longiflorum* and *Olea europaea* by SDS-PAGE coordinated with the LC-MS/MS. However, given the un-completely genome annotation of the two plants, a database search algorithm (Mascot) was employed and identified 51 proteins in Lily and 57 in olive, of which only 13 were present in both SEs. In-depth analysis of these proteins showed that more than half of the proteins contain a signal peptide, and it was predicted that the SE might participate in at least 80 different biological processes and 97 molecular functions, of which included the carbohydrate metabolism, cell signaling and response to the biotic and abiotic stresses. During pollination, the catabolic enzymes disintegrate large polysaccharides and lipids into smaller units to regulate pollen tube growth by selective degradation of cell-wall polysaccharides. Two Stigma/stylar cysteine-rich adhesion (SCA) isoforms, a chemotropic protein and a fasciclin-like domain (FAS) protein were identified in the lily SE with the role of pollen tube adhesion [38]. By proteomics, a comprehensive map of proteins can be built to discover their biological function within pollination.

When the male and female gametes are prepared for the next journey, the recognition and interaction will induce the tube growth via compatible signal and chemical gradients. The extension keeps until the tube enters into the embryo sac, where the molecules and receptors of tube response with the guidance from the female gametophyte [39]. Comparing to angiosperm, the pollination droplet is a unique and conservative pollination mechanism in gymnosperm. Proteomics analysis and identification taken on the mechanism revealed the proteins participating in the drop were related to the pathogen defense and pollen development [40]. Although there have not been a complete network for the interaction between the tube and pistil, some key regulators and receptors have been identified by multidisciplinary approaches including biochemistry, molecular genetics and functional genomics. On the contrary, in *Arabidopsis*, the female gametophytic guidance is divided into two stages: the funicular guidance, implying that the tubes extend from the septum to the funiculus, and the micropylar guidance, implying that the tube navigates from the funiculus to the female gametophyte through the micropyle [41]. For example, MYB98, as a transcription factor, expresses in the synergid cell as micropyle guidance [42,43]. AtLURE1 peptides belong to the defensin-like (DEFL) peptides, expressed in synergid cells and secreted toward the funicular surface to guide the tube growth as especially guidance [44]. Looking back on how the receptors on the pollen tube interact with the female guidance, the ion gradient regulated by the transmembrane proteins and channels are critical for the tube growth, such as the cyclic nucleotide-gated channel 18 (CNGC18), working as a channel to regulate the Ca^{2+} concentration, and a GABA transaminase, which is coded by the *pollen pistil 2* (*POP2*) and has the function to increase content of GABA in the style and ovary [39,45]. Besides, lost in pollen tube guidance 1 (LIP1) and 2 (LIP2) were verified as a receptor complex to respond to AtLURE1 [46]. Especially, the mitogen-activated protein kinase 3 (MPK3) and the mitogen-activated protein kinase 6 (MPK6) were verified with the function of guiding the direction to the right towards the mutational plants of mpk3 and mpk4 [47].

Although much research had been conducted on several model species around compatible and even incompatible pollination and fertilization, there was a common theory among the types of molecular regulation on the pollen-pistil interaction [48]. Researchers suggested the absence was as a result of rapid evolution and change of proteins regulating sexual reproductive process [49]. For instance, arabinogalactan proteins (AGPs), a family of hydroxyproline-rich glycoproteins (HRGPs), showed an increased expression during pollination in the olive pistil (*Olea europaea* L.) compared to the non-pollinated pistil, while the expression of AGPs decreased after pollination. AGPs were localized predominantly in the cell wall of secretory cell of the stigma, as well as in the transmitting tissue of the pistil during the pollination period by means of immunofluorescence localization. These results proved that proteins play roles of supporting pollen performance and tube growth during the pollination stage in olive, which corresponds to the conclusion that AGPs had roles in vegetative, reproductive, and cellular growth and development, previously [50,51].

3. Proteomic Analysis of Pollen-Pistil Interaction in SI Response

If one plant is hermaphrodite, both stamen and pistil from the same flower, it has more chance to self-fertilize. Self-fertilization means that the offspring is obtained within a short time and is identical genetically to the parent plant, which is in favor of maintaining stability with low genetic diversity. However, the genetic stability makes it difficult to adapt to the variable environments so that species can achieve optimized continuation. Therefore, it naturally led to the generation of cross-fertilization and evolution of various mechanisms for preventing self-fertilization [52]. To enhance diversity and obtain more chances to survive in such a changeable and complex world, the evolution of several morphologic and genetic barriers of self-fertilization has occurred in the life history of plants. For instance, dichogamy, namely the condition of anthers and pistils maturing at different times, can prevent self-fertilization. Except physical and suited isolation, discrimination between genetically related (self) and unrelated (non-self) pollen grains that then inhibit the related grains is one major genetic mode. That is self-incompatibility (SI), the failure of fertilization between the pollen and pistil from the same flower to produce zygote and endosperm [31]. To date, more than half of species in angiosperms have self-sterility and SI.

Dissecting the mechanism of the barrier system contributes to understand the series of events for pollen-pistil interaction. In genetics, SI is controlled by a single S (Sterility) locus with high polymorphism, which encodes proteins in pistil and pollen, respectively, as the basis of recognition and reaction during reproduction. There are two principal genetic forms of SI, gametophytic (GSI) and sporophytic (SSI), which are distinguished by the decider for the phenotype of S-locus in pollen [53]. The representative family for SSI is Brassicaceae, while Solanaceae and Papaceraceae are model plants for GSI. The mechanisms of the two categories are diverged with different controlled genes within a special signal pathway and regulatory network [54].

3.1. Proteomic Analysis of the Gametophytic Self-Incompatibility Response

In GSI, the S phenotype of the pollen is determined by its own haploid genome and the mechanism of GSI is mainly related to the inhibition of the growth of pollen tube. In addition there are two mechanisms of the GSI incompatibility system, where Solanaceae, Rosaceae and Plantaginaceae depend on the S-RNase-based rejection system and diversely Papaveraceae undergoes the S-glycoprotein mediated Ca^{2+} signaling system [55]. The female determinant of SI in Papaveraceae is PrsS (*P. rhoeas* style S) secreted by the stigma. When SI is trigged with the self-pollen, programmed cell death and pollen inhibition begin with increased concentration of Ca^{2+} [56,57]. Here, the review focuses on the S-RNase-based self-incompatibility.

The S-RNase-based rejection system is mediated through an interaction between S-locus ribonuclease (S-RNase, female determinant) and pollen tube-borne F-box proteins, S-locus F-box (SLF)/S-haplotype-specific F-box (SFB) (SLF/SFB male determinant) [58] (Figure 1).

The first S-specific proteins were found in *Nicotiana alata* in the stigma and style by isoelectric focusing [59]. Then, researchers confirmed that the S-specific proteins from different S-haplotypes in *Nicotiana alata* were ribonucleases with the RNase activity of inhibition of pollen tube growth in SI response [60].The similar approaches, including 2-D GE, were used in other Solanaceae and Rosaceae family to isolate and analysis of the style S-glycoprotein, such as *Petunia hybrida* [61,62] and Japanese pear [63,64]. The S-specific proteins were renamed S-RNases. The special and basic glycoproteins secreted into the extracellular matrix of the stigma, transmitting tract, and the inner epidermis of the ovary after pollination, which are cytotoxin contributing to RNA degradation during the extension of the pollen tube ending in the rejection of "self" pollen. The sequence analysis shows S-RNase has five conserved regions, C1 to C5, and two specially hyper-variable regions, HVa and HVb, which all contribute to the S-specific recognition [5]. However, there are not enough evidences to prove which domain determines the recognition.

Figure 1. A simplified diagram of the S-RNase-based self-incompatible system. There was more than one type of S-locus F-box complex (SLF) to recognize all non-self S-Rnase. If the "self" S-RNases were compartmentalized in the vacuole for the degradation of HT-B, a protein indispensable to the release of S-RNase, the type of S-RNase may be cytotoxin contributing to RNA degradation during the extension of the pollen tube ending in the rejection of "self" pollen. On the contrary, the "non-self" S-RNase may be ubiquitinated thus degraded through the proteasomal pathway by the SLF complex.

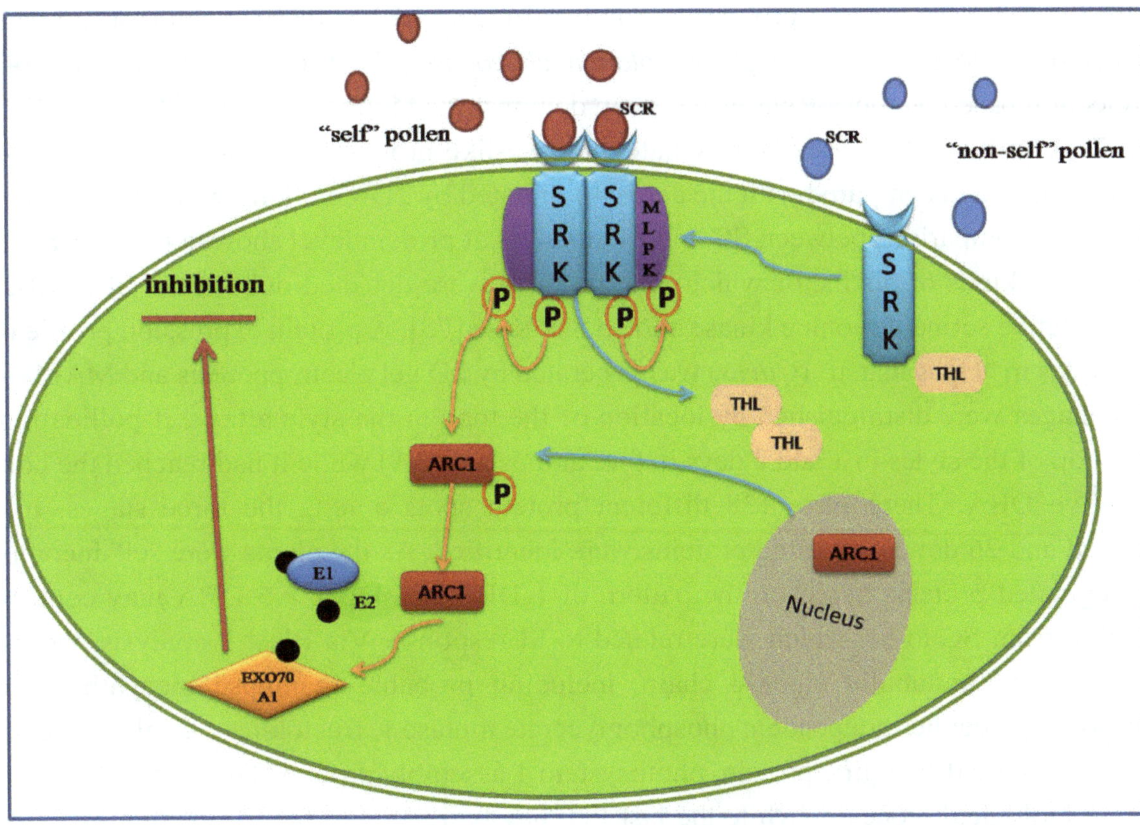

To understand the mechanism of GSI, the male determinant related to S-RNase should be fully unfolded. The S-determinant should fulfill three rules as follows: (i) linkage to S-RNase gene; (ii) variable sequences comparing different S-haplotypes; (iii) expression in pollen, specifically. Based on the above characteristics, the male S-specific genes were surfaced. In *Antirrhinum*, *AhSLF-S2* was identified as a novel F-box gene closely involved in the interaction of SI, which is polymorphic and expressed in *tapetum*, microspores and pollen grains, specifically. The protein encoded by *AhSLF-*S_2 had a conserved F-box domain [65]. Similarly, SLF was shown to be the pollen S-determinant in *Petunia inflate* and *Prunus mume* [66,67]. Further studies predicted that Pi SLF was one component of an E3 ligase complex, cooperating with PI CUL1-G and SBP1, to mediate ubiquitination of non-self S-RNases, which may be degraded by the 26s proteasome [68]. The structure of SLF also confirmed the result with an ubiquitin-binding domain (UBD) in C-terminal region [69]. The transgenic functional assay was used to examine the interaction between the S_2-SLF of *P. inflate* and non-self S-RNase, which proved that S_2-SLF interacted with S_7-RNase and S_{13}-RNase but not with S_5-RNase and S_{11}-RNase. Through microRNA expression assays, it was also confirmed that there was more than one type of SLF protein to recognize all non-self S-RNase, of which each type of SLF would interact with a subset of non-self S-RNase [70].

In addition to inhibition and rejection, there is a theory of "compartmentalization", whereby S-RNaseis compartmentalized in the vacuole for the degradation of HT-B, a protein indispensable to the release of S-RNase, and eventually promotes the compatible pollination [71]. Except the S-specific determinants, there are some modifier factors during GSI. The SBP1 and SCF-like E3 pitching in pollen could form a complex with SLF to allow the degradation or ubiquitination of non-self S-RNase, while the 120 kD a glycoprotein, the pistil extension-like protein III (PELPIII) and the transmitting tract-specific glycoprotein (TTS) are the S-RNase binding proteins in the pistil [52]. eEF1A, the instigator of the actin binding activity, could bind to S-RNase during SI in *Solanum chacoense* [72]. The exhaustive network of S-specific pollen rejection based on the S-RNases has been demonstrated in the reviews of McClure [71,73].

Actin-7, actin-8 and fructose bisphosphate aldolase-like protein were found only in self-pollinated pistil in apricots when compared with the across-pollinated by 2D-GE along with MS/MS analysis [74]. In apricots, the comparison between SC and SI cultivars at protein level showed there were 15 different proteins in two kinds of pistils, of which nine proteins were detected only in the SI pistils, including actin-7, a putative serine/threonine kinase and an S-RNase [75]. A protein expression profile in different stages of styles in "Hyuganatsu" (*Citrus*) was generated by 2D gel electrophoresis and MALDI-TOF/MS. The three stages were distinguished by location of the tube in the style after self-pollination: the tube was at the top of the styles in 1 and 3 days before anthesis (DBA) while it had reached the bottom of the styles in five DBA. There were 138 different protein spots among the three stages, in which 17 up-regulated and 26 down-regulated proteins were identified. As the plants were self-incompatible, the nine up-regulated proteins based on the pattern of 1 DBA > 3 DBA > 5 DBA may contribute to the transmission from SC to SI, which were related to SI response. Via Blast P analysis, there were nine proteins involved in tubulin alpha-4 chain, including probable rhamnose biosynthetic enzyme 1, 2,3-bisphosphoglycerate-independent phosphoglycerate mutase 1, fructokinase-2, allene oxide synthase, chloroplastic, luminal-bingding protein, photosystem 1 assembly protein. The first eight all participated in different biological processes while the last PG may be related to the SI reaction. The comparative studies in proteomics may be helpful in isolating the key proteins related to SI and eventually in explaining the reproduction process [76]. These studies uncover the special proteins in SC and/or SI process by proteomic approaches, which may also expand our insights on the possible pathways related to the interactions during both compatible and incompatible pollination.

3.2. Proteomic Analysis of the Sporophytic Self-Incompatibility Response

Sporophytic self-incompatibility (SSI) is triggered by the interaction between a polymorphic stigma receptor and its pollen ligand, if the expressed genotypes of S allele are alike. The restriction ends in the failure of the development of the pollen tube.

In Brassicaceae, the genetic factors encoded by *S-locus* are S-locus receptor kinase (SRK), which is a transmembrane Ser/Thr receptor kinase on the epidermis of the stigma papilla cells and functions as the female S-determinant, and S-locus protein 11 (SP11, or S-locus cysteine-rich protein, SCR). Its expression level in pollen is as high as that in the male determinant. The interaction of SRKn–SCRn results in the compatibility or incompatibility based on the polymorphism of the two proteins [77]. Even if the intracellular signal pathway is still not acquainted entirely, a model of ubiquitin-proteasomal degradation based on the interaction between armadillo-repeat-containing 1 (ARC1) and EXO70A1, a substrate of ARC1, has been demonstrated (Figure 2) [78,79]. The degradation of these proteins may lead to the generation of SI, in other words, the predicted proteins represent various compatibility factors required for pollen germination and growth [80]. Around the SI pathway, the positive regulator M locus protein kinase (MLPK), which can recognize and phosphorylate ARC1, and the negative regulators THL1 and THL2, which belong to the thioredoxin-h family, were all discovered participating in the incompatibility response in Brassicaceae [81,82].

Figure 2. Amode of sporophytic self-incompatibility signaling pathway regulated by the ARCI. When the cognate S-locus cysteine-rich protein (SCR) binds to the extracellular domain of SRK, its intracellular domain is activated and relieves inhibition from THL proteins. A phosphorylation cascades transfer from the MLPK to ARC1. The phosphorylated ARC1 E3 ubiquitin ligase, supported by the E1 and E2 ligase, may activate the ubiquitination and degradation of EXO70A1. The ARC1-mediated degradation of EXO70A1 leads to the inhibition secretion of "compatible" factors to reject the "self" pollens.

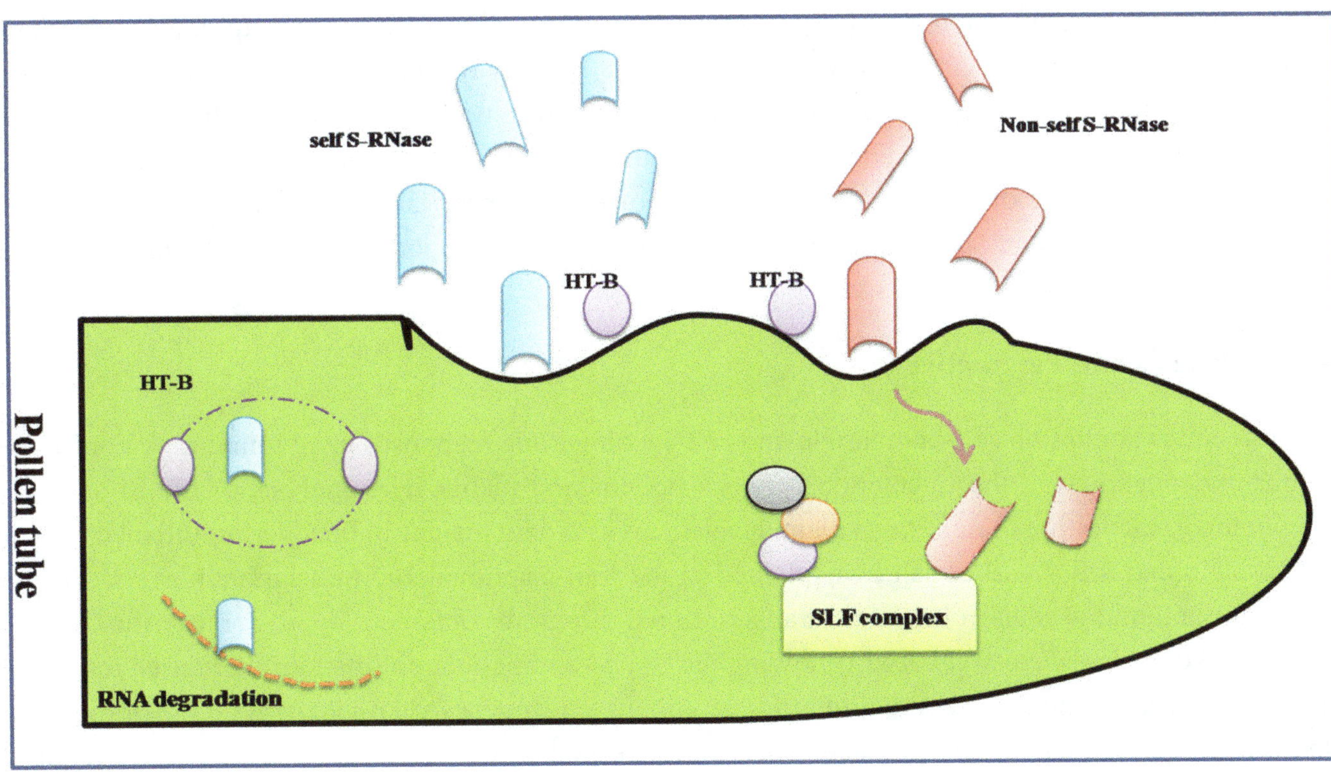

Based on the above, an accurate view of alternation in abundance of proteins within SI is a way to understand pollination and fertilization, comparing the similar but direct studies on pollen and/or pistil. The change of proteins can either be up-regulated or down-regulated. The purpose of this research was to obtain the comparative protein profile of SI reproductive tissues, and the differentially expressed proteins were predicted by functional classification, which may provide the mechanism of SI system. To explore this point, a changeable protein list on SI response was acquired by 2D-DIGE and mass spectrometry. Nineteen decreased proteins were identified and predicted to be involved in various pathways including biosynthetic pathways, signaling transduction pathway and cytoskeletal organization.

Based on GO annotation, the 19 protein candidates were classified into six groups: metabolism (SLR glycoprotein, RUBISCO and so on), trafficking (annexin), development (actin), translation (GTP and several putative chloroplast translation), chaperone (protein disulphide isomerase), and structural (alpha 2–4 tubulin). Analyzing the various functions of these proteins can help to understand their mechanisms for down-regulation, consistent with the SI response. Compatibility between pharmacology and cell biological techniques—including using mutant lines and depolymerization of alpha 2–4 tubulin and microtubule (MT) in the stigmatic papilla contribution to the process of compatible pollen acceptance, which is likely mediated by EXO70A1—has been shown. Combing the down-regulated tubulin in SI, it indicates that the alteration of MT dynamics cannot affect the SI response [83]. Similar selection of candidate proteins involved in the SI system were reported in non-heading Chinese cabbage, of which traces of pistils were found at 0 h and 2 h after pollination in SI and SC lines. Among 22 potential proteins, the UDP-sugar pyrophosphorylase (USPase) and DNADP-dependent glyceraldehyde-3-phosphate dehydrogenase, two proteins classified in the energy metabolism, were down-regulated at 2 h after incompatible pollination, which showed the existence of sucrose degradation and ATP supply during the SI response. Similar analysis in the up-regulated methionine synthase (METS), involving the protein methylation, implied that DNA methyltransferases might have a role in SI response, which is related to post-translational modification in SI [84]. Contrary to incompatible pollen germination, the dynamics of proteins' file established in *Brassica napus* showed that the enzymes involved in glycolysis, TCA cycle and electron transport chain were up-regulated in germinating pollen compared to the mature pollen, such as glyceraldehydes-3-phosphate dehydrogenase, malate dehydrogenase and cytochrome b5 reductases [24]. This indicated that the circulation of energy materials was necessary during normal pollination, while the down-regulated proteins in SI may provide opportunity to reject and suppress the "self" pollen.

4. Conclusions and Perspectives

Proteomics studies on self-compatible and self-incompatible responses have immensely contributed to our understanding of the interaction between pollens and pistils. The analysis of the SI system, a phenomenon exhibiting in numerous plant species, gives a new insight in uncovering the mechanisms. The protein and mRNA level can be explained in the transcriptional and translational levels. The two fields are mutually reinforcing. The transcriptional program obtained by a comparison between 0–30 min stigmas of *Brassica napus* following the SI and compatible pollinations were used to uncover the genes participating in the compatible and incompatible responses via microarrays. One of the results showed that the absence of un-regulated genes in SI response was consistent with the down-regulated proteins following SI. This study's results are consistent with the findings of previous study [85]. However, 33% of genes experienced different variation trends in mRNA levels compared with the protein levels in the pistils of non-heading Chinese cabbage [84]. The difference between the two levels may result from post-transcriptional regulation or post-translational modification [83]. Post-translational modifications include glycosylation, phosphorylation and ubiquitination [7]. So far, these modifications were identified in the reproductive process, involving the protein interactions, signaling transductions, protein degradations, and so on. For example, the SRK would be auto-phosphorylated for the homoplastic SCR from the pollen in Brassica. The phosphorylation cascades pass on to MLPK to transmit incompatible signals, which can activate ARC1, holding the E3 ligase, and leading to the ubiquitination and degradation of protein substrates [79,82]. In addition to the post-translational modification, the interaction partners and the sub-cellular localization also make sure that the transcript levels cannot reveal the dynamic protein properties. So, to remedy the imperfection of transcriptomics, the dynamics of proteins involved in the pollen and/or pistil (pre-pollination and after-pollination) should be tested at multiple time-points based on the development of MS technologies, which can be used by virtue of the peptide sequencing and isotope labeling methods to play a great role in obtaining the dynamic profile during the interaction between pollens and pistils.

However, the coverage of proteins during the SI and SC processes is limited for different reasons, such as the protein expressions in space-time effect and the low-abundance of proteins. To overcome these limitations, multiple, more sensitive technologies should be developed. Therein, multi-dimensional protein identification technology (Mud PIT) can help to capture the proteins at multiple time-points [6]. Besides, the proteomics screen the candidates especially, which requires verifying their special functions in the pollen-pistil interaction via various molecular and genetic methods. Given the difficulties in the collection of reproductive tissues, the multidisciplinary fields and approaches within transcriptomics, proteomics and metabolomics should be combined to reveal the global network involved in the process of pollination.

Acknowledgments

The work was supported by 100 Talents Project of Chinese Academy of Sciences.

Author Contributions

All the authors have contributed to the writing and editing of the paper.

References

1. Kenrick, P.; Crane, P.R. The origin and early evolution of plants on land. *Nature* **1997**, *389*, 33–39.
2. Hiscock, S.J.; Allen, A.M. Diverse cell signalling pathways regulate pollen-stigma interactions: The search for consensus. *New Phytol*. **2008**, *179*, 286–317.
3. Dutta, S.K.; Srivastav, M.; Rymbai, H.; Chaudhary, R.; Singh, A.K.; Dubey, A.K.; Lal, K. Pollen-pistil interaction studies in mango (*Mangifera indica* L.) cultivars. *Sci. Horticult*. **2013**, *160*, 213–221.
4. Takayama, S.; Shiba, H.; Iwano, M.; Shimosato, H.; Che, F.-S.; Kai, N.; Watanabe, M.; Suzuki, G.; Hinata, K.; Isogai, A. The pollen determinant of self-incompatibility in *Brassica campestris*. *Proc. Natl. Acad. Sci. USA* **2000**, *97*, 1920–1925.
5. Kao, T.-H.; Tsukamoto, T. The molecular and genetic bases of S-RNase-based self-incompatibility. *Plant Cell Online* **2004**, *16*, S72–S83.

6. Ahmad, Y.; Lamond, A.I. A perspective on proteomics in cell biology. *Trends Cell Biol.* **2014**, *24*, 257–264.
7. Sankaranarayanan, S.; Jamshed, M.; Samuel, M.A. Proteomics approaches advance our understanding of plant self-incompatibility response. *J. Proteome Res.* **2013**, *12*, 4717–4726.
8. Panchaud, A.; Affolter, M.; Moreillon, P.; Kussmann, M. Experimental and computational approaches to quantitative proteomics: Status quo and outlook. *J. Proteomics* **2008**, *71*, 19–33.
9. Lee, J.-H.; Cho, J.-Y. Proteomics approaches for the studies of bone metabolism. *BMB Rep.* **2014**, *47*, 141–148.
10. Wiese, S.; Reidegeld, K.A.; Meyer, H.E.; Warscheid, B. Protein labeling by iTRAQ: A new tool for quantitative mass spectrometry in proteome research. *Proteomics* **2007**, *7*, 340–350.
11. Becker, C.H.; Bern, M. Recent developments in quantitative proteomics. *Mutat. Res. Genet. Toxicol. Environ. Mutagenes.* **2011**, *722*, 171–182.
12. McCormick, S. Male gametophyte development. *Plant Cell* **1993**, *5*, 1265–1275.
13. McCormick, S. Control of male gametophyte development. *Plant Cell* **2004**, *16*, S142–S153.
14. Taylor, L.P.; Hepler, P.K. Pollen germination and tube growth. *Annu. Rev. Plant Biol.* **1997**, *48*, 461–491.
15. Honys, D.; Twell, D. Comparative analysis of the arabidopsis pollen transcriptome. *Plant Physiol.* **2003**, *132*, 640–652.
16. Becker, J.D.; Boavida, L.C.; Carneiro, J.; Haury, M.; Feijó, J.A. Transcriptional profiling of Arabidopsis tissues reveals the unique characteristics of the pollen transcriptome. *Plant Physiol.* **2003**, *133*, 713–725.
17. Holmes-Davis, R.; Tanaka, C.K.; Vensel, W.H.; Hurkman, W.J.; McCormick, S. Proteome mapping of mature pollen of *Arabidopsis thaliana. Proteomics* **2005**, *5*, 4864–4884.
18. Grobei, M.A.; Qeli, E.; Brunner, E.; Rehrauer, H.; Zhang, R.; Roschitzki, B.; Basler, K.; Ahrens, C.H.; Grossniklaus, U. Deterministic protein inference for shotgun proteomics data provides new insights into *Arabidopsis* pollen development and function. *Genome Res.* **2009**, *19*, 1786–1800.
19. Mayfield, J.A.; Fiebig, A.; Johnstone, S.E.; Preuss, D. Gene families from the *Arabidopsis thaliana* pollen coat proteome. *Science* **2001**, *292*, 2482–2485.
20. Dai, S.; Li, L.; Chen, T.; Chong, K.; Xue, Y.; Wang, T. Proteomic analyses of *Oryza sativa* mature pollen reveal novel proteins associated with pollen germination and tube growth. *Proteomics* **2006**, *6*, 2504–2529.
21. Suen, D.F.; Wu, S.S.H.; Chang, H.C.; Dhugga, K.S.; Huang, A.H.C. Cell wall reactive proteins in the coat and wall of maize pollen: Potential role in pollen tube growth on the stigma and through the style. *J. Biol. Chem.* **2003**, *278*, 43672–43681.
22. Kerim, T.; Imin, N.; Weinman, J.J.; Rolfe, B.G. Proteome analysis of male gametophyte development in rice anthers. *Proteomics* **2003**, *3*, 738–751.
23. Dai, S.; Chen, T.; Chong, K.; Xue, Y.; Liu, S.; Wang, T. Proteomics identification of differentially expressed proteins associated with pollen germination and tube growth reveals characteristics of germinated *Oryza sativa* pollen. *Mol. Cell. Proteomics* **2007**, *6*, 207–230.
24. Sheoran, I.S.; Pedersen, E.J.; Ross, A.R.; Sawhney, V.K. Dynamics of protein expression during pollen germination in canola (*Brassica napus*). *Planta* **2009**, *230*, 779–793.

25. Fernando, D.D. Characterization of pollen tube development in *Pinus strobus* (Eastern white pine) through proteomic analysis of differentially expressed proteins. *Proteomics* **2005**, *5*, 4917–4926.
26. Chen, Y.; Chen, T.; Shen, S.; Zheng, M.; Guo, Y.; Lin, J.; Baluska, F.; Samaj, J. Differential display proteomic analysis of *Picea meyeri* pollen germination and pollen-tube growth after inhibition of actin polymerization by latrunculin B. *Plant J. Cell Mol. Biol.* **2006**, *47*, 174–195.
27. Zaidi, M.A.; O'Leary, S.; Wu, S.; Gleddie, S.; Eudes, F.; Laroche, A.; Robert, L.S. A molecular and proteomic investigation of proteins rapidly released from triticale pollen upon hydration. *Plant Mol. Biol.* **2012**, *79*, 101–121.
28. Han, B.; Chen, S.; Dai, S.; Yang, N.; Wang, T. Isobaric tags for relative and absolute quantification-based comparative proteomics reveals the features of plasma membrane-associated proteomes of pollen grains and pollen tubes from *Lilium davidii*. *J. Integr. Plant Biol.* **2010**, *52*, 1043–1058.
29. Gasser, C.S.; Robinson-Beers, K. Pistil development. *Plant Cell* **1993**, *5*, 1231.
30. Faure, J.-E. Double fertilization in flowering plants: Discovery, study methods and mechanisms. *Comptes Rendus de l'Académie des Sciences-Series III-Sciences de la Vie* **2001**, *324*, 551–558.
31. Rea, A.C.; Nasrallah, J.B. Self-incompatibility systems: Barriers to self-fertilization in flowering plants. *Int. J. Dev. Biol.* **2008**, *52*, 627–636.
32. Tung, C.W.; Dwyer, K.G.; Nasrallah, M.E.; Nasrallah, J.B. Genome-wide identification of genes expressed in *Arabidopsis* pistils specifically along the path of pollen tube growth. *Plant Physiol.* **2005**, *138*, 977–989.
33. Boavida, L.C.; Borges, F.; Becker, J.D.; Feijo, J.A. Whole genome analysis of gene expression reveals coordinated activation of signaling and metabolic pathways during pollen-pistil interactions in *Arabidopsis*. *Plant Physiol.* **2011**, *155*, 2066–2080.
34. Ohyanagi, H.; Sakata, K.; Komatsu, S. Soybean proteome database 2012: Update on the comprehensive data repository for soybean proteomics. *Front. Plant Sci.* **2012**, *3*, e110.
35. Haerizadeh, F.; Wong, C.E.; Bhalla, P.L.; Gresshoff, P.M.; Singh, M.B. Genomic expression profiling of mature soybean (*Glycine max*) pollen. *BMC Plant Biol.* **2009**, *9*, e25.
36. Li, M.; Sha, A.; Zhou, X.; Yang, P. Comparative proteomic analyses reveal the changes of metabolic features in soybean (*Glycine max*) pistils upon pollination. *Sex. Plant Reprod.* **2012**, *25*, 281–291.
37. Verhoeven, T.; Feron, R.; Wolters-Arts, M.; Edqvist, J.; Gerats, T.; Derksen, J.; Mariani, C. STIG1 controls exudate secretion in the pistil of petunia and tobacco. *Plant Physiol.* **2005**, *138*, 153–160.
38. Rejon, J.D.; Delalande, F.; Schaeffer-Reiss, C.; Carapito, C.; Zienkiewicz, K.; de Dios Alche, J.; Rodriguez-Garcia, M.I.; van Dorsselaer, A.; Castro, A.J. Proteomics profiling reveals novel proteins and functions of the plant stigma exudate. *J. Exp. Bot.* **2013**, *64*, 5695–5705.
39. Takeuchi, H.; Higashiyama, T. Attraction of tip-growing pollen tubes by the female gametophyte. *Curr. Opin. Plant Biol.* **2011**, *14*, 614–621.
40. Wagner, R.E.; Mugnaini, S.; Sniezko, R.; Hardie, D.; Poulis, B.; Nepi, M.; Pacini, E.; Aderkas, P. Proteomic evaluation of gymnosperm pollination drop proteins indicates highly conserved and complex biological functions. *Sex. Plant Reprod.* **2007**, *20*, 181–189.
41. Márton, M.-L.; Dresselhaus, T. Female gametophyte-controlled pollen tube guidance. *Biochem. Soc. Trans.* **2010**, *38*, 627–630.

42. Kasahara, R.D.; Portereiko, M.F.; Sandaklie-Nikolova, L.; Rabiger, D.S.; Drews, G.N. MYB98 is required for pollen tube guidance and synergid cell differentiation in *Arabidopsis*. *Plant Cell* **2005**, *17*, 2981–2992.
43. Kessler, S.A.; Grossniklaus, U. She's the boss: Signaling in pollen tube reception. *Curr. Opin. Plant Biol.* **2011**, *14*, 622–627.
44. Takeuchi, H.; Higashiyama, T. A species-specific cluster of defensin-like genes encodes diffusible pollen tube attractants in *Arabidopsis*. *PLoS Biol.* **2012**, *10*, e1001449.
45. Sanchez, A.M.; Bosch, M.; Bots, M.; Nieuwland, J.; Feron, R.; Mariani, C. Pistil factors controlling pollination. *Plant Cell* **2004**, *16*, S98–S106.
46. Liu, J.; Zhong, S.; Guo, X.; Hao, L.; Wei, X.; Huang, Q.; Hou, Y.; Shi, J.; Wang, C.; Gu, H.; *et al.* Membrane-bound RLCKs LIP1 and LIP2 are essential male factors controlling male-female attraction in *Arabidopsis*. *Curr. Biol.* **2013**, *23*, 993–998.
47. Guan, Y.; Lu, J.; Xu, J.; McClure, B.; Zhang, S. Two Mitogen-activated protein kinases, MPK3 and MPK6, are required for funicular guidance of pollen tubes in *Arabidopsis*. *Plant Physiol.* **2014**, *165*, 528–533.
48. Swanson, R.; Edlund, A.F.; Preuss, D. Species specificity in pollen-pistil interactions. *Annu. Rev. Genet.* **2004**, *38*, 793–818.
49. Swanson, W.J.; Vacquier, V.D. The rapid evolution of reproductive proteins. *Nat. Rev. Genet.* **2002**, *3*, 137–144.
50. Showalter, A. Arabinogalactan-proteins: Structure, expression and function. *Cell. Mol. Life Sci.* **2001**, *58*, 1399–1417.
51. Suárez, C.; Zienkiewicz, A.; Castro, A.J.; Zienkiewicz, K.; Majewska-Sawka, A.; Rodríguez-García, M.I. Cellular localization and levels of pectins and arabinogalactan proteins in olive (*Olea europaea* L.) pistil tissues during development: Implications for pollen-pistil interaction. *Planta* **2012**, *237*, 305–319.
52. Cruz-Garcia, F.; Hancock, C.N.; McClure, B. S-RNase complexes and pollen rejection. *J. Exp. Bot.* **2003**, *54*, 123–130.
53. De Nettancourt, D. Incompatibility in angiosperms. *Sex. Plant Reprod.* **1997**, *10*, 185–199.
54. McCubbin, A.G.; Kao, T.-H. Molecular recognition and response in pollen and pistil interactions. *Annu. Rev. Cell Dev. Biol.* **2000**, *16*, 333–364.
55. Meng, X.; Sun, P.; Kao, T.H. S-RNase-based self-incompatibility in *Petunia inflata*. *Ann. Bot.* **2011**, *108*, 637–646.
56. Wu, J.; Wang, S.; Gu, Y.; Zhang, S.; Publicover, S.J.; Franklin-Tong, V.E. Self-Incompatibility in *Papaver rhoeas* activates nonspecific cation conductance permeable to Ca^{2+} and K^{+}. *Plant Physiol.* **2010**, *155*, 963–973.
57. Wheeler, M.J.; de Graaf, B.H.; Hadjiosif, N.; Perry, R.M.; Poulter, N.S.; Osman, K.; Vatovec, S.; Harper, A.; Franklin, F.C.; Franklin-Tong, V.E. Identification of the pollen self-incompatibility determinant in *Papaver rhoeas*. *Nature* **2009**, *459*, 992–995.
58. Ushijima, K.; Sassa, H.; Dandekar, A.M.; Gradziel, T.M.; Tao, R.; Hirano, H. Structural and transcriptional analysis of the self-incompatibility locus of almond: Identification of a pollen-expressed F-box gene with haplotype-specific polymorphism. *Plant Cell Online* **2003**, *15*, 771–781.

59. Bredemeijer, G.; Blaas, J. S-Specific proteins in styles of self-incompatible *Nicotiana alata*. *Theor. Appl. Genet.* **1981**, *59*, 185–190.
60. McClure, B.A.; Haring, V.; Ebert, P.R.; Anderson, M.A.; Simpson, R.J.; Sakiyama, F.; Clarke, A.E. Style self-incompatibility gene products of *Nicotlana alata* are ribonucleases. *Nature* **1989**, *342*, 955–957.
61. Broothaerts, W.J.; van Laere, A.; Witters, R.; Préaux, G.; Decock, B.; van Damme, J.; Vendrig, J.C. Purification and N-terminal sequencing of style glycoproteins associated with self-incompatibility in *Petunia hybrida*. *Plant Mol. Biol.* **1990**, *14*, 93–102.
62. Singh, A.; Ai, Y.; Kao, T.-H. Characterization of ribonuclease activity of three S-allele-associated proteins of *Petunia inflata*. *Plant Physiol.* **1991**, *96*, 61–68.
63. Hiratsuka, S. Characterization of an S-allele associated protein in Japanese pear. *Euphytica* **1992**, *62*, 103–110.
64. Ishimizu, T.; Sato, Y.; Saito, T.; Yoshimura, Y.; Norioka, S.; Nakanishi, T.; Sakiyama, F. Identification and partial amino acid sequences of seven S-RNases associated with self-incompatibility of Japanese pear, *Pyrus pyrifolia* Nakai. *J. Biochem.* **1996**, *120*, 326–334.
65. Lai, Z.; Ma, W.; Han, B.; Liang, L.; Zhang, Y.; Hong, G.; Xue, Y. An F-box gene linked to the self-incompatibility (S) locus of *Antirrhinum* is expressed specifically in pollen and tapetum. *Plant Mol. Biol.* **2002**, *50*, 29–41.
66. Entani, T.; Iwano, M.; Shiba, H.; Che, F.S.; Isogai, A.; Takayama, S. Comparative analysis of the self-incompatibility (S-) locus region of *Prunus mume*: Identification of a pollen-expressed F-box gene with allelic diversity. *Genes Cells* **2003**, *8*, 203–213.
67. Sijacic, P.; Wang, X.; Skirpan, A.L.; Wang, Y.; Dowd, P.E.; McCubbin, A.G.; Huang, S.; Kao, T.-H. Identification of the pollen determinant of S-RNase-mediated self-incompatibility. *Nature* **2004**, *429*, 302–305.
68. Hua, Z.; Kao, T.-H. Identification and characterization of components of a putative *Petunia* S-locus F-box–containing E3 ligase complex involved in S-RNase-based self-incompatibility. *Plant Cell Online* **2006**, *18*, 2531–2553.
69. Chen, G.; Zhang, B.; Liu, L.; Li, Q.; Zhang, Y.; Xie, Q.; Xue, Y. Identification of a ubiquitin-binding structure in the S-locus F-box protein controlling S-RNase-based self-incompatibility. *J. Genet. Genomics (Yi Chuan Xue Bao)* **2012**, *39*, 93–102.
70. Sun, P.; Kao, T.H. Self-Incompatibility in *Petunia inflata*: The relationship between a self-incompatibility locus F-box protein and its non-self S-RNases. *Plant Cell* **2013**, *25*, 470–485.
71. McClure, B.; Cruz-Garcia, F.; Romero, C. Compatibility and incompatibility in S-RNase-based systems. *Ann. Bot.* **2011**, *108*, 647–658.
72. Soulard, J.; Boivin, N.; Morse, D.; Cappadocia, M. eEF1A Is an S-RNase binding factor in self-incompatible solanum chacoense. *PLoS One* **2014**, *9*, e90206.
73. McClure, B. Darwin's foundation for investigating self-incompatibility and the progress toward a physiological model for S-RNase-based SI. *J. Exp. Bot.* **2009**, *60*, 1069–1081.
74. Feng, J.; Chen, X.; Yuan, Z.; He, T.; Zhang, L.; Wu, Y.; Liu, W.; Liang, Q. Proteome comparison following self- and across-pollination in self-incompatible apricot (*Prunus armeniaca* L.). *Protein J.* **2006**, *25*, 328–335.

75. Feng, J.; Chen, X.; Yuan, Z.; Zhang, L.; Ci, Z.; Liu, X.; Zhang, C. Primary molecular features of self-incompatible and self-compatible F1 seedling from apricot (*Prunus armeniaca* L.) Katy× Xinshiji. *Mol. Biol. Rep.* **2009**, *36*, 263–272.
76. Uchida, A.; Takenaka, S.; Sakakibara, Y.; Kurogi, S. Comprehensive analysis of expressed proteins in the different stages of the style development of self-incompatible "Hyuganatsu" (*Citrus tamurana* hort. ex Tanaka). *J. Jpn. Soc. Horticult. Sci.* **2012**, *81*, 150–158.
77. Tantikanjana, T.; Nasrallah, M.E.; Nasrallah, J.B. Complex networks of self-incompatibility signaling in the Brassicaceae. *Curr. Opin. Plant Biol.* **2010**, *13*, 520–526.
78. Gu, T.; Mazzurco, M.; Sulaman, W.; Matias, D.D.; Goring, D.R. Binding of an arm repeat protein to the kinase domain of the S-locus receptor kinase. *Proc. Natl. Acad. Sci. USA* **1998**, *95*, 382–387.
79. Stone, S.L.; Anderson, E.M.; Mullen, R.T.; Goring, D.R. ARC1 is an E3 ubiquitin ligase and promotes the ubiquitination of proteins during the rejection of self-incompatible *Brassica* pollen. *Plant Cell Online* **2003**, *15*, 885–898.
80. Samuel, M.A.; Chong, Y.T.; Haasen, K.E.; Aldea-Brydges, M.G.; Stone, S.L.; Goring, D.R. Cellular pathways regulating responses to compatible and self-incompatible pollen in *Brassica* and *Arabidopsis* stigmas intersect at Exo70A1, a putative component of the exocyst complex. *Plant Cell Online* **2009**, *21*, 2655–2671.
81. Bower, M.S.; Matias, D.D.; Fernandes-Carvalho, E.; Mazzurco, M.; Gu, T.; Rothstein, S.J.; Goring, D.R. Two members of the thioredoxin-h family interact with the kinase domain of a *Brassica* S locus receptor kinase. *Plant Cell Online* **1996**, *8*, 1641–1650.
82. Kakita, M.; Murase, K.; Iwano, M.; Matsumoto, T.; Watanabe, M.; Shiba, H.; Isogai, A.; Takayama, S. Two distinct forms of M-locus protein kinase localize to the plasma membrane and interact directly with S-locus receptor kinase to transduce self-incompatibility signaling in *Brassica rapa*. *Plant Cell Online* **2007**, *19*, 3961–3973.
83. Samuel, M.A.; Tang, W.; Jamshed, M.; Northey, J.; Patel, D.; Smith, D.; Siu, K.M.; Muench, D.G.; Wang, Z.-Y.; Goring, D.R. Proteomic analysis of *Brassica* stigmatic proteins following the self-incompatibility reaction reveals a role for microtubule dynamics during pollen responses. *Mol. Cell. Proteomics* **2011**, *10*, doi:10.1074/mcp.M111.011338.
84. Wang, L.; Peng, H.; Ge, T.; Liu, T.; Hou, X.; Li, Y. Identification of differentially accumulating pistil proteins associated with self-incompatibility of non-heading Chinese cabbage. *Plant Biol.* **2013**, *16*, 49–57.
85. Sankaranarayanan, S.; Jamshed, M.; Deb, S.; Chatfield-Reed, K.; Kwon, E.-J.G.; Chua, G.; Samuel, M.A. Deciphering the stigmatic transcriptional landscape of compatible and self-incompatible pollinations in *Brassica napus* reveals a rapid stigma senescence response following compatible pollination. *Mol. Plant* **2013**, doi:10.1093/mp/sst066.

11

Quantitative Proteomics of the Root of Transgenic Wheat Expressing *TaBWPR-1.2* Genes in Response to Waterlogging

Emdadul Haque †, Fumitaka Abe *, Masahiko Mori ‡, Yohei Nanjo, Setsuko Komatsu, Atsushi Oyanagi and Kentaro Kawaguchi

NARO Institute of Crop Science (NICS), National Agriculture and Food Research Organization (NARO), 2-1-18 Kannondai, Tsukuba, Ibaraki 305-8518, Japan;
E-Mails: haque@yokohama-cu.ac.jp (E.H.); mmmori@obihiro.ac.jp (M.M.); ynanjo@affrc.go.jp (Y.N.); skomatsu@affrc.go.jp (S.K.); oyanagi@affrc.go.jp (A.O.); kentaro@affrc.go.jp (K.K.)

† Current address: Kihara Institute for Biological Research, Yokohama City University, Maioka-cho 641-12, Totsuka-ku, Yokohama 244-0813, Japan

‡ Current address: Department of Agro-Environmental Science, Division of Plant Production Science, Obihiro University of Agriculture and Veterinary Medicine, Inada-cho, Nishi 2-11, Obihiro, Hokkaido 080-8555, Japan.

* Author to whom correspondence should be addressed; E-Mail: fabe@affrc.go.jp

External Editor: Dr. Jacek R. Wisniewski

Abstract: Once candidate genes are available, the application of genetic transformation plays a major part to study their function in plants for adaptation to respective environmental stresses, including waterlogging (WL). The introduction of stress-inducible genes into wheat remains difficult because of low transformation and plant regeneration efficiencies and expression variability and instability. Earlier, we found two cDNAs encoding WL stress-responsive wheat pathogenesis-related proteins 1.2 (*TaBWPR-1.2*), *TaBWPR-1.2#2* and *TaBWPR-1.2*#13. Using microprojectile bombardment, both cDNAs were introduced into "Bobwhite". Despite low transformation efficiency, four independent T_2 homozygous lines for each gene were isolated, where transgenes were ubiquitously and variously expressed. The highest transgene expression was obtained in *Ubi:TaBWPR-1.2#2* L#11a and

Ubi:TaBWPR-1.2#13 L#4a. Using quantitative proteomics, the root proteins of L#11a were analyzed to explore possible physiological pathways regulated by TaBWPR-1.2 under normal and waterlogged conditions. In L#11a, the abundance of proteasome subunit alpha type-3 decreased under normal conditions, whereas that of ferredoxin precursor and elongation factor-2 increased under waterlogged conditions in comparison with normal plants. Proteomic results suggest that L#11a is one of the engineered wheat plants where TaBWPR-1.2#2 is most probably involved in proteolysis, protein synthesis and alteration in the energy pathway in root tissues via the above proteins in order to gain metabolic adjustment to WL.

Keywords: transgenic wheat; pathogenesis-related protein-1.2; proteomics; seminal roots; waterlogging

1. Introduction

In the last two decades, genetic transformation has become a powerful tool to transfer new genes into crop plants. This approach offers an attractive alternative to conventional breeding, because specific traits can be transferred into selected genotypes without adverse effects on desirable genetic backgrounds. Wheat (*Triticum aestivum* L.) is one of the most important crops that feeds the growing world population. Its production is predicted to decline (along with that of other cereals) due to adverse environments. Among cereals, wheat was the last to be genetically modified, because of inherent difficulties associated with gene delivery into regenerable explants and recovery of transformants; wheat, particularly hexaploid, has a larger genome than other cereals [1–3]. Transgenic wheat lines producing some proteins involved in development have been obtained [2–5], and corresponding genes, proteins or metabolites have been analyzed; yet, this approach is still a challenge for stress-inducible genes [6–8].

Transcription of stress-inducible genes depends on the strength and duration of stimuli. These genes can be divided into early and late responsive [9]. To the best of our knowledge, no well-characterized wheat-derived promoters for constitutive, tissue-specific or stress-inducible expression are available. In wheat, the maize ubiquitin promoter and intron (*Ubi*) [10] and the rice actin promoter with the 5' intron (*Act1*) [11] appear to result in the highest and most stable constitutive expression. Position effects, the developmental stage [12] and, rarely, stress [13] may affect *Ubi* activity in transgenic wheat lines. Recently, promising stress-inducible promoters, such as *Arabidopsis rd29A* [8], maize *Rab17* [6] and barley *HvDhn4s* [7], have been used to study the effect of drought, but these promoters may be not efficient in heterologous systems. To circumvent these problems, some wheat genes, particularly stress-inducible ones, have been overexpressed in other plants [14,15].

Proteomic techniques in conjunction with mass spectrometry (MS), including gel-based and gel-free proteomics, enable comparative quantitative protein profiling. Because of the disadvantages of gel-based proteomics (labor intensiveness, low sensitivity and reproducibility and the inability to characterize complete proteomes), gel-free proteomics has become a valuable tool for functional analyses of particular biological processes or responses to the environment [5,16,17].

Plants' ability to tolerate water stresses, such as drought and waterlogging (WL), is crucial for agricultural production worldwide. Stress environments trigger a wide variety of plant responses through sensing, signaling and adaption. Soil WL has been a serious environmental stress that imposes on plant growth and productivity [18]. To design molecular mechanism for WL tolerance, elucidation of cellular systems involved in responses and adaptations have been required to efficiently discover key genes to be applied to engineer its tolerance. For this purpose, here, we focus on response and tolerance systems against WL in wheat plants.

The hexaploid wheat genome contains 23 pathogenesis-related (PR) protein-1-like genes, designated as TaPR-1.1 to 20 [19]. Among them, the deduced TaPR-1.20 protein sequence was highly identical to that encoded by the *TaPR-1.2* gene [20]. TaPR-1.2 (TaPR-1.20) is not a marker for systemic acquired resistance [19], but a stress (aluminum, humidity)-responsive gene [19,21]. Although little is known about *PR-1.2* gene expression and protein production in root in response to environmental stresses, relevant information has been obtained for other PR families. For example, PR-10 proteins in rice and maize were found to respond to drought and cold, respectively [22–24]. In a previous study on morphological adaptation to WL in the seminal roots of hexaploid spring wheat "Bobwhite SH 98 26" [25], we found that levels of a TaPR-1.2 significantly increased during lysigenous aerenchyma formation [26]. We thought that there was a relationship between TaPR-1.2 and WL response and/or aerenchyma tissue formation in wheat seminal roots. Very recently, we identified two TaPR-1.2 cDNAs, *TaBWPR-1.2#2* (AB711115) and *TaBWPR-1.2#13* (AB711116), from the seminal root of Bobwhite as WL-responsive at the RNA and protein levels [27]. These clones differ by the presence or absence of two amino acids (FA) at positions 164–165 and one amino acid, lysine ("K"; *i.e.*, a positive charge), at the C-terminal end. However, the functional differences between these two TaBWPR-1.2 clones in wheat were unknown. Moreover, wheat plants transformed with WL stress-responsive gene(s) are not yet available.

In the present work, we used the biolistic approach to transform wheat cultivar "Bobwhite SH 98 26" and produced homozygous transgenic lines overexpressing *TaBWPR-1.2#2* or *TaBWPR-1.2#13* under the control of the *Ubi* promoter. To explore the physiological pathway of *TaBWPR-1.2*, we compared protein abundance in control and transgenic wheat seminal roots under control and WL conditions by gel-free proteomics. This work may be useful for those who attempt to produce transgenic wheat plants and for those interested in the role of PR-1.2 proteins in wheat.

2. Experimental Section

2.1. Construct Preparation

The original TaBWPR-1.2#2 and TaBWPR-1.2#13 cDNAs were cloned into the pCRII-TOPO vector (Invitrogen, Carlsbad, CA, USA) and were described previously [27]. The coding regions were then amplified with a primer set containing the BamHI sites, and the fragments were inserted into the BamHI site of the plasmid, pAHC17 [28]. The two constructs were designated as pUbi:TaBWPR-1.2#2 and pUbi:TaBWPR-1.2#13.

2.2. Plant Material Preparation for Transformation

Spring wheat (*Triticum aestivum* L. cv. Bobwhite 98 26) [29] was used in all experiments. Seeds were sown (four seeds per 18-cm pot) in a 2:1 mixture of Sakata Soil Mix (Sakata Seed Corp., Yokohama, Japan) and Kureha fertilized granulated soil (Kureha Corp., Tokyo, Japan). Plants were grown in a greenhouse at 17 °C (day)/10 °C (night) with an 8-h photoperiod for 12 weeks and then transferred into a controlled-environment chamber at 20/13 °C with a 16-h photoperiod (750 μmol·m^{-2}·s) and 55%–65% relative humidity. Tillers were harvested 13–15 days after anthesis by cutting below the third node of the tiller with three leaves retained and kept at 5 °C without water supply for 5–7 days.

2.3. Isolation of Scutellar Tissues from Immature Embryos

Immature caryopses were collected from the spikelets 10–12 days after anthesis, rinsed with 70% ethanol, surface-sterilized in sodium hypochlorite solution (0.5% v/v available chlorine) containing 0.1% v/v Tween 20 for 15 min and then rinsed three times with sterile distilled water. Immature embryos were isolated aseptically under a dissecting microscope, and the entire axis of the embryos was removed by a fine blade. Isolated scutellar tissues were cultured, scutellum side up, on callus induction medium containing 0.2 M mannitol (CI-0.2Man) at 25 °C in the dark for 3–4 h before bombardment.

2.4. Biolistic Transformation

Scutellar tissues were bombarded with each plasmid. Plasmid DNA was prepared using a Qiagen Maxi Kit (Qiagen, Hilden, Germany). The plasmid pUba [30], carrying the *bar* gene that confers resistance to the herbicide, phosphinothricin, was co-bombarded with each plasmid, pUbi:TaBWPR-1.2#2 or pUbi:TaBWPR-1.2#13, in a 1:1.5 molar ratio. Plasmid DNA (total 5 μg) was precipitated onto gold particles (2 mg; 1.0 μm in diameter) in the presence of 1 M $CaCl_2$ and 16 mM spermidine, and then, the DNA-gold particles were washed twice with ethanol and resuspended in 100 μL of ethanol. For each bombardment, 5 μL of the DNA-gold suspension was used (100 μg particles per shot). Particles were bombarded with a PDS 1000/He particle delivery system (Bio-Rad, Hercules, CA, USA). The target tissues were placed 5.5 cm from the stopping screen at a helium pressure of 6.2 MPa.

2.5. Tissue Culture and Selection of Transgenic Plants

The composition of tissue culture media is listed in Supplementary Table 1. At 2 days after bombardment, scutellar tissues (16 per 90-mm plate) were cultured on callus maintenance medium containing 3 mg·L^{-1} phosphinothricin (CM-3P) for 3 weeks. The explants were transferred to shoot growth medium containing 1 mg·L^{-1} phosphinothricin (SG-1P) (8 calluses per plate) under illumination for 3 weeks for shoot regeneration and then to root growth medium containing 3 mg·L^{-1} phosphinothricin (RG-3P) for a further 3 weeks for root regeneration. Plants resistant to phosphinothricin were transferred to soil.

2.6. PCR Analysis of Transgenic Plants

DNA-PCR and reverse-transcription PCR (RT-PCR) were used to screen transgenic plants. Genomic DNA was isolated from leaf tips as described by [31]. Total RNA was isolated from leaf tips as described by [27]. The forward primer for DNA-PCR was designed within the *Ubi* promoter (5'-ttagccctgccttcatacgc-3'). That for RT-PCR was designed between *Ubi*, and the sequence was identical in *TaBWPR-1.2#2* and *TaBWPR-1.2#13* region (5'-actctagaggatccccatgg-3'). The reverse primers for DNA-PCR and RT-PCR corresponded to the unique sequences of *TaBWPR-1.2#2* (5'-ttgttgtcccatgccacgg-3') and *TaBWPR-1.2#13* (5'-ctgttgtcccacgtcacag-3').

2.7. Analysis of Gene Expression in Different Organs by RT-PCR

Seeds germinated on wet filter paper in a glass Petri dish for 4 days were raised in either big glass Petri dishes (height 6 cm × diameter 9 cm; As One Stock, Tokyo, Japan) for another 4 days, then the leaf, the root base (1 cm), the middle part of the root (3–5 cm) and the root tip (1 cm) were collected; or 30 cm-long well-drained pots [25] in the phytotron chamber; the whole leaf and root were collected 15 days later, and spikes were collected 90 days later (before anthesis). Total RNA was extracted from wheat organs as described previously [27]. One-step PCR was performed with the PrimeScript RT reagent kit (Takara, Kyoto, Japan) in a 10-μL reaction volume (200 ng of total RNA). One-step RT-PCR was performed using a PCR System (Takara) under conditions of 50 °C for 30 m followed by 33 cycles of 95 °C for 30 s, 60 °C for 30 s and 72 °C for 30 s with gene-specific primers as above.

2.8. Gene Expression Analysis by qRT-PCR in Homozygous Transformants under WL

Homozygous transformants and their null-segregants were grown in pots in a phytotron chamber for 7 days, followed by 5 days of WL [26]. Whole-root samples were prepared from normal and waterlogged 12-day-old plants and stored at −80 °C. qRT-PCR was performed according to [32], as slightly modified by [27]. The primer design and amplification efficiency are also described in detail in these two publications [27,32]. To detect the transgenes, we used the gene-specific primers described above. To detect endogenous genes, we used the primers 5'-cttgacgccgaagcctagta-3' (forward) and 5'-gccggaatgtgtgcttattt-3' (reverse) for *TaBWPR-1.2#2* and 5'-cgcactggtcatagtcatgg-3' (forward) and 5'-ctgttgtcccacgtcacag-3' (reverse) for *TaBWPR-1.2#13*. An actin gene was used as an internal control. All RT- and qRT-PCR experiments were performed in biological triplicates and technical triplicates.

2.9. Protein Extraction and Immunoblot Analysis with Rice Anti-PR-1 Antibody

Homozygous lines and control (12-day-old plants) were subjected to 5-day WL, and whole roots were collected as samples. Samples were ground in SDS sample buffer consisting of 60 mM Tris-HCl (pH 6.8), 2% SDS, 10% glycerol and 5% 2-mercaptoethanol. After centrifugation, supernatant was separated on a 12% SDS polyacrylamide electrophoresis gel. Immunoblot analysis was performed according to [27] with an anti-rice PR-1 antibody [33].

2.10. Preparation of Proteins for Mass Spectrometry (MS)

Protein concentration in the extracts was estimated by a Pierce 660 nm Protein Assay Kit with the Ionic Detergent Compatibility Reagent (Thermo Fisher Scientific, Waltham, MA, USA). Detergent was removed from the extracted proteins (100 μg) by chloroform-methanol extraction as follows. Samples (adjusted to 100 μL) were mixed consecutively with methanol (400 μL), chloroform (100 μL) and water (300 μL) and centrifuged at 20,000× *g* for 5 min for phase separation. The upper (aqueous) phase was discarded, and methanol (300 μL) was added to the organic phase. The samples were centrifuged again at 20,000× *g* for 5 min; the supernatants were discarded and the pellets dried. Proteins were reduced with 50 mM dithiothreitol for 1 h at 56 °C, alkylated with 50 mM iodoacetamide for 1 h at 37 °C in the dark and digested with trypsin and lysyl endopeptidase at a 1:100 enzyme/protein ratio for 16 h at 37 °C.

The resulting peptides were acidified with formic acid to pH < 3, desalted with a C18-pipette tip and analyzed by MS.

2.11. Data Acquisition by Nano-Liquid Chromatography (LC) MS/MS

Peptides were analyzed on a nanospray LTQ XL Orbitrap mass spectrometer (Thermo Fisher Scientific) operated in data-dependent acquisition mode with Xcalibur software (version 2.0.7, Thermo Fisher Scientific). Using an Ultimate 3000 nanoLC system (Dionex, Germering, Gemany), peptides in 0.1% formic acid were loaded onto a C18 PepMap trap column (300 μm ID × 5 mm, Dionex), eluted and separated on a C18 Tip column (75 μm ID × 120 mm nano-HPLC capillary column NTTC-360/75-3; Nikkyo Technos, Tokyo, Japan) in a linear acetonitrile gradient (8%–30% in 120 min) in 0.1% formic acid at a flow rate of 200 nL/min. A spray voltage of 1.5 kV was used. Full-scan mass spectra were acquired over a mass range of 400–1500 *m/z* with a resolution of 30,000.

The lock mass function was used to obtain high mass accuracy. The ten most intense precursor ions were selected for collision-induced fragmentation in the linear ion trap at a normalized collision energy of 35%. Dynamic exclusion was used within 90 s to prevent repetitive selection of the same peptides.

2.12. Protein Identification

Proteins were identified by the Mascot search engine (version 2.3.0.2, Matrix Science, London, U.K.) through Mascot Daemon client software (version 2.3.2, Matrix Science) using a customized *T. aestivum* database containing 21,690 protein sequences. The protein sequences were obtained from the Triticeae Full-Length CDS database (6146 sequences) [34], NCBI database (10,690 sequences) [35] and UniProt database (4854 sequences) [36]. The parameters used in Mascot searches were as follows: cysteine carbamidomethylation was set as a fixed modification, and methionine oxidation was set as a variable modification.

Trypsin was specified as the proteolytic enzyme, and one missed cleavage was allowed. Peptide mass tolerance was set at 5 ppm. Fragment mass tolerance was set at 0.5 Da, and peptide charge was set at +2, +3 or +4. An automatic decoy database search was performed as part of the search. Mascot results were filtered with Mascot Percolator to improve the accuracy and sensitivity of peptide identification. False discovery rates for peptide identification were <1.0% in all searches. The Mascot results were exported in XML format for SIEVE (version 2.0, Thermo Fisher Scientific) analysis.

2.13. Analysis of Differential Protein Abundance Using Acquired MS Data

Analysis of protein abundance was performed by using the label-free quantification package, SIEVE (Thermo Fisher Scientific), to compare the relative abundance of peptides and proteins in the control and experimental groups, as previously described by [37]. It is important to note that we performed this study in a phytotron chamber where the stress level is mild (low light intensity and low temperature compared to a greenhouse). Therefore, the thresholds for fold changes in protein quantities in transgenic *vs.* non-transgenic samples were set at >1.4 or <0.6 with a significant difference ($p < 0.05$).

3. Results

3.1. Regeneration and Establishment of Homozygous Lines

Shoots regenerated six weeks after bombardment are shown in Figure 1A, and roots regenerated nine weeks after bombardment are shown in Figure 1B. Fewer plantlets were regenerated after bombardment with *Ubi:TaBWPR-1.2#2* than with *Ubi:TaBWPR-1.2#13* [38]. Plantlets were adapted to a room environment for seven days by removing paraffin with surgical tape and then transplanted into soil. Transgenic plants were established in soil in a plastic pocket tray after 30 days (Figure 1C, left) and subsequently transferred to plastic pots (Figure 1C, right) after 45 days of acclimatization.

Figure 1. Regeneration of transgenic Bobwhite SH 98 26 after biolistic transformation of immature embryos with *Ubi:TaBWPR-1.2#13* (as a representative of both transgenes). (**A**) Shoot differentiation from calluses; (**B**) rooting of differentiated shoots; (**C**) transgenic wheat at 30 days (**left**) and 45 days (**right**).

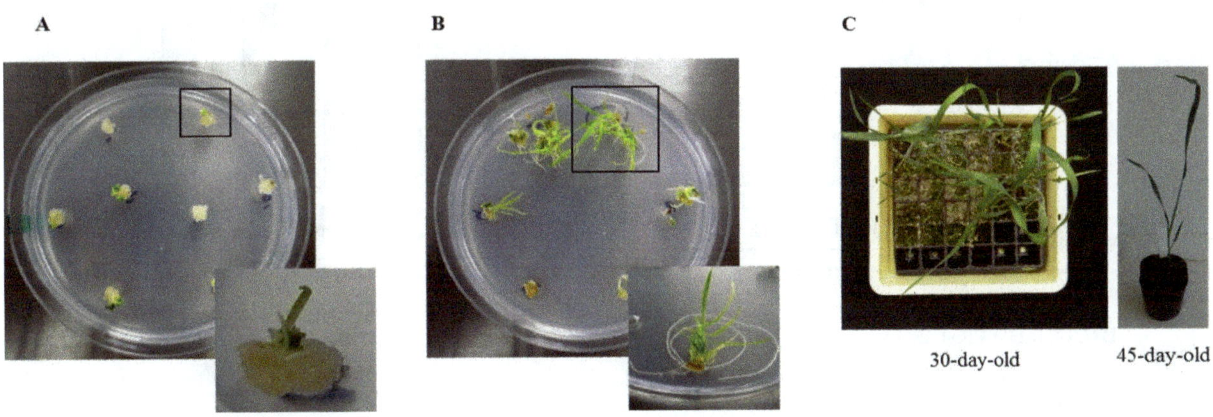

Transgenic callus-derived plants (Figure 1A) were considered as T_0 plants. The presence of transgenes was confirmed by DNA-PCR and RT-PCR analysis (Supplementary Figure 1, upper panels). T_1 seeds were obtained from all T_0 plants expressing the genes of interest. Sixteen seeds of each T_1 plant were sown and segregation of transgenes in the leaf tip was confirmed by DNA-PCR and RT-PCR analysis (Supplementary Figure 1, lower panels). Nine out of twenty plants for *Ubi:TaBWPR-1.2#13* and seven out of 12 for *Ubi:TaBWPR-1.2#13* showed a Mendelian 3:1 ratio of transgene-positiveness. For each independent T_1 plant, seven randomly selected transgene-positive plants and one transgene-negative plant were propagated to obtain T_2 seeds and null-segregants. T_2 seeds of all spikes from each individual positive T_1 plant were bulked and further propagated to check homozygosity. Four independent, homozygous T_2 lines were obtained for each, *Ubi:TaBWPR-1.2#2* (4, 11a, 11b', 13a) and *Ubi:TaBWPR-1.2#13* (2, 4a, 4b', 5b').

3.2. mRNA and Protein Levels in Ubi:TaBWPR-1.2 Transformants

To examine protein production in *Ubi:TaBWPR-1.2#2* and *Ubi:TaBWPR-1.2#13* transformants, we used immunoblotting of total protein from whole seminal roots of plants grown for five days under normal or waterlogged conditions. *Ubi:TaBWPR-1.2#2* transformants showed a slight, but not significant, increase in protein levels under normal conditions, but a significant increase under waterlogged conditions

(Figure 2A). The increase in line L#11a was the largest among all lines. In *Ubi:TaBWPR-1.2#13* transformants, the protein level increased (but not significantly) under normal and waterlogged conditions; the increase was somewhat higher in L#4a than in other lines (Figure 2A).

Figure 2. The levels of TaBWPR-1.2 proteins in the roots of homozygous transformants under control conditions and five days of waterlogging (WL). Immunoblotting was performed with anti-rice PR-1 antibody. (**A**) Four transgenic TaBWPR-1.2#2 lines; (**B**) four transgenic TaBWPR-1.2#13 lines. E, endogenous proteins; T, transgene proteins. The data are the means of three independent biological samples; error bars represent ±SEM. ns, not significant; * $p < 0.05$; ** $p < 0.01$ by a two-sample *t*-test.

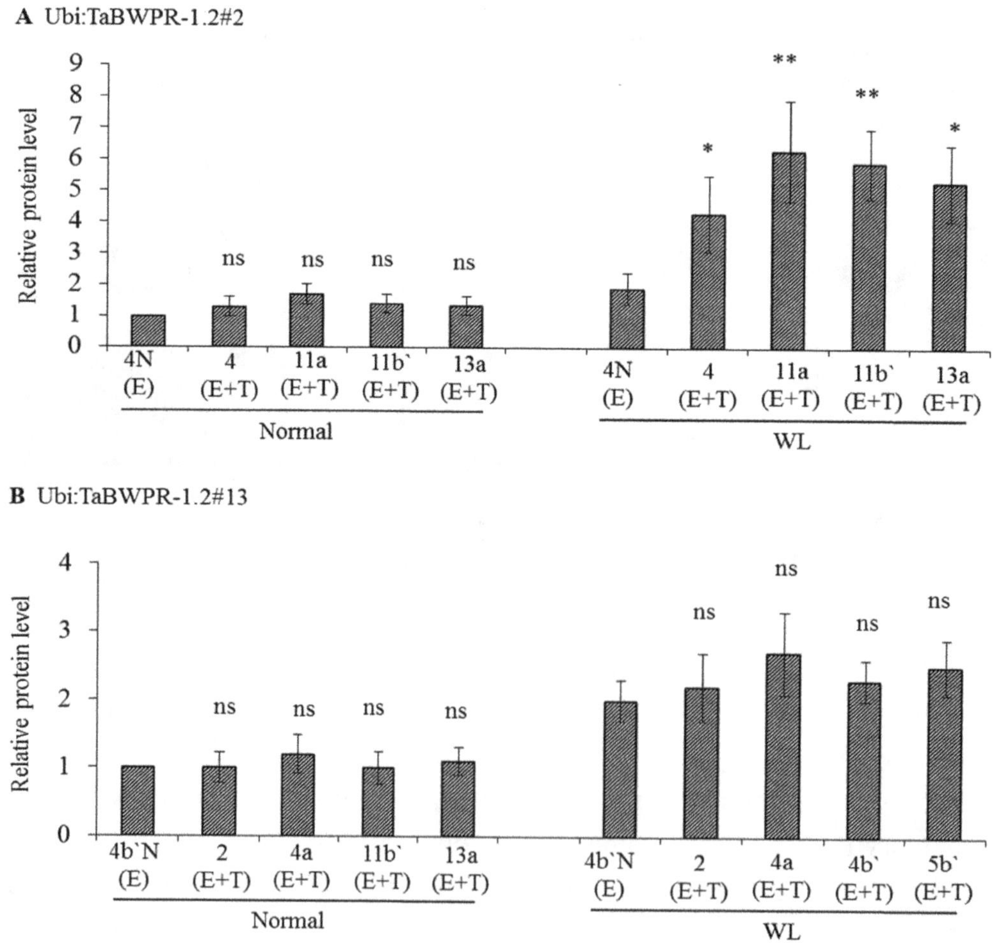

We compared the RNA levels of transgenes in the *Ubi:TaBWPR-1.2#2* (line L#11a) and *Ubi:TaBWPR-1.2#13* (line L#4a) transformants with the levels of corresponding endogenous RNA in the same samples by qRT-PCR. In L#11a, the level of the transgenic RNA was increased dramatically under both conditions, although the increase was less pronounced under waterlogged conditions (Figure 3B). The level of the endogenous RNA in this line was also increased and responded similarly to WL (Figure 3C). In L#4a, the level of the transgenic RNA was also dramatically increased, but the increase was higher under waterlogged than under normal conditions (Figure 3C). The level of the endogenous RNA was slightly elevated in L#4a under both conditions (Figure 3C). As shown in Figure 3A, when compared with Bobwhite SH 98 26, L#11a and L#4a transgenic lines showed only slightly longer roots under WL. L#11a and its null-segregant 4N were selected for proteome analysis.

Figure 3. Expression of *TaBWPR-1.2* transgenes and respective endogenous genes in wheat roots under control conditions and after five days of waterlogging. (**A**) Phenotypes of control (Bobwhite SH 98 26) and the best transgenic lines (L#11a for *Ubi:TaBWPR-1.2#2*; L#4a for *TaBWPR-1.2#13*), showing no differences, except slightly longer roots under WL when compared to Bobwhite SH 98 26. *Ubi:TaBWPR-1.2#2* L#11a (**B**) and *TaBWPR-1.2#13* L#4a (**C**) were compared with the respective negative lines (N). Transcript levels were normalized to an actin gene as an internal control. The relative mRNA levels of the E and T from negative plants were set to one. E, relative RNA levels of endogenous genes; T, relative RNA levels of transgenes. WL, waterlogged conditions. Data for endogenous and transgenes of transgenic seedlings are presented as solid black and hatched columns, respectively, whereas those for the negative controls are presented as white columns. The data are the means of three independent biological samples; error bars represent ±SEM. ** Significant differences ($p < 0.01$) by a two-sample *t*-test.

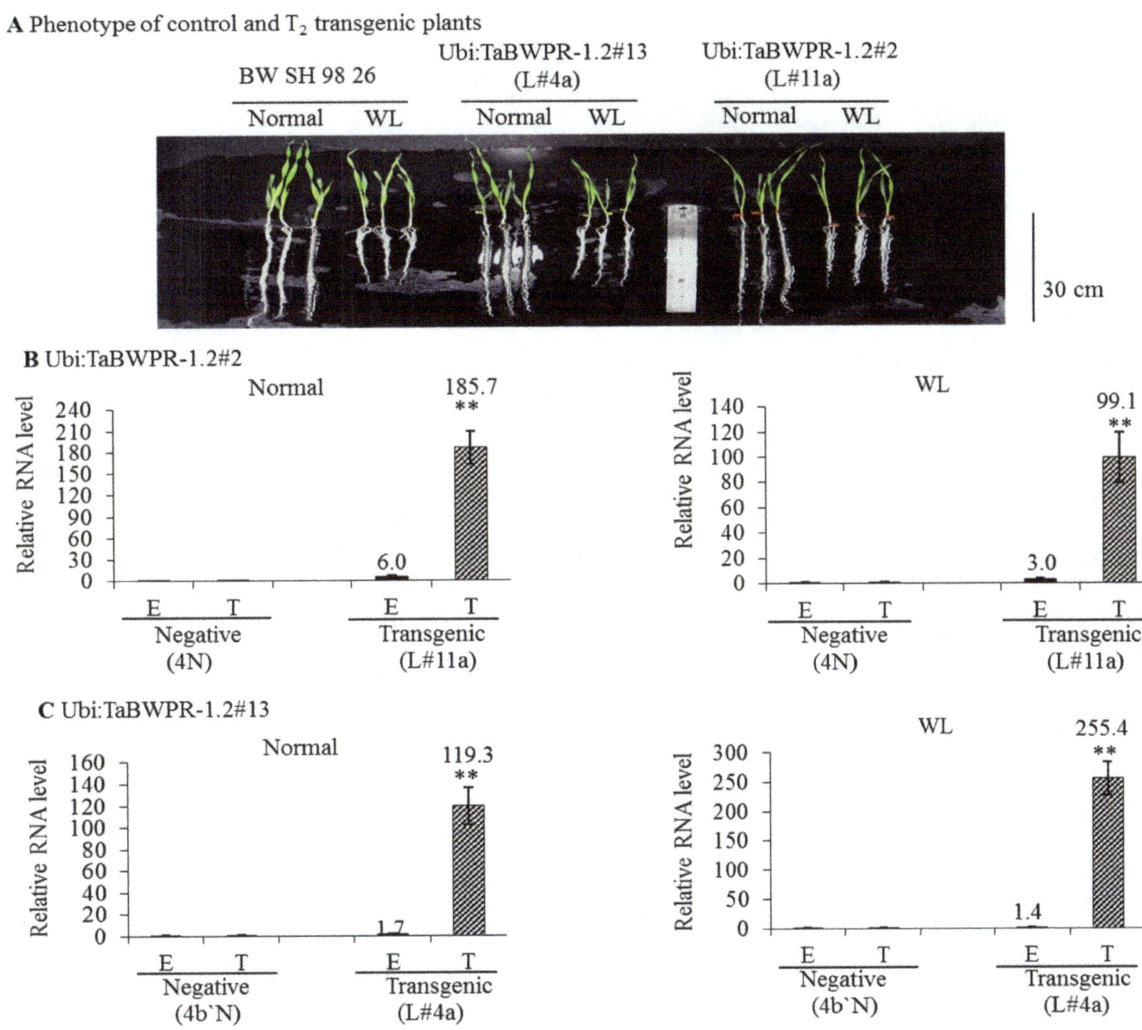

3.3. *Changes in Protein Levels in Seminal Roots of TaBWPR-1.2#2-Overexpressor Transgenic Plants*

To gain insights into the physiological role of TaBWPR-1.2s in wheat root, we analyzed the total soluble proteome using gel-free proteomics technique in whole seminal roots of L#11a from *TaBWPR-1.2#2* and its null-segregant 4N grown under normal and waterlogged conditions. Under normal conditions, we detected two upregulated proteins (PR-1.17 and -1.14) and three downregulated proteins (PR, PR10 and unknown/proteasome subunit alpha type-3) (Table 1). When compared under waterlogged condition, we found four upregulated proteins (PR-1.6, ferredoxin precursor, elongation factor-2 (EF-2) and one unknown peptide (contig 2626) and one downregulated protein (PR) (Table 1). The upregulation of some PRs reconfirms that the *TaBWPR-1.2#2* transgene is at least translated into protein in L#11a. However, there are some other pathogenesis-related candidate proteins found to be up- and down-regulated in response to *TaBWPR-1.2#2* transgenic plants under both normal as well as waterlogged conditions. Results showed that there is a characteristic reverse tendency among the *PR-1* gene family. Silencing of tobacco *PR-1a* leads to silencing of other *PR-1* genes, but under certain treatments, some *PR-1* genes were upregulated in *PR-1a*-silenced tobacco plants [39]. There are 23 *TaPR-1* genes [19], and the expression of one of them may affect the expression of other *TaPR-1* genes. The functions of these new members of TaBWPR-1 proteins that are up- and down-regulated in L#11a are still unknown. In this study, a discussion was done on the possible application of the above four differentially expressed partner proteins of TaBWPR-1.2#2 in root physiology in response to soil WL.

Table 1. Changes in protein abundance in the seminal roots of the wheat transgenic line overexpressing *TaBWPR-1.2#2*.

	Protein Name	Accession No. [a]	Organism	MP [b]	Ratio [c]	SD [d]
		Normal conditions (11a/4N)				
1	Pathogenesis-related protein 1_17	F8S6U7	*T. aestivum*	4	3.0	0.45
2	Pathogenesis-related protein 1_14	F8S6U4	*T. aestivum*	2	1.6	0.5
3	Pathogenesis-related protein	H2KXF7	*T. aestivum*	4	0.44	0.02
4	Pathogenesis-related protein 10	B5B3P8	*T. aestivum*	4	0.46	0.08
5	Unknown Proteasome subunit alpha type-3	AK332255 * ACN10361 *	*T. aestivum* *Salmo salar*	5	0.58	0.05
		WL conditions (11a/4N)				
1	Pathogenesis-related protein 1_6	F8S6T6	*T. aestivum*	2	2.0	0.6
2	Ferredoxin precursor	Q8S3J5	*T. aestivum*	2	2.0	1.2
3	Elongation factor-2	Q9M7S5	*T. aestivum*	2	1.9	1.1
4	Unknown (contig 2626)	AK331943 *	*T. aestivum*	2	1.4	0.1
5	Pathogenesis-related protein	H2KXF7	*T. aestivum*	4	0.6	0.09

Protein hits were validated if identified with $p < 0.05$. [a] Accession numbers are from specific wheat databases (see main text) and from the NCBI database. * cDNA clones. [b] MP, the number of query-matched peptides (cutoff value: <3). [c] The ratio was calculated by dividing the protein level in transgenic wheat to that in wild-type wheat. [d] SD, standard deviation ($n = 3$).

4. Discussion

We produced homozygous transgenic *Ubi:TaBWPR-1.2* wheat and examined the RNA and proteins of seminal roots responsive to transgenes overexpression under normal and WL conditions. To the best of our knowledge, this is the first report of successful wheat transformation with *TaBWPR-1.2* constructs. Our transgenic *TaBWPR-1.2#2* line stably produced the RNA and protein of interest.

In comparison with the reported efficiency of biolistic transformation of wheat (1%) [40], the transformation efficiency in our study was low (approximately 0.2% for *Ubi:TaBWPR-1.2#2* and 0.4%

for *Ubi:TaBWPR-1.2#13*), and it took us approximately two years to produce four homozygous lines for each transgene. The difference between the two transgenes might be due to the specific effects of these genes. We analyzed the expression of transgenes in various organs of four homozygous lines and one null-segregant from *Ubi:TaBWPR-1.2#13* in the absence of stress. The *Ubi:TaBWPR-1.2#13* transgene was ubiquitously expressed in germinating embryo and in all tested organs of 8- and 15-day-old seedlings (Supplementary Figure 2). We detected variations in both RNA expression (Figures 2 and 3) and protein abundance (Supplementary Figure 2) among these lines. Studies in *Drosophila melanogaster* [41], *Saccharomyces cerevisiae* [42] and wheat [4] showed that the positions of the introduced genes on chromosomes may influence their expression. Thus, different insertion positions of the transgenes in the genome may have resulted in variations in their RNA expression. Under waterlogged conditions, the level of *TaBWPR-1.2#2* mRNA decreased (Figure 3B), whereas that of *TaBWPR-1.2#13* mRNA increased (Figure 3C). These differences in the stress response of transgene expression may also be caused by the insertion positions of the transgenes.

Line L#11a had the highest RNA expression and consistently produced the protein of interest under waterlogged conditions (Figure 3, Supplementary Figure 2). Although the effect of protein degradation and the difference in the detection of TaBWPR-1.2#2 and TaBWPR-1.2#13 by the extraction method and antibody used in this study cannot be excluded, we believe that *TaBWPR-1.2* mRNAs, particularly *TaBWPR-1.2#13* mRNA, are highly unstable. *TaBWPR-1.2* mRNAs reached its maximum at Day 1 after the onset of WL and then started to decline, which comes close to the baseline after Day 10 [26]. To bypass the adverse effects of constitutive overexpression, the early stress-responsive nature of *TaBWPR-1.2* genes requires suitable stress-inducible or root-specific promoters [6,18] and, thereby, sufficient activation for adequate translation in transgenic plants. The development of a WL-inducible promoter is urgently needed and is currently under way in our laboratory. Taking into account the difficulties in wheat transformation, L#11a and probably L#4a are good candidate lines with which to study the role of TaBWPR-1.2 proteins in wheat seminal roots.

The level of PSMA3 was lower in line L#11a than in control plants under normal conditions, but not under WL. Proteasome-mediated proteolysis plays a key role in plant responses to several environmental stresses [43]. In soybean roots, accumulation of proteasome and COP9 signalosome proteins increases in response to flooding stress and returns close to baseline upon de-submergence [44]. Thus, it is suggested that PSMA3 in non-transgenic wheat increases upon WL. Based on the present result and previous information, the decreased amount of PSMA3 caused by TaBWPR-1.2#2 overproduction in L#11a may return close to baseline levels through the increase of its endogenous level upon WL, because transgene expression under normal and waterlogged conditions is the same as that of the *Ubi* promoter. This may be why PSMA3 is downregulated under normal conditions, but not under waterlogged conditions. Furthermore, Haque *et al.* [26] reported that the proteasome subunits did not increase in waterlogged wheat roots; the reason for the apparent discrepancy is that we previously used a more stringent threshold of two-fold differences in protein abundance. Although the mechanism of the PSMA3 decrease in L#11a is unclear and the decrease is moderate, it is indicated that it should be taken into account. It will be interesting to test whether TaBWPR-1.2#2 inhibits PSMA3 synthesis.

Ferredoxin (Fd) was increased in line L#11a under waterlogged conditions, but not under control conditions. Ferredoxins are iron-sulfur proteins that transfer electrons in a wide variety of metabolic reactions. In higher plants, distinct Fd isoforms are detected in photosynthetic and non-photosynthetic organs [45,46]. In non-photosynthetic root plastids, Fd-dependent enzymes need Fd reduced with NADPH (Fd:NADP$^+$); one such enzyme is Fd:NADP$^+$ oxidoreductase (FNR) [46,47]. Onda *et al.* [46] demonstrated that the interaction between root FNR and Fds was stronger than between leaf FNRs and Fds, which is crucial for efficient electron allocation and flux from NADH to Fd in the NADH-FNR-Fd cascade. Here, Fd increased in L#11a only under waterlogged conditions. It could be that a certain amount of PR-1.2 is needed to interact with Fd, which was not sufficient by overloaded TaBWPR-1.2#2 protein under control conditions, but together with elevated endogenous protein, the total TaBWPR-1.2#2 was sufficient under waterlogged conditions. We also found that Fd increases only under WL *vs.* control conditions in L#11, but is absent in the wild-type under WL *vs.* control conditions [38], suggesting that Fd is undetectable in the wild-type and responds only upon TaBWPR-1.2#2 expression. It is suggested that TaBWPR-1.2#2 may play an important role in a higher rate of electron flux in metabolic reactions mediated by Fd in wheat roots under the limited energy conditions caused by WL.

EF-2 is an essential protein catalyzing ribosomal translocation during protein synthesis [48], and EF accumulates in soybean under flooding stress [49]. Because protein synthesis needs to continue in plant roots under WL conditions [26], the increase in EF-2 in L#11a may regulate the synthesis of some proteins in wheat seminal roots. However, no EF-2 increase was found in either transgenic or non-transgenic plants compared between WL and control conditions (Supplementary Tables 2 and 3), suggesting that further studies are required to reconfirm that EF-2 is a responding protein to TaBWPR-1.2#2. Another potential responsive protein of TaBWPR-1.2#2 is encoded by contig 2626. Like Fd, this protein was present only in transgenic, but not in non-transgenic plants. Further studies are needed to elucidate the role of TaBWPR-1.2#2 in wheat roots, which involves the protein encoded by contig 2626. This study was performed in a phytotron chamber under mild WL stress conditions; hence the identification of more TaBWPR-1.2-responsive proteins can be expected under much more severe stress conditions, such as waterlogged conditions in a greenhouse [26]. These results suggest that TaBWPR-1.2#2 appears to be an inhibitor of the proteasome under normal conditions and an inducer of Fd and EF-2 under WL, and TaBWPR-1.2#2 might be a potential candidate root protein that mitigates the effects of WL.

5. Conclusions

We developed transgenic wheat lines overexpressing two *TaBWPR-1.2* genes and obtained some evidence regarding the physiological pathways possibly affected by TaBWPR-1.2#2 in wheat roots under WL. Further studies are needed to develop transgenic wheat lines expressing *TaBWPR-1.2* genes under the control of root-specific or WL-inducible promoters to examine the phenotypic responses under more natural waterlogged conditions.

Acknowledgments

We thank Shunshuke Oda (National Agriculture and Food Research Organization-NARO Institute of Crop Science) and Toyoaki Anai (Saga University) for suggestions and critically reviewing the manuscript. We also thank Alessandro Pellegrineschi (Applied Biotechnology Center, International Maize and Wheat Improvement Center-CIMMYT, Mexico) for kindly providing Bobwhite 98 26 seeds. This work was supported by a Grant-in-Aid from the Bio-oriented Technology Research Advancement Institution (Promotion of Basic Research Activities for Innovative Biosciences, No. H20/seeds-01-01), Japan.

Author Contributions

The conception of the experimental design was performed by E.H. Transgenic wheat development were performed by F.A., M.M. and E.H. Protein preparation, western blotting, gel-free proteomics and sample analysis were performed by S.K., Y.N. and E.H.. K.K. and A.O. contributed to technical suggestions and manuscript preparation.

References

1. Abdul, R.; Ma, Z.; Wang, H. Genetic transformation of wheat (*Triticum aestivum* L): A review. *Triticeae Genome Genet.* **2010**, *1*, 1–7.
2. Bhalla, P.L.; Ottenhof, H.H.; Singh, M.B. Wheat transformation—An update of recent progress. *Euphytica* **2006**, *149*, 353–366.
3. Jones, H.D. Wheat transformation: Current technology and applications to grain development and composition. *J. Cereal Sci.* **2005**, *41*, 137–147.
4. Zeller, S.L.; Kalinina, O.; Brunner, S.; Keller, B.; Schmid, B. Transgene environment interactions in genetically modified wheat. *PLoS One* **2010**, *5*, e11405.
5. Guo, H.; Zhang, H.; Li, Y.; Ren, J.; Wang, X.; Niu, H.; Yin, J. Identification of changes in wheat (*Triticum aestivum* L.) seeds proteome in response to anti-*trxs* gene. *PLoS One* **2011**, *6*, e22255.
6. Morran, S.; Eini, O.; Pyvovarenko, T.; Parent, B.; Singh, R.; Ismagul, A.; Eliby, S.; Shirley, N.; Langridge, P.; Lopato, S. Improvement of stress tolerance of wheat and barley by modulation of expression of DREB/CBF factors. *Plant Biotechnol. J.* **2011**, *9*, 230–249.
7. Xue, G.P.; Way, H.M.; Richardson, T.; Drenth, J.; Joyce, P.A.; McIntyre, C.L. Overexpression of *TaNAC69* leads to enhanced transcript levels of stress up-regulated genes and dehydration tolerance in bread wheat. *Mol. Plant* **2011**, *4*, 697–712.
8. Pellegrineschi, A.; Reynolds, M.; Pacheco, M.; Brito, R.M.; Almeraya, R.; Yamaguchi-Shinozaki, K.; Hoisington, D. Stress-induced expression in wheat of the *Arabidopsis thaliana DREB1A* gene delays water stress symptoms under greenhouse conditions. *Genome* **2004**, *47*, 493–500.
9. Mahajan, S.; Tuteja, N. Cold, salinity and drought stresses: An overview. *Arch. Biochem. Biophys.* **2005**, *444*, 139–158.
10. Christensen, A.H.; Sharrock, R.A.; Quail, P.H. Maize polyubiquitin genes: Structure, thermal perturbation of expression and transcript splicing, and promoter activity following transfer to protoplasts by electroporation. *Plant Mol. Biol.* **1992**, *18*, 675–689.
11. McElroy, D.; Blowers, A.D.; Jenes, B.; Wu, R. Construction of expression vectors based on the rice actin-1 (Act1) 5' region for use in monocot transformation. *Mol. Gen. Genet.* **1991**, *231*, 150–160.
12. Rooke, L.; Bekes, F.; Fido, R.; Barro, F.; Gras, P.; Tatham, A.S.; Barcelo, P.; Lazzeri, P.; Shewry, P.R. Overexpression of a gluten protein in transgenic wheat results in greatly increased dough strength. *J. Cereal Sci.* **2000**, *30*, 115–120.
13. Stoger, E.; Williams, S.; Keen, D.; Christou, P. Constitutive *versus* seed specific expression in transgenic wheat: Temporal and spatial control. *Transgenic Res.* **1999**, *8*, 73–82.
14. Kovalchuk, N.; Jia, W.; Eini, O.; Morran, S.; Pyvovarenko, T.; Fletcher, S.; Bazanova, N.; Harris, J.;

Beck-Oldach, K.; Shavrukov, Y.; Langridge, P.; Lopato, S. Optimization of *TaDREB3* gene expression in transgenic barley using cold-inducible promoters. *Plant Biotechnol. J.* **2013**, *11*, 659–670.

15. Tang, Y.; Liu, M.; Gao, S.; Zhang, Z.; Zhao, X.; Zhao, C.; Zhang, F.; Chen, X. Molecular characterization of novel *TaNAC* genes in wheat and overexpression of *TaNAC2a* confers drought tolerance in tobacco. *Physiol. Plant.* **2012**, *144*, 210–224.
16. Eldakak, M.; Milad, S.I.M.; Nawar, A.; Rohila, J.S. Proteomics: A biotechnology tool for crop improvement. *Front. Plant Sci.* **2013**, *4*, doi:10.3389/fpls.2013.00035.
17. Jayaraman, D.; Forshey, K.L.; Grimsrud, P.A.; Ane, J.M. Leveraging proteomics to understand plant-microbe interactions. *Front. Plant Sci.* **2012**, *3*, doi:10.3389/fpls.2012.00044.
18. Redillas, M.C.F.R.; Jeong, J.S.; Kim, Y.S.; Jung, H.; Bang, S.W.; Choi, Y.D.; Ha, S.H.; Reuzeau, C.; Kim, J.K. The overexpression of *OsNAC9* alters the root architecture of rice plants enhancing drought resistance and grain yield under field conditions. *Plant Biotechnol. J.* **2012**, *10*, 792–805.
19. Lu, S.; Friesen, T.L.; Faris, J.D. Molecular characterization and genomic mapping of the pathogenesis-related protein 1 (PR-1) gene family in hexaploid wheat (*Triticum aestivum* L.). *Mol. Genet. Genomics* **2011**, *285*, 485–503.
20. Molina, A.; Gorlach, J.; Volrath, S.; Ryals, J. Wheat genes encoding two types of PR-1 proteins are pathogen inducible, but do not respond to activators of systemic acquired resistance. *Mol. Plant Microbe Interact.* **1999**, *12*, 53–58.
21. Milla, M.A.R.; Butler, E.D.; Huete, A.R.; Wilson, C.F.; Anderson, O.; Gustafson, J.P. Expressed sequence tag-based gene expression analysis under aluminum stress in rye. *Plant Physiol.* **2002**, *130*, 1706–1716.
22. Hashimoto, M.; Kisseleva, L.; Sawa, S.; Furukawa, T.; Komatsu, S.; Koshiba, T. A novel rice PR10 protein, RSOsPR10, specifically induced in roots by biotic and abiotic stresses, possibly via the jasmonic acid signaling pathway. *Plant Cell Physiol.* **2004**, *45*, 550–559.
23. Xie, Y.R.; Chen, Z.Y.; Brown, R.L.; Bhatnagar, D. Expression and functional characterization of two pathogenesis-related protein 10 genes from *Zea mays*. *J. Plant. Physiol.* **2010**, *167*, 121–130.
24. Takeuchi, K.; Gyohda, A.; Tominaga, M.; Kawakats, M.; Atakeyama, A.; Ishii, N.; Shimaya, K.; Nishimura, T.; Riemann, M.; Nick, P.; Hashimoto, M.; *et al.* RSOsPR10 expression in response to environmental stresses is regulated antagonistically by jasmonate/ethylene and salicylic acid signaling pathways in rice roots. *Plant Cell Physiol.* **2011**, *52*, 1686–1696.
25. Haque, M.E.; Abe, F.; Kawaguchi, K. Formation and extension of lysigenous aerenchyma in seminal root cortex of spring wheat (*Triticum aestivum* cv. Bobwhite line SH 98 26) seedlings under different strengths of waterlogging. *Plant Root* **2010**, *4*, 31–39.
26. Haque, M.E.; Kawaguchi, K.; Komatsu, S. Analysis of proteins in aerenchymatous seminal roots of wheat grown in hypoxic soils under waterlogged conditions. *Protein Pept. Lett.* **2011**, *18*, 912–924.
27. Haque, M.E.; Abe, F.; Mori, M.; Oyanagi, A.; Komatsu, S.; Kawaguchi, K. Characterization of a wheat pathogenesis related protein-1.2, TaBWPR-1.2, expressed in seminal roots in response to waterlogging. *J. Plant Physiol.* **2014**, *171*, 602–609.
28. Christensen, A.H.; Quail, P.H. Ubiquitin promoter-based vectors for high-level expression of selectable and/or screenable marker genes in monocotyledonous plants. *Transgenic Res.* **1996**, *5*, 213–218.

29. Pellegrineschi, A.; Noguera, L.M.; Skovmand, B.; Brito, R.M.; Velazquez, L.; Salgado, M.M.; Hernandez, R.; Warburton, M.; Hoisington, D. Identification of highly transformable wheat genotypes for mass production of fertile transgenic plants. *Genome* **2002**, *45*, 421–430.
30. Toki, S.; Takamatsu, S.; Nojiri, C.; Ooba, S.; Anzai, H.; Iwata, M.; Christensen, A.H.; Quail, P.H.; Uchimiya, H. Expression of a maize ubiquitin gene promoter-*bar* chimeric gene in transgenic rice plants. *Plant Physiol.* **1992**, *100*, 1503–1507.
31. Taniguchi, Y.; Kawata, M.; Ando, I.; Shimizu, T.; Ohshima, M. Selecting genetic transformants of *indica* and *indica*-derived rice cultivars using bispyribac sodium and a mutated ALS gene. *Plant Cell Rep.* **2010** *29*, 1287–1295.
32. Nakamura, S.; Abe, F.; Kawahigashi, H.; Nakazono, K.; Tagiri, A.; Matsumoto, T.; Utsugi, S.; Ogawa, T.; Handa, H.; Ishida, H.; *et al.* A Wheat Homolog of MOTHER OF FT AND TFL1 Acts in the Regulation of Germination. *Plant Cell* **2011**, *23*, 3215–3229.
33. Rakwal, R.; Komatsu, S. Role of jasmonate in the rice (*Oryza sativa* L.) self-defense mechanism using proteome analysis. *Electrophoresis* **2000**, *21*, 2492–2500.
34. TriFLDB: Triticeae Full-Length CDS DataBase. Available online: http://trifldb.psc.riken.jp/download.pl (accessed on 20 February 2013).
35. NCBI: National Center for Biotechnology Information. Available online: http://www.ncbi.nlm.nih.gov (accessed on 20 February 2013).
36. UniProt. Available online: http://www.uniprot.org (accessed on 20 February 2013).
37. Komatsu, S.; Han, C.; Nanjo, Y.; Altaf-Un-Nahar, M.; Wang, K.; He, D.; Yang, P. Label-free quantitative proteomic analysis of abscisic acid effect in early-stage soybean under flooding. *J. Proteome Res.* **2013**, *12*, 4769–4784.
38. Haque, M.E.; Abe, F.; Mori, M.; Oyanagi, A.; Komatsu, S.; Kawaguchi, K. NARO Institute of Crop Science, Tsukuba, Japan. Unpublished data, 2012.
39. Rivière, M.P.; Marais, A.; Ponchet, M.; Willats, W.; Galiana, E. Silencing of acidic pathogenesis-related PR-1 genes increases extracellular b-(1/3)-glucanase activity at the onset of tobacco defence reactions. *J. Exp. Bot.* **2008**, *59*, 1225–1239.
40. Altpeter, F.; Vasil, V.; Srivastava, V.; Stoger, E.; Vasil, I.K. Accelerated production of transgenic wheat (*Triticum aestivum* L.) plants. *Plant Cell Rep.* **1996**, *16*, 12–17.
41. Henikoff, S. Position effects and variegation enhancers in an autosomal region of *Drosophila melanogaster*. *Genetics* **1979**, *93*, 105–115.
42. Gottschling, D.E.; Aparicio, O.M.; Billington, B.L.; Zakian, V.A. Position effect at *S. cerevisiae* telomeres: Reversible repression of Pol II transcription. *Cell* **1990**, *63*, 751–762.
43. Kurepa, J.; Wang, S.; Li, Y.; Smalle, J. Proteasome regulation, plant growth and stress tolerance. *Plant Signal Behav.* **2009**, *4*, 924–927.
44. Yanagawa, Y.; Komatsu, S. Ubiquitin/proteasome-mediated proteolysis is involved in the response to flooding stress in soybean roots, independent of oxygen limitation. *Plant Sci.* **2012**, *185–186*, 250–258.
45. Alonso, J.M.; Chamarro, J.; Granell, A. A nonphotosynthetic ferredoxin gene is induced by ethylene in citrus organs. *Plant. Mol. Biol.* **1995** *29*, 1211–1221.

46. Onda, Y.; Matsumura, T.; Kimata-Ariga, Y.; Sakakibara, H.; Sugiyama, T.; Hase, T. Differential interaction of maize root Ferredoxin: NADP1 oxidoreductase with photosynthetic and non-photosynthetic ferredoxin isoproteins. *Plant Physiol.* **2000**, *123*, 1037–1046.
47. Ritchie, S.W.; Redinbaugh, M.G.; Shiraishi, N.; Vrba, J.M.; Campbell, W.H. Identification of a maize root transcript expressed in the primary response to nitrate: Characterization of a cDNA with homology to ferredoxin-NADP1 oxidoreductase. *Plant Mol. Biol.* **1994**, *26*, 678–690.
48. Justice, M.C.; Hsu, M.J.; Tse, B.; Ku, T.; Balkovec, J.; Schmatz, D.; Nielsen, J. Elongation factor 2 as a novel target for selective inhibition of fungal protein synthesis. *J. Biol. Chem.* **1998**, *273*, 3148–3151.
49. Komatsu, S.; Nanjo, Y.; Nishimura, M. Proteomic analysis of the flooding tolerance mechanism in mutant soybean. *J. Proteomics* **2013**, *79*, 231–250.

12

Plant Cell Wall Proteins: A Large Body of Data, but what about Runaways?

Cécile Albenne [1,2], Hervé Canut [1,2], Laurent Hoffmann [1,2] and Elisabeth Jamet [1,2,]*

[1] Université de Toulouse; UPS, UMR 5546, Laboratoire de Recherche en Sciences Végétales, BP 42617 Auzeville, F-31326 Castanet-Tolosan, France; E-Mails: albenne@lrsv.ups-tlse.fr (C.A.); canut@lrsv.ups-tlse.fr (H.C.); hoffmann@lrsv.ups-tlse.fr (L.H.)

[2] CNRS, UMR 5546, BP 42617, F-31326 Castanet-Tolosan, France

* Author to whom correspondence should be addressed; E-Mail: jamet@lrsv.ups-tls.fr

---

Abstract: Plant cell wall proteomics has been a very dynamic field of research for about fifteen years. A full range of strategies has been proposed to increase the number of identified proteins and to characterize their post-translational modifications. The protocols are still improving to enlarge the coverage of cell wall proteomes. Comparisons between these proteomes have been done based on various working strategies or different physiological stages. In this review, two points are highlighted. The first point is related to data analysis with an overview of the cell wall proteomes already described. A large body of data is now available with the description of cell wall proteomes of seventeen plant species. CWP contents exhibit particularities in relation to the major differences in cell wall composition and structure between these plants and between plant organs. The second point is related to methodology and concerns the present limitations of the coverage of cell wall proteomes. Because of the variety of cell wall structures and of the diversity of protein/polysaccharide and protein/protein interactions in cell walls, some CWPs can be missing either because they are washed out during the purification of cell walls or because they are covalently linked to cell wall components.

Keywords: cell wall; plant; proteomics

---

1. Introduction

Plant cell wall proteomics is a tricky field of research, since proteins are not only minor components of plant cell walls, but are also trapped in complex networks of polysaccharides with which they can interact. Plant cell walls are mainly composed of cellulose microfibrils wrapped in and connected with hemicelluloses and inserted into a complex pectin gel [1]. At the end of growth, secondary walls are formed [2]. Such walls are more rigid and may contain lignin. The structure and composition of cell walls are constantly modified to allow plant growth and development, and to contribute to the adaptation of plants to their changing environment [3–5]. All these processes involve *de novo* assembly and/or remodeling of wall components as well as signaling processes [6].

Cell wall proteins (CWPs) are the "blue collar workers," modifying cell wall components and customizing them to confer appropriate properties to cell walls [6]. They also contribute to signaling by interacting with plasma membrane receptors or by releasing signal molecules such as peptides or oligosaccharides [7–9]. Thus, a large variety of proteins are present in cell walls [10]. They have different physico-chemical properties, they may interact with other cell wall components and their relative abundance is variable. Proteomics strategies should allow the full inventory of proteins in a tissue, an organ or an organelle at a given stage of development or in response to an external stimulus. However, in the case of cell walls, these strategies are particularly difficult to establish [11]. The three main drawbacks are: (i) cell walls constitute open compartments, (ii) proteins are trapped in a complex polysaccharide matrix with which they interact and (iii) most CWPs are modified at the post-translational level. Two types of flowcharts have been designed and used: non-destructive or non-disruptive ones elute proteins outside the cells without disrupting plasma membranes; destructive or disruptive ones start with the purification of cell walls followed by the elution of proteins with various solutions. Each of them has advantages and drawbacks which have been previously reviewed [10,12]. The combination of these strategies has led to the identification of hundreds of proteins in various plants and in different organs. *Arabidopsis thaliana* has been the most studied plant with 500 CWPs identified at present, representing about one fourth of the expected CWPs. In *Oryza sativa* and *Brachypodium distachyon*, the second and third most studied plants, 314 and 270 CWPs have been identified so far respectively.

Comparisons between different cell wall proteomes have been done using two criteria. In a few cases, different strategies have been used to analyze the same organs. For example, *Populus deltoides* CWPs have been identified either after separation by 1D-electrophoresis followed by LC-MS/MS analysis or after direct analysis by LC-MS/MS [13]. Two partly overlapping sets of proteins have been identified showing that different technologies are required to enlarge the coverage of cell wall proteomes. In other cases, organs at different stages of development or different organs have been analyzed using the same strategies. Cell wall proteomes of *A. thaliana* etiolated hypocotyls have been analyzed 5 or 11 days after germination [14]. In the same way, cell wall proteomes have been studied in growing and mature leaf and stems of *B. distachyon* [15], and in apical and basal stems of

Medicago sativa [16]. Such experiments have allowed the identification of candidate proteins possibly involved in cell wall extension or in cell wall strengthening at the end of growth. Finally, a quantitative approach has allowed the identification of the *A. thaliana* GLIP1 GDSL lipase as a contributor to plant defense against *A. brassicicola* infection [17].

Despite the accumulation of data, well-known CWPs are still under-represented in cell wall proteomes, like structural proteins forming covalent networks, *i.e.*, Proline-Rich Proteins (PRPs) and extensins (EXTs), or highly glycosylated proteins, like ArabinoGalactan Proteins (AGPs). In addition, the analysis of the content of the buffers used during the washings steps of cell walls during their purification has shown that some proteins are lost at that step. In this review, we focus on two points: (i) an overview of the existing cell wall proteomics data highlighting differences between monocots and dicots in relation to differences in cell wall composition and structure or between cell wall proteomes of different organs and (ii) the limitations to the full coverage of cell wall proteomes.

2. A Large Body of Data

With 53 papers reporting plant cell wall proteomes, much data has been accumulated during the last 15 years (Table 1). Seventeen plant species have been the subject of investigations among which 13 dicots and 4 monocots. As previously reviewed, different plant organs, mainly roots, hypocotyls, stems, leaves, ovules and fruits, as well as suspension cultures and seedlings grown in liquid medium have been studied using different strategies [10,18].

Xylem sap proteomes have been considered in this analysis because they contain many secreted proteins which could originate from root stele cells or from dying xylem cells [19]. Altogether, 2170 CWPs encoded by distinct genes have been identified. Classifications into functional classes have been proposed to get overviews of cell wall proteomes [10,20].

It is noteworthy that the same classes have been found in all proteomes: proteins acting on polysaccharides (PAC, e.g., glycoside hydrolases, carbohydrate esterases and lyases, expansins), oxido-reductases (OR, e.g., peroxidases, multicopper oxidases, blue copper binding proteins and multicopper oxidases), proteases (P, e.g., Asp proteases, Cys proteases, Ser proteases, Ser carboxypeptidases), proteins having interacting domains (ID) with polysaccharides (e.g., lectins) or proteins (e.g., enzyme inhibitors, leucine-rich repeats proteins), proteins possibly involved in lipid metabolism (LM, e.g., lipases GDSL, lipid transfer proteins), proteins possibly involved in signaling (S, e.g., arabinogalactan proteins), structural proteins (SP, e.g., leucine-rich repeat extensins, glycine-rich proteins) and proteins of yet unknown function (UF). Proteins with predicted function which are not falling into these categories have been grouped into the miscellaneous class (M, e.g., purple acid phosphatases, phosphate-inducible (phi) proteins, germin and germin-like proteins).

Table 1. Plant cell wall proteomics (CWPs) studies.

Plant species	Type of proteome	Number of identified CWPs [a]	References
Dicots			
Arabidopsis thaliana	cell wall	913	[14,17,21–36]
	N-glycoproteome	200	[37,38]
		495	
Brassica napus/oleracea	xylem sap	147	[19,39]
	N-glycoproteome	92	[19]
		162	
Cicer arietinum	cell wall	nd	[40–42]
Glycine max	cell wall	nd	[43]
Gossypium hirsutum	*N*-glycoproteome	**116**	[44]
Helianthus annuus	cell wall	nd	[45]
Linum usitatissimum	cell wall	**106**	[46]
Medicago sativa	cell wall	**199**, nd	[16,47]
Nicotiana benthamiana	cell wall	nd	[48]
Nicotiana tabacum	cell wall	nd	[34,49–51]
Populus deltoides	cell wall	144	[13]
P. trichocarpa x	xylem sap	33	[52]
P. deltoides (hybrid poplar)		**142**	
Solanum lycopersicum	cell wall	nd, 60	[34,53]
	N-glycoproteome	104	[20]
		161	
Solanum tuberosum	cell wall	nd, **136**	[54,55]
Monocots			
Brachypodium distachyon	cell wall	689	[15]
		314	
Oryza sativa	cell wall	381	[56–60]
		270	
Saccharum officinarum	cell wall	**69**	[61]
Zea mays	cell wall,	nd	[62,63]
	xylem sap	nd	[64]

[a] All these proteomes are in the WallProtDB database (See Supplementary Material). Only proteins having a predicted signal peptide are considered (see Supplementary Material). The number of identified proteins is only mentioned when the identification has been done using homologous sequences. Otherwise, nd means that this number could not be calculated. Numbers in black correspond to the total number of proteins identified whereas numbers in bold blue correspond to numbers of different proteins identified in each species.

To date, the overall distribution of CWPs into these functional classes is similar between dicot and monocot cell wall proteomes with three major classes (Figure 1a,b): PAC (around 26%), oxido-reductases (around 17%), and proteases (around 13%). These average proteomes contain data (i) originating from different kinds of plant organs or from cell suspension cultures, (ii) obtained using various methods of extraction and (iii) identified using different mass spectrometry techniques [10]. They give an overview of the types of proteins which can be identified using the variety of available strategies. Although xylem sap proteomes contain CWPs [19,52], their distribution into functional classes is very different from that of CWPs extracted from plant organs (Figure 1c), with a higher proportion of PAC, oxido-reductases and proteases.

Figure 1. Distribution of CWPs into functional classes. All the proteins have been annotated according to the presence of functional domains (see Supplementary Material), thus providing homogeneous annotations. (**a**) Pool of dicot proteomes; (**b**) Pool of monocot proteomes; (**c**) Pool of xylem sap proteomes.

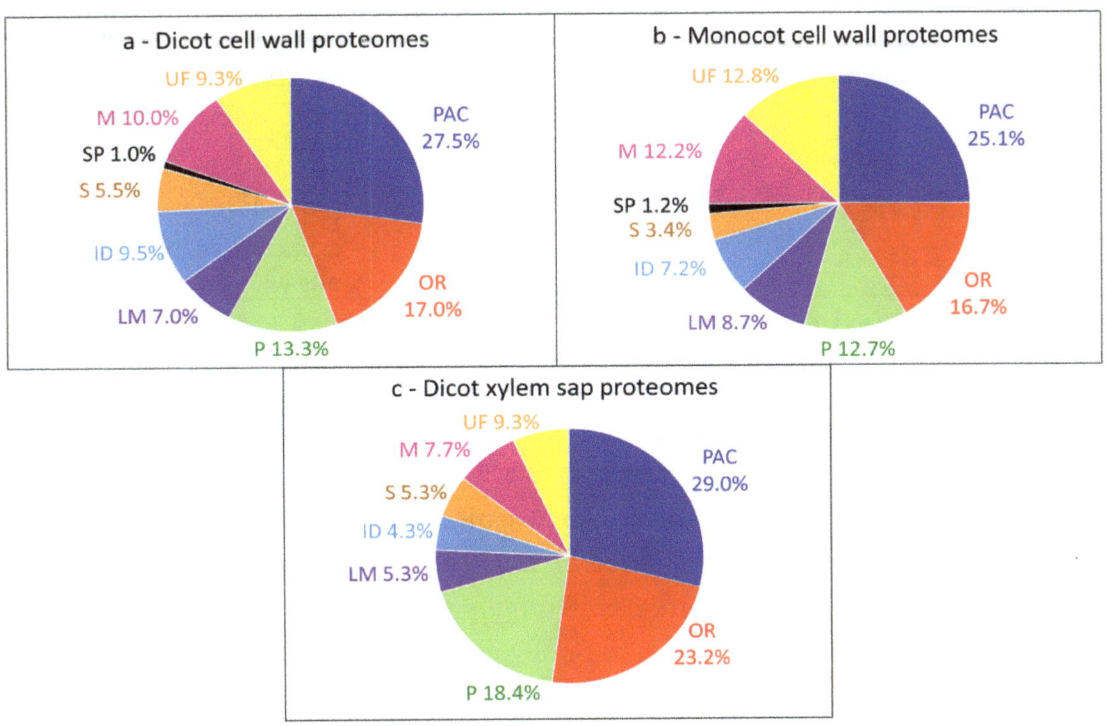

Interestingly, differences can be highlighted when comparisons of cell wall proteomes obtained in similar conditions are done between different tissues or organs of the same plant (Table 2). The comparison of the cell wall proteomes of *Solanum lycopersicum* fruit pericarp [20] and cuticle [53] shows striking changes in the relative importance of PAC (32.4% *vs.* 10.0%), oxido-reductases (9.3% *vs.* 16.7%), proteases (24.1% *vs.* 6.7%), proteins related to lipid metabolism (7.4% *vs.* 15%), proteins having interacting domains (7.4% *vs.* 26.7%) and miscellaneous proteins (7.4% *vs.* 20.0%) (Figure 2a,b). It is not surprising that the proportion of PAC is lower in the cuticle proteome than in the pericarp cell wall proteome and that the proportion of proteins related to lipid metabolism is higher. Indeed, the biogenesis of the cuticle composed of waxes and cutin occurs at the plant surface [53]. In the same way, major differences are found between cell wall proteomes of mature leaves and basal internodes of *Brachypodium distachyon* [15]: 26.5% *vs.* 19.4% PAC and 15.1% *vs.* 21.2% oxido-reductases (Figure 2c,d). Although both organs are mature, basal internodes are more lignified than mature leaves and the presence of more oxido-reductases and less PAC is probably required for lignin monomer polymerization.

Table 2. Information about the cell wall or xylem sap proteomes used for overall comparisons.

	Stems	Leaves	Fruit pericarp	Fruit cuticle	Xylem sap	Protocols	Ref.
Dicots							
B. napus/oleracea					x	xylem sap	[19]
L. usitatissimum	x					- cell wall preparation - extraction of proteins from cell walls with $CaCl_2$, LiCl	[46]
M. sativa	x					- cell wall preparation - extraction of proteins from cell walls with EGTA, LiCl	[16]
P. deltoides					x	xylem sap	[13]
S. lycopersicum				x		chloroform extraction	[53]
S. lycopersicum			x			*N*-glycoproteome (total protein extraction followed by ConA affinity chromatography	[20]
S. tuberosum		x				- cell wall preparation - extraction of proteins from cell walls with $CaCl_2$	[55]
Monocots							
B. distachyon	x	x				- cell wall preparation - extraction of proteins from cell walls with $CaCl_2$, LiCl	[15]

Figure 2. Comparisons of cell wall proteomes of different plant tissues or organs. **(a)** *L. esculentum* fruit pericarp; **(b)** *L. esculentum* fruit cuticle; **(c)** *B. distachyon* mature leaves; **(d)** *B. distachyon* basal internodes. All the proteins have been annotated according to the presence of functional domains (see Supplementary Material).

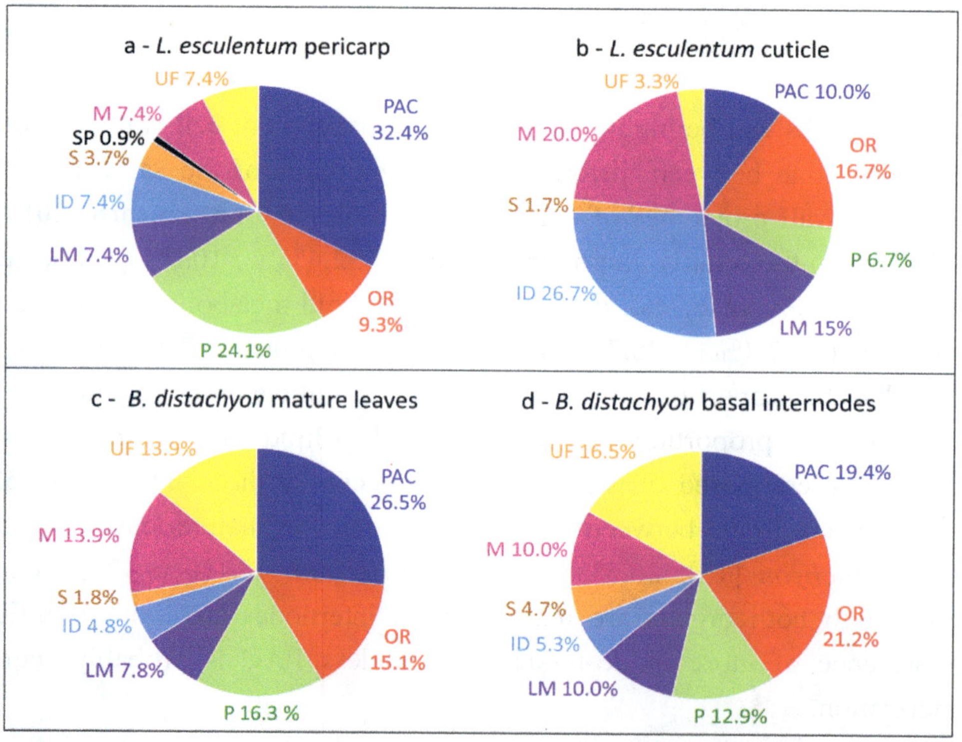

Comparisons of cell wall proteomes between similar organs of monocots and dicots show differences related to the composition of their cell walls [1]. For example, cell wall proteomes of leaves of *B. distachyon* [15] and *Solanum tuberosum* [55] show differences in the relative proportions of PAC (25.6% *vs.* 33.6%), oxido-reductases (15.5% *vs.* 9.9%), proteins related to lipid metabolism (7.7% *vs.* 5.3%) and proteins having interacting domains (4.8% *vs.* 9.2%) (Figure 3a,b). In both cases, proteins have been extracted from purified cell walls using salt solutions. Such differences have been discussed [15]. It was suggested that the presence of aromatic compounds in monocot primary cell walls could explain the higher proportion of oxido-reductases. The higher proportion of proteins related to lipid metabolism has been related to the presence of a cuticle on both sides of monocot leaves. Finally, only a few enzyme inhibitors have been identified in the *B. distachyon* leaf proteome as well as no lectin. A similar comparison between cell wall proteomes of stems such as those of *B. distachyon* [15], *Linum usitatissimum* [46] and *Medicago sativa* [16] does not show striking differences between monocots and dicots probably because both contain lignified secondary walls.

Figure 3. Comparisons of cell wall proteomes of mature leaves between a monocot and a dicot. **(a)** *B. distachyon*; **(b)** *S. tuberosum*. All the proteins have been annotated according to the presence of functional domains (see Supplementary Material).

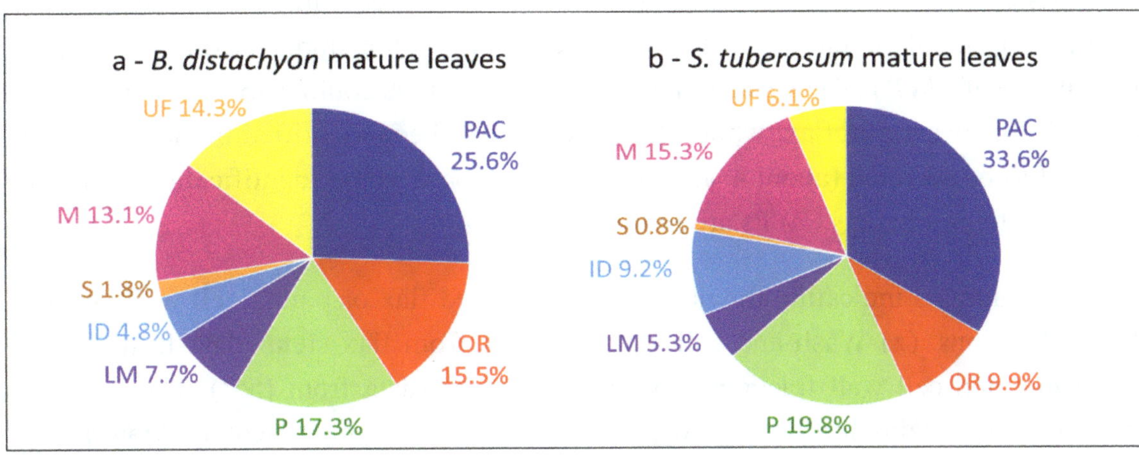

All these comparisons are qualitative ones based on presence/absence of proteins in cell wall proteomes. Inside each functional class, the comparison of protein families can be refined to look for candidate proteins possibly involved in cell wall remodeling in specific organs, during particular stages of development, or in response to changes in environmental conditions. Such results are discussed in detail in experimental papers (see Table 1). Quantitative data are still scarce and the limitations of the available protocols to completely extract CWPs from cell walls do not allow getting fully reliable information as for transcriptomes. However, transcriptomic data do not provide any information about post-transcriptional levels of gene regulation, and both types of data are complementary [65].

3. The Limitations for Full Coverage of Cell Wall Proteomes

Although well-documented, plant cell wall proteomes are probably missing proteins lost during the purification of cell walls and important protein families such as structural proteins are still lacking. These limitations will be examined in the following paragraphs [30].

3.1. Loss of Proteins during the Purification of Cell Walls

It is difficult to obtain a high coverage of the complete set of proteins present in cell walls because of the lack of surrounding membrane which can result in the loss of CWPs during the isolation procedure [66]. CWPs can have little or no interactions with cell wall components and thus move freely in the extracellular space. Non-destructive techniques such as vacuum infiltration [25], or recovery of liquid culture media from cell suspension cultures or seedlings [23,27] were developed to overcome this obstacle. Large sets of "labile CWPs" have been identified. Most of them have acidic pI ranging from 2 to 6 while CWPs are mainly basic proteins [67].

Two recent studies using destructive methods to isolate cell walls of flax stems or potato leaves have considered the loss of proteins during the cell wall purification steps [46,55]. Starting with ground plant material, the isolation procedures retained a differential centrifugation approach to separate cell wall and cytoplasmic fractions [55]. Several washing steps were performed to exclude cytoplasmic and membrane proteins [46]. Figure 4 shows the number of CWPs identified in the different fractions, *i.e.*, wash *vs.* cell wall fractions (flax stem) and cytoplasmic *vs.* cell wall fractions (potato leaves). Surprisingly, about 15% of the CWPs identified in these studies were only present in the wash or in the cytoplasmic fractions. These CWPs did not show any distinctive features, e.g., their pIs are in the basic range in contrast to the "labile CWPs" identified with non-destructive methods and no particular protein family could be found [67]. The isolation procedures used to purify cell walls led to a significant loss of CWPs. The wash and cytoplasmic fractions could also be considered in cell wall proteomic studies. However, in flax, while 958 proteins have been identified in the wash fraction, only 42 are predicted to be secreted (about 4%). The main drawback is the identification of a large number of intracellular proteins whereas CWPs are in the minority.

Figure 4. Diagrams indicating the number of identified flax or potato cell wall proteins in different fractions. (**a**) Wash and cell wall fractions from flax stem (data from [46]); (**b**) Cytoplasmic and cell wall fractions from potato leaves (data from [55]). The sub-cellular localization of proteins has been predicted as described in Supplementary Material. Only proteins having a predicted signal peptide and no known intracellular retention signal are considered as CWPs.

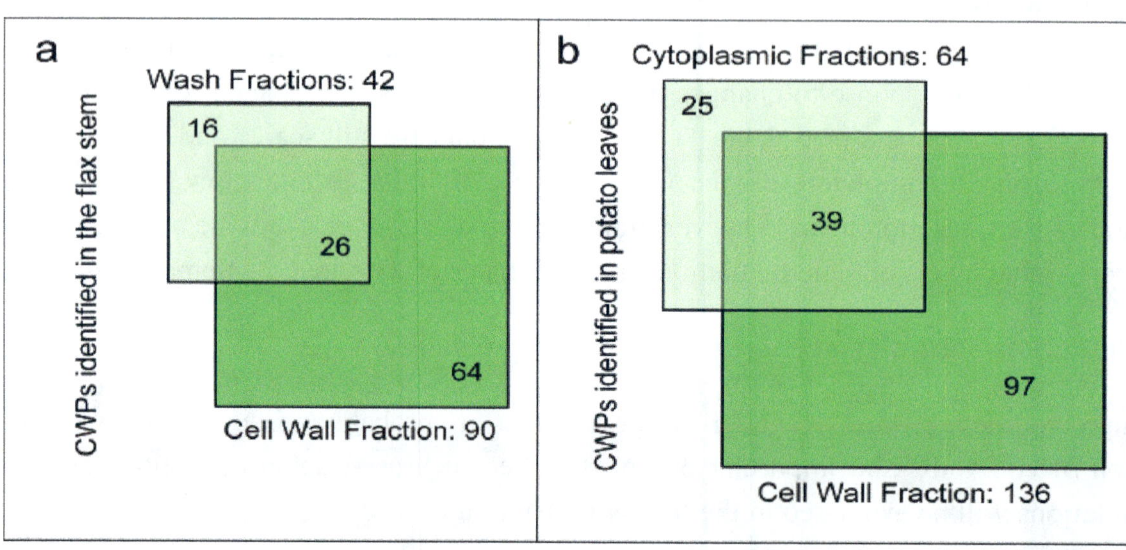

3.2. Extraction of Proteins by Salt Solutions

Most plant cell wall proteomic studies use salts to release CWPs from cell walls using non-destructive strategies or to extract proteins from purified cell walls [10]. Different types of salt solutions have been used, but $CaCl_2$ solutions appeared to be among of the most efficient ones [25]. In the case of destructive methods, there are doubts with regard to the release of *bona fide* CWPs since the intracellular content is released at the time of tissue grinding. Actually, two kinds of proteins are identified, those having predicted signal peptides which are considered as CWPs in this review, and those having no signal peptide. This point has been discussed in previous reviews [10,68].

To illustrate the efficiency of CWP extraction from purified call walls using salt solutions, we have examined the cell wall localization of a protein identified in numerous cell wall proteomic studies, namely At5g11420. This is one of the so-called DUF642 (Domain of Unknown function) proteins which all have a predicted signal peptide [69]. In addition, since the observation of fluorescent chimeric proteins by confocal microscopy offers the opportunity to explore the effect of exogenous treatments on the protein localization dynamic at the cellular scale, we show the release of At5g11420 after a salt solution treatment.

The plant cell wall is an acidic compartment and the sub-cellular localization of protein of interest labeled with a fluorescent protein (FP) is challenging in a low pH environment. The spectral properties of Green FP (GFP) are influenced by pH, and the fluorescence of GFP variants (e.g., monomeric Enhanced GFP, mEGFP and Yellow FP, YFP) decreases at a pH below 6. In this study, we have chosen the tagRFP as a fluorescent reporter taking advantage of its low pKa (3.1) [70].

The *N. benthamiana* leaf epidermal cells, transiently expressing the p35S::At5g11420::tagRFP construct, produced a red fluorescent signal at the cell periphery (Figure 5c). In non-plasmolyzed and glycerol-plasmolyzed cells, the At5g11420::tagRFP protein co-localized with the calcofluor labelling, a specific cell wall marker (data not shown, Figure 5b,e). Under plasmolysis condition with glycerol, the plasma membrane labeled by the pm::YFP marker was progressively loosened from the cell wall, while the At5g11420::tagRFP fluorescence was maintained into the cell wall (Figure 5d,f). These data indicate that At5g11420 is specifically targeted to the cell wall.

When plasmolysis was induced by $CaCl_2$, the detachment of the plasma membrane from the cell wall was accompanied by a new At5g11420::tagRFP labelling pattern (Figure 5h). After a few minutes of incubation, the At5g11420::tagRFP fluorescence diffused from the cell wall into the apoplastic compartment delimited by the plasma membrane (Figure 5k). This experiment illustrates how proteins can be released from cell walls using salt solutions. It should be noted that they can be released together with other cell wall components like pectins.

The efficiency of CWP extraction by salt solutions depends on the type of interactions between CWPs and cell wall components. This is also the reason why different extraction methods have been used in cell wall proteomic studies. Alternatively, glycoproteins have been captured by lectin affinity chromatography, starting from total extracts of proteins [20,37,44]. This strategy has proved to be very efficient since CWPs are synthesized in the secretory pathway. However, care should be taken to distinguish glycoproteins which are resident in the secretory pathway from those which are targeted to the extracellular space.

Figure 5. The At5g11420 protein is localized in the cell wall (see Supplementary Material for methods). (**a–f**): *N. benthamiana* leaf epidermal cells plasmolyzed by incubation with glycerol. Cell wall localization of the At5g11420::tagRFP protein; (**g–k**): *N. benthamiana* leaf epidermal cells plasmolyzed by incubation with $CaCl_2$. Under $CaCl_2$ treatment the At5g11420::tagRFP protein partially relocalizes to the apoplasm; (**a**, **g**) Bright field; (**b**) Calcofluor labelling of the cell wall; (**c**, **h**) RFP labelling; At5g11420::tagRFP was used to observe At5g11420 protein localization. (**d**, **i**) YFP labelling; aquaporin::YFP allows plasma membrane visualization; (**j**) Chloroplast labeling; (**e**) Merge of (**b**) and (**c**); (**f**, **k**) Merge of (**c**) and (**d**) and (**h**) and (**i**), respectively. (**l**) Merge of (**i**) and (**j**).

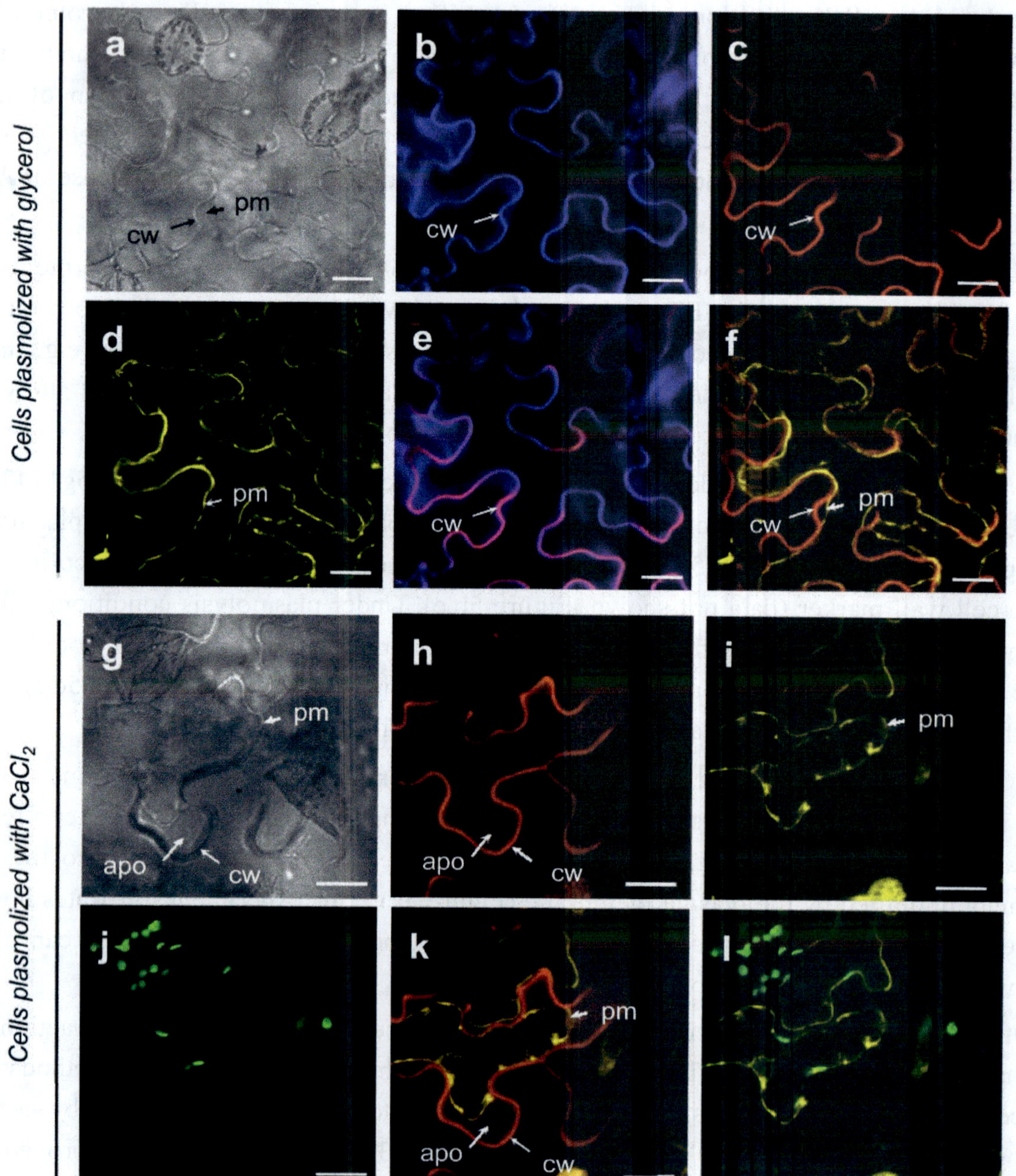

cw, cell wall; pm, plasma membrane; apo, apoplasm. Bars = 20 µm.

3.3. Difficulties to Extract Structural Proteins

As mentioned above, cell wall proteomic studies mentioned in this review rely on protein extraction methods using salt extractions. However, these strategies were shown to be inefficient to solubilize covalently-linked proteins, like structural proteins. To date, only a few PRPs, Leucine-Rich repeat Extensin (LRXs), Glycine-Rich Proteins (GRPs) or Thr/Hyp-rich GlycoProteins (THRGPs) have been identified (Table 3). Structural proteins are thus under-represented in cell wall proteomes, *i.e.*, 3 PRPs and no EXT out of the 18 and 32 respectively predicted in *A. thaliana* [71]. Main features concerning these families are reported below to pinpoint the bottlenecks preventing their extraction.

Table 3. Structural proteins identified in cell wall proteomes.

Protein family	Plant	References
PRP (At5g09530; At5g14920, AtGASA14)	*A. thaliana*	[14]
AGP/PRP (At1g28290, AtAGP31)	*A. thaliana*	[14,30]
LRX (At1g62440, AtLRX2; At4g13340; At3g24480; AtLRX3, AtLRX4; At4g18670, AtLRX5)	*A. thaliana*	[14,22,38]
GRP (At2g05580)	*A. thaliana*	[14]
LRX (Os01g0594300, Os05g0180300, Os06g0704500, Os02g0138000	*O. sativa*	[56]
GRP (Os07g0688700, Os07g0440100)	*O. sativa*	[57]
THRGP (Os03g0676300, Os04g0418800)	*O. sativa*	[56,57]
AGP/PRP (Lus10015434)	*L. usitatissimum*	[46]
LRX (Medtr8g103700.1, Medtr6g086120.1)	*M. sativa*	[16]
LRX (Solyc11g005150.1)	*L. esculentum*	[20]

EXTs belong to the superfamily of hydroxyproline-rich glycoproteins (HRGPs) and are involved in cell wall assembly, cell shape and growth [72–74]. They have been widely studied since the sixties and constitute one of the best known CWP family [75]: (i) they are basic proteins, (ii) they contain repetitive sequence with contiguous Hyp *O*-glycosylated with short arabino-oligosaccharides, (iii) they adopt a polyproline II helical structure, (iv) they can be crossed-linked through isodityrosine or di-isodityrosine links [76] and (v) they interact with pectins. The molecular bases of their insolubilization have been highlighted recently. It was shown by atomic force microscopy (AFM) analysis that the purified *A. thaliana* AtEXT3 self-assemble to form dendritic structures, consistent with cross-linking by peroxidases observed *in vitro* [77]. Similar network structures were observed by AFM for a maize THRGP, but peroxidases were not involved in their cross-linking [78]. AFM observations corroborate previously reported electronic microscopy data showing intramolecular and short intermolecular cross-links [79]. It was proposed that self-assembled extensins form positively charged scaffolds in the cell plate, able to react with negatively charged pectins through ionic interactions. Besides, covalent cross-links between extensins and pectins were also suggested [80,81].

EXT-like chimeras and hybrid-EXTs also exist in cell walls [72,73]. They are assumed to be insolubilized *in muro* but the presence of other protein domains may modify their behavior. For instance, the *A. thaliana* LRX1 is insolubilized in the cell wall, but this does not involve Tyr cross-links [82]. However, Tyr residues are required for LRX1 function in root hair formation [82].

PRPs are highly basic, mostly lowly glycosylated proteins, and they display specific repetitive motifs [83,84]. PRPs are probably covalently cross-linked in the cell wall, but direct evidence is still lacking [85–87].

GRPs are characterized by a high content in glycine residues (up to 70%) [88,89]. Several studies using immunocytochemistry have shown that they are associated with the protoxylem, suggesting a function in a repair system during the stretching phase [88]. It is assumed that the repetitive nature of the glycine-rich domains leads to the formation of β-pleated sheet structures allowing hydrophobic interactions. Interestingly, *in vitro* cross-linking experiments carried out in presence of peroxidase suggested the formation of networks only in Tyr-containing GRPs [90]. However, further experimental data should be obtained to characterize with more details intra- and inter-molecular networks involving GRPs *in muro*.

Finally, some AGPs were shown to bind covalently to the cell wall. They constitute a category of HRGPs *O*-hyperglycosylated by arabinogalactans at non-contiguous Hyp, playing essential roles in a wide range of plant growth and development processes [91]. AGPs have been assumed to form complexes with pectins and xylans [91]. The first experimental evidence for covalent attachment between an *A. thaliana* AGP and hemicellulosic and pectic polysaccharides, forming a complex called Arabinoxylan Pectin Arabinogalactan Protein1 (APAP1), has been recently reported [92]. Interestingly, the *apap1* mutant showed an increased extractability of pectin and xylan, supporting the structural role proposed for APAP1 [92]. This result indicates that some AGPs may serve as cross-linker in cell walls, corroborating previous reports where AGPs were described as pectic plasticizers [93,94].

Alternative extraction strategies using SDS buffer to extract structural proteins have been tried but they were inefficient [30]. The question of the extraction of covalently bound CWPs thus remains unanswered and further research is necessary to improve their identification by proteomics.

4. Concluding Remarks

The knowledge of plant cell wall proteomes has been greatly enlarged through the numerous studies performed during the last fifteen years. Thanks to various complementary strategies, it is possible to get an overview of proteins present in the cell walls of numerous plant organs and in cell suspension cultures. However, the full coverage of plant cell wall proteomes remains challenging since some proteins are lost during the purification of cell walls and cross-linked proteins are not extracted. Global approaches avoiding cell wall purification such as direct capture of glycoproteins on lectin affinity columns did not allow to significantly enlarge cell wall proteomes [20,37,44]. It can be anticipated that a better coverage of cell wall proteomes will require strategies adapted to protein families of interest as for AGPs which have been specifically targeted by the Yariv reagent [35].

A major drawback for the use of cell wall proteomic data is the heterogeneity of protein functional annotation which limits relevant interpretation of data and comparisons between proteomes [95]. In this regard, WallProtDB is a useful tool since all the proteins are annotated in the same way. At present, it is probable that the identified proteins are the most abundant and the most accessible within the intricate extracellular polysaccharide networks. Besides, reliable quantitative information is now required to better describe CWP profiles and correlate them to plant physiological state.

Acknowledgements

This work was supported by the French Centre National de la Recherche Scientifique (CNRS) and the Paul Sabatier University of Toulouse. This work has been done at LRSV, part of the "Laboratoire d'Excellence" (LABEX) entitled TULIP (ANR -10-LABX-41; ANR-11-IDEX-0002-02). The authors wish to thank A. Le Ru and A. Jauneau at the *Toulouse Réseau Imagerie* (TRI-GenoToul) for providing access to microscopy equipments. They are thankful to D. Roujol for technical assistance. They also thank G. Lomonossoff (John Innes Center, Norwich, UK) and Plant Bioscience Limited (Norwich, UK) for providing the pEAQ vector.

Author Contributions

All the authors have participated to the writing of the paper. Each of them has specifically contributed to the following parts: proteins strongly associated to cell walls (CA); runaways (HC); cell wall localization of At5g11420 (LH); collect of proteomic data, comparison of proteomes and coordination of the writing (EJ).

References

1. Carpita, N.C.; Gibeaut, D.M. Structural models of primary cell walls in flowering plants, consistency of molecular structure with the physical properties of the walls during growth. *J. Plant* **1993**, *3*, 1–30.
2. Wang, H.; Dixon, R. On-off switches for secondary cell wall biosynthesis. *Mol. Plant* **2012**, *5*, 297–303.
3. Knox, J.P. Revealing the structural and functional diversity of plant cell walls. *Curr. Opin. Plant Biol.* **2008**, *11*, 308–313.
4. Lee, K.; Marcus, S.; Knox, J. Cell wall biology: perspectives from cell wall imaging. *Mol. Plant* **2011**, *4*, 212–219.
5. Roppolo, D.; Geldner, N. Membrane and walls: Who is master, who is servant? *Curr. Opin. Plant Biol.* **2012**, *15*, 608–617.
6. Fry, S.C. Primary cell wall metabolism: Tracking the careers of wall polymers in living plant cells. *New Phytol.* **2004**, *161*, 641–675.
7. Seifert, G.; Roberts, K. The biology of arabinogalactan proteins. *Annu. Rev. Plant Biol.* **2007**, *58*, 137–161.
8. Murphy, E.; Smith, S.; de Smet, I. Small signaling peptides in *Arabidopsis* development: How cells communicate over a short distance. *Plant Cell* **2012**, *24*, 3198–3217.
9. Ferrari, S.; Savatin, D.; Sicilia, F.; Gramegna, G.; Cervone, F.; Lorenzo, G. Oligogalacturonides: Plant damage-associated molecular patterns and regulators of growth and development. *Front. Plant Sci.* **2013**, *4*, e49.

10. Albenne, C.; Canut, H.; Jamet, E. Plant cell wall proteomics: The leadership of *Arabidopsis thaliana*. *Front. Plant Sci.* **2013**, *4*, e111.
11. Jamet, E.; Albenne, C.; Boudart, G.; Irshad, M.; Canut, H.; Pont-Lezica, R. Recent advances in plant cell wall proteomics. *Proteomics* **2008**, *8*, 893–908.
12. Lee, S.J.; Saravanan, R.S.; Damasceno, C.M.; Yamane, H.; Kim, B.D.; Rose, J.K. Digging deeper into the plant cell wall proteome. *Plant Physiol. Biochem.* **2004**, *42*, 979–988.
13. Pechanova, O.; Hsu, C.-Y.; Adams, J.; Pechan, T.; Vandervelde, L.; Drnevich, J.; Jawdy, S.; Adeli, A.; Suttle, J.; Lawrence, A.; *et al.* Apoplast proteome reveals that extracellular matrix contributes to multistress response in poplar. *BMC Genomics* **2010**, *11*, e674.
14. Irshad, M.; Canut, H.; Borderies, G.; Pont-Lezica, R.; Jamet, E. A new picture of cell wall protein dynamics in elongating cells of *Arabidopsis thaliana*: Confirmed actors and newcomers. *BMC Plant Biol.* **2008**, *8*, e94.
15. Douché, T.; San Clemente, H.; Burlat, V.; Roujol, D.; Valot, B.; Zivy, M.; Pont-Lezica, R.; Jamet, E. *Brachypodium distachyon* as a model plant toward improved biofuel crops: Search for secreted proteins involved in biogenesis and disassembly of cell wall polymers. *Proteomics* **2013**, *13*, 2438–2454.
16. Verdonk, J.; Hatfield, R.; Sullivan, M. Proteomic analysis of cell walls of two developmental stages of alfalfa stems. *Front. Plant Sci.* **2012**, *3*, e279.
17. Oh, I.S.; Park, A.R.; Bae, M.S.; Kwon, S.J.; Kim, Y.S.; Lee, J.E.; Kang, N.Y.; Lee, S.; Cheong, H.; Park, O.K. Secretome analysis reveals an *Arabidopsis* lipase involved in defense against *Alternaria brassicicola*. *Plant Cell* **2005**, *17*, 2832–2847.
18. Ruiz-May, E.; Rose, J. Progress toward the tomato fruit cell wall proteome. *Front. Plant Sci.* **2013**, *4*, e159.
19. Ligat, L.; Lauber, E.; Albenne, C.; San Clemente, H.; Valot, B.; Zivy, M.; Pont-Lezica, R.; Arlat, M.; Jamet, E. Analysis of the xylem sap proteome of *Brassica oleracea* reveals a high content in secreted proteins. *Proteomics* **2011**, *11*, 1798–1813.
20. Catalá, C.; Howe, K.; Hucko, S.; Rose, J.; Thannhauser, T. Towards characterization of the glycoproteome of tomato (*Solanum lycopersicum*) fruit using Concanavalin A lectin affinity chromatography and LC-MALDI-MS/MS analysis. *Proteomics* **2011**, *11*, 1530–1544.
21. Basu, U.; Francis, J.L.; Whittal, R.W.; Stephens, J.L.; Wang, Y.; Zaiane, O.R.; Goebel, R.; Muench, D.G.; Good, A.G.; Taylor, G.J. Extracellular proteomes of *Arabidopsis thaliana* and *Brassica napus* roots: Analysis and comparison by MUdPIT and LC-MS/MS. *Plant Soil* **2006**, *286*, 357–376.
22. Bayer, E.M.; Bottrill, A.R.; Walshaw, J.; Vigouroux, M.; Naldrett, M.J.; Thomas, C.L.; Maule, A.J. *Arabidopsis* cell wall proteome defined using multidimensional protein identification technology. *Proteomics* **2006**, *6*, 301–311.
23. Borderies, G.; Jamet, E.; Lafitte, C.; Rossignol, M.; Jauneau, A.; Boudart, G.; Monsarrat, B.; Esquerré-Tugayé, M.T.; Boudet, A.; Pont-Lezica, R. Proteomics of loosely bound cell wall proteins of *Arabidopsis thaliana* cell suspension cultures: A critical analysis. *Electrophoresis* **2003**, *24*, 3421–3432.
24. Borner, G.H.; Lilley, K.S.; Stevens, T.J.; Dupree, P. Identification of glycosylphosphatidylinositol-anchored proteins in *Arabidopsis*. A proteomic and genomic analysis. *Plant Physiol.* **2003**, *132*, 568–577.

25. Boudart, G.; Jamet, E.; Rossignol, M.; Lafitte, C.; Borderies, G.; Jauneau, A.; Esquerré-Tugayé, M.-T.; Pont-Lezica, R. Cell wall proteins in apoplastic fluids of *Arabidopsis thaliana* rosettes: Identification by mass spectrometry and bioinformatics. *Proteomics* **2005**, *5*, 212–221.
26. Casasoli, M.; Spadoni, S.; Lilley, K.; Cervone, F.; de Lorenzo, G.; Mattei, B. Identification by 2-D DIGE of apoplastic proteins regulated by oligogalacturonides in *Arabidopsis thaliana*. *Proteomics* **2008**, *8*, 1042–1054.
27. Charmont, S.; Jamet, E.; Pont-Lezica, R.; Canut, H. Proteomic analysis of secreted proteins from *Arabidopsis thaliana* seedlings: Improved recovery following removal of phenolic compounds. *Phytochemistry* **2005**, *66*, 453–461.
28. Cheng, F.-Y.; Blackburn, K.; Lin, Y.-M.; Goshe, M.; Wiliamson, J. Absolute protein quantification by LC/MS for global analysis of salicylic acid-induced plant protein secretion responses. *J. Proteome Res.* **2009**, *8*, 82–93.
29. Chivasa, S.; Ndimba, B.K.; Simon, W.J.; Robertson, D.; Yu, X.-L.; Knox, J.P.; Bolwell, P.; Slabas, A.R. Proteomic analysis of the *Arabidopsis thaliana* cell wall. *Electrophoresis* **2002**, *23*, 1754–1765.
30. Feiz, L.; Irshad, M.; Pont-Lezica, R.F.; Canut, H.; Jamet, E. Evaluation of cell wall preparations for proteomics: A new procedure for purifying cell walls from *Arabidopsis hypocotyls*. *Plant Methods* **2006**, *2*, e10.
31. Haslam, R.P.; Downie, A.L.; Raventon, M.; Gallardo, K.; Job, D.; Pallett, K.E.; John, P.; Parry, M.A.J.; Coleman, J.O.D. The assessment of enriched apoplastic extracts using proteomic approaches. *Ann. Appl. Biol.* **2003**, *143*, 81–91.
32. Kwon, H.-K.; Yokoyama, R.; Nishitani, K. A proteomic approach to apoplastic proteins involved in cell wall regeneration in protoplasts of *Arabidopsis* suspension-cultured cells. *Plant Cell Physiol.* **2005**, *46*, 843–857.
33. Ndimba, B.K.; Chivasa, S.; Hamilton, J.M.; Simon, W.J.; Slabas, A.R. Proteomic analysis of changes in the extracellular matrix of *Arabidopsis* cell suspension cultures induced by fungal elicitors. *Proteomics* **2003**, *3*, 1047–1059.
34. Robertson, D.; Mitchell, G.P.; Gilroy, J.S.; Gerrish, C.; Bolwell, G.P.; Slabas, A.R. Differential extraction and protein sequencing reveals major differences in patterns of primary cell wall proteins from plants. *J. Biol. Chem.* **1997**, *272*, 15841–15848.
35. Schultz, C.J.; Ferguson, K.L.; Lahnstein, J.; Bacic, A. Post-translational modifications of arabinogalactan-peptides of *Arabidopsis thaliana*. Endoplasmic reticulum and glycosylphosphatidylinositol-anchor signal cleavage sites and hydroxylation of proline. *J. Biol. Chem.* **2004**, *279*, 45503–45511.
36. Tran, H.; Plaxton, W. Proteomic analysis of alterations in the secretome of *Arabidopsis thaliana* suspension cells subjected to nutritional phosphate deficiency. *Proteomics* **2008**, *8*, 4317–4326.
37. Minic, Z.; Jamet, E.; Negroni, L.; der Garabedian, P.A.; Zivy, M.; Jouanin, L. A sub-proteome of *Arabidopsis thaliana* trapped on Concanavalin A is enriched in cell wall glycoside hydrolases. *J. Exp. Bot.* **2007**, *58*, 2503–2512.
38. Zhang, Y.; Giboulot, A.; Zivy, M.; Valot, B.; Jamet, E.; Albenne, C. Combining various strategies to increase the coverage of the plant cell wall glycoproteome. *Phytochemistry* **2011**, *72*, 1109–1123.
39. Kehr, J.; Buhtz, A.; Giavalisco, P. Analysis of xylem sap proteins from *Brassica napus*. *BMC*

Plant Biol. **2005**, *5*, e11.

40. Bhushan, D.; Pandey, A.; Chattopadhyay, A.; Choudhary, M.; Chakraborty, S.; Datta, A.; Chakraborty, N. Extracellular matrix proteome of chickpea (*Cicer arietinum* L.) illustrates pathway abundance, novel protein functions and evolutionary perspect. *J. Proteome Res.* **2006**, *5*, 1711–1720.
41. Bhushan, D.; Pandey, A.; Choudhary, M.; Datta, A.; Chakraborty, S.; Chakraborty, N. Comparative proteomics analysis of differentially expressed proteins in chickpea extracellular matrix during dehydration stress. *Mol. Cell. Proteomics* **2007**, *6*, 1868–1884.
42. Bhushan, D.; Jaiswal, D.; Ray, D.; Basu, D.; Datta, A.; Chakraborty, S.; Chakraborty, N. Dehydration-responsive reversible and irreversible changes in the extracellular matrix: Comparative proteomics of chickpea genotypes with contrasting tolerance. *J. Proteome Res.* **2011**, *10*, 2027–2046.
43. Komatsu, S.; Kobayashi, Y.; Nishizawa, K.; Nanjo, Y.; Furukawa, K. Comparative analysis of differentially expressed proteins in soybean cell wall during flooding stress. *Amino Acids* **2010**, *39*, 1435–1449.
44. Kumar, S.; Kumar, K.; Pandey, P.; Rajamani, V.; Padmalatha, K.; Dhandapani, G.; Kanakachari, M.; Leelavathi, S.; Kumar, P.; Reddy, V. Glycoproteome of elongating cotton fibre cells. *Mol. Cell. Proteomics* **2013**, *12*, 3777–3789.
45. Pinedo, M.; Regenten, M.; Elizalde, M.; Quiroga, I.; Pagnussat, L.; Jorrin-Novo, J.; Maldonado, A.; de la Canal, L. Extracellular sunflower proteins: Evidence on non-classical secretion of a jacalin-related lectin. *Protein Pept. Lett.* **2012**, *19*, 270–276.
46. Day, A.; Fénart, S.; Neutelings, G.; Hawkins, S.; Rolando, C.; Tokarski, C. Identification of cell wall proteins in the flax (*Linum usitatissimum*) stem. *Proteomics* **2013**, *13*, 812–825.
47. Watson, B.S.; Lei, Z.; Dixon, R.A.; Sumner, L.W. Proteomics of *Medicago sativa* cell walls. *Phytochemistry* **2004**, *65*, 1709–1720.
48. Goulet, C.; Goulet, C.; Goulet, M.; Michaud, D. 2-DE proteome maps for the leaf apoplast of *Nicotiana benthamiana*. *Proteomics* **2010**, *10*, 2336–2344.
49. Dani, V.; Simon, W.J.; Duranti, M.; Croy, R.R. Changes in the tobacco leaf apoplast proteome in response to salt stress. *Proteomics* **2005**, *5*, 737–745.
50. Delannoy, M.; Alves, G.; Vertommen, D.; Ma, J.; Boutry, M.; Navarre, C. Identification of peptidases in *Nicotiana tabacum* leaf intercellular fluid. *Proteomics* **2008**, *8*, 2285–2298.
51. Millar, D.; Whitelegge, J.; Bindschedler, L.; Rayon, C.; Boudet, A.; Rossignol, M.; Borderies, G.; Bolwell, G. The cell wall and secretory proteome of a tobacco cell line synthesising secondary wall. *Proteomics* **2009**, *9*, 2355–2372.
52. Dafoe, N.; Constabel, P. Proteomic analysis of hybrid poplar xylem sap. *Phytochemistry* **2009**, *70*, 856–863.
53. Yeats, T.; Howe, K.; Matas, A.; Buda, G.; Thannhauser, T.; Rose, J. Mining the surface proteome of tomato (*Solanum lycopersicum*) fruit for proteins associated with cuticle biogenesis. *J. Exp. Bot.* **2010**, *61*, 3759–3771.
54. Fernandez, M.; Pagano, M.; Daleo, G.; Guevara, M. Hydrophobic proteins secreted into the apoplast may contribute to resistance agaisnt *Phytophtora infestans* in potato. *Plant Physiol. Biochem.* **2012**, *60*, 59–66.

55. Lim, S.; Chisholm, K.; Coffin, R.; Peters, R.; Al-Mughrabi, K.; Wang-Pruski, G.; Pinto, D. Protein profiling in potato (*Solanum tuberosum* L.) leaf tissues by differential centrifugation. *J. Proteome Res.* **2012**, *11*, 2594–2601.
56. Chen, X.; Kim, S.; Cho, W.; Rim, Y.; Kim, S.; Kim, S.; Kang, K.; Park, Z.; Kim, J. Proteomics of weakly bound cell wall proteins in rice calli. *J. Plant Physiol.* **2008**, *166*, 665–685.
57. Cho, W.; Chen, X.; Chu, H.; Rim, Y.; Kim, S.; Kim, S.; Kim, S.-W.; Park, Z.-Y.; Kim, J.-Y. The proteomic analysis of the secretome of rice calli. *Physiol. Plant.* **2009**, *135*, 331–341.
58. Jung, Y.-H.; Jeong, S.-H.; Kim, S.; Singh, R.; Lee, J.-E.; Cho, Y.-S.; Agrawal, G.; Rakwal, R.; Jwa, N.-S. Systematic secretome analyzes of rice leaf and seed callus suspension-cultured cells: Workflow development and establishment of high-density two-dimensional gel reference maps. *J. Proteome Res.* **2008**, *7*, 5187–5210.
59. Zhou, L.; Bokhari, S.; Dong, C.; Liu, J. Comparative proteomics analysis of the root apoplasts of rice seedlings in response to hydrogen peroxide. *PLoS One* **2011**, *6*, e16723.
60. Kim, S.; Wang, Y.; Lee, K.; Park, Z.; Park, J.; Wu, J.; Kwon, S.; Lee, Y.; Agrawal, G.; Rakwal, R.; *et al*. In-depth insight into *in vivo* apoplastic secretome of rice-*Magnaporthe oryzae* interaction. *J. Proteomics* **2013**, *78*, 58–71.
61. Calderan-Rodrigues, M.; Jamet, E.; Calderan Rodrigues Bonassi, M.; Guidetti-Gonzalez, S.; Carmanhanis Begossi, A.; Vaz Setem, L.; Franceschini, L.; Guimarães Fonseca, J.; Labate, C. Cell wall proteomics of sugarcane cell suspension cultures. *Proteomics* **2014**, *14*, 738–749.
62. Zhu, J.; Chen, S.; Alvarez, S.; Asirvatham, V.S.; Schachtman, D.P.; Wu, Y.; Sharp, R.E. Cell wall proteome in the maize primary root elongation zone. I. Extraction and identification of water-soluble and lightly ionically bound proteins. *Plant Physiol.* **2006**, *140*, 311–325.
63. Zhu, J.; Alvarez, S.; Marsh, E.; Lenoble, M.; Cho, I.; Sivaguru, M.; Chen, S.; Nguyen, H.; Wu, Y.; Schachtman, D.; *et al*. Cell wall proteome in the maize primary root elongation zone. II. Region-specific changes in water soluble and lightly ionically bound proteins under water deficit. *Plant Physiol.* **2007**, *145*, 1533–1548
64. Alvarez, S.; Goodger, J.Q.; Marsh, E.L.; Chen, S.; Asirvatham, V.S.; Schachtman, D.P. Characterization of the maize xylem sap proteome. *J. Proteome Res.* **2006**, *5*, 963–972.
65. Jamet, E.; Roujol, D.; San Clemente, H.; Irshad, M.; Soubigou-Taconnat, L.; Renou, J.-P.; Pont-Lezica, R. Cell wall biogenesis of *Arabidopsis thaliana* elongating cells: Transcriptomics complements proteomics. *BMC Genomics* **2009**, *10*, e505.
66. Jamet, E.; Boudart, G.; Borderies, G.; Charmont, S.; Lafitte, C.; Rossignol, M.; Canut, H.; Pont-Lezica, R. Isolation of plant cell wall proteins. In *Sample Preparation and Fractionation for 2-D PAGE/Proteomics*; Posch, A., Ed.; Humana Press: Totowa, NJ, USA, 2007.
67. Jamet, E.; Canut, H.; Boudart, G.; Pont Lezica, R. Cell wall proteins: A new insight through proteomics. *Trends Plant Sci.* **2006**, *11*, 33–39.
68. Rose, J.K.C.; Lee, S.-J. Straying off the highway: Trafficking of secreted plant proteins and complexity in the plant cell wall proteome. *Plant Physiol. Biochem.* **2010**, *153*, 433–436.
69. Vázquez-Lobo, A.; Roujol, D.; Zuñiga-Sánchez, E.; Albenne, C.; Piñero, D.; Gamboa de Buen, A.; Jametb, E. The highly conserved spermatophyte cell wall DUF642 protein family: Phylogeny and first evidence of interaction with cell wall polysaccharides *in vitro*. *Mol. Phylogenet. Evol.* **2012**, *63*, 510–520.

70. Shaner, N.; Lin, M.; McKeown, M.; Steinbach, P.; Hazelwood, K.; Davidson, M.; Tsien, R.Y. Improving the photostability of bright monomeric orange and red fluorescent proteins. *Nat. Methods* **2008**, *5*, 545–551.
71. Showalter, A.; Keppler, B.; Lichtenberg, J.; Gu, D.; Welch, L. A bioinformatics approach to the identification, classification, and analysis of hydroxyproline-rich glycoproteins. *Plant Physiol.* **2010**, *153*, 485–513.
72. Hall, Q.; Cannon, M.C. The cell wall hydroxyproline-rich glycoprotein RSH is essential for normal embryo development in *Arabidopsis*. *Plant Cell* **2002**, *14*, 1161–1172.
73. Velasquez, S.; Ricardi, M.; Dorosz, J.; Fernandez, P.; Nadra, A.; Pol-Fachin, L.; Egelund, J.; Gille, S.; Harholt, J.; Ciancia, M.; *et al.* *O*-Glycosylated cell wall proteins are essential in root hair growth. *Science* **2011**, *332*, 1401–1403.
74. Lamport, D.; Northcote, D. Hydroxyproline in primary cell walls of higher plants. *Nature* **1960**, *188*, 665–666.
75. Lamport, D.; Kieliszewski, M.; Chen, Y.; Cannon, M. Role of the extensin superfamily in primary cell wall architecture. *Plant Physiol.* **2011**, *156*, 11–19.
76. Brady, J.; Sadler, I.; Fry, S. Pulcherosine, an oxidatively coupled trimer of tyrosine in plant cell walls: Its role in cross-link formation. *Phytochemistry* **1998**, *47*, 349–353.
77. Cannon, M.; Terneus, K.; Hall, Q.; Tan, L.; Wang, Y.; Wegenhart, B.; Chen, L.; Lamport, D.; Chen, Y.; Kieliszewski, M. Self-assembly of the plant cell wall requires an extensin scaffold. *Proc. Natl. Acad. Sci. USA* **2008**, *105*, 2226–2231.
78. Schnabelrauch, L.S.; Kieliszewski, M.J.; Upham, B.L.; Alizedeh, H.; Lamport, D.T.A. Isolation of pI 4.6 extensin peroxidase from tomato cell suspension cultures and identification of Val-Tyr-Lys as putative intermolecular cross-link site. *Plant J.* **1996**, *9*, 477–489.
79. Stafstrom, J.P.; Staehelin, L.A. The role of carbohydrate in maintaining extensin in an extended conformation. *Plant Physiol.* **1986**, *81*, 242–246.
80. Qi, X.Y.; Behrens, B.X.; West, P.R.; Mort, A.J. Solubilization and partial characterization of extensin fragments from cell walls of cotton suspension-cultures, evidence for a covalent cross-link between extensin and pectin. *Plant Physiol.* **1995**, *108*, 1691–1701.
81. Nuñez, A.; Fishman, M.; Fortis, L.; Cooke, P.; Hotchkiss, A.J. Identification of extensin protein associated with sugar beet pectin. *J. Agric. Food Chem.* **2009**, *57*, 10951–10958.
82. Ringli, C. The hydroxyproline-rich glycoprotein domain of the *Arabidopsis* LRX1 requires Tyr for function but not for insolubilization in the cell wall. *Plant J.* **2010**, *63*, 662–669.
83. Cassab, G.I. Plant cell wall proteins. *Annu. Rev. Plant Physiol. Plant Mol. Biol.* **1998**, *49*, 281–309.
84. Datta, K.; Schmidt, A.; Marcus, A. Characterization of two soybean repetitive proline-rich proteins and a cognate cDNA from germinated axes. *Plant Cell* **1989**, *1*, 945–952.
85. Bradley, D.J.; Kjellbom, P.; Lamb, C.J. Elicitor-induced and wound induced oxidative cross-linking of a proline rich plant cell wall protein: a novel, rapid defense response. *Cell* **1992**, *70*, 21–30.
86. Brisson, L.F.; Tenhaken, R.; Lamb, C. Function of oxidative cross linking of cell wall structural proteins in plant disease resistance. *Plant Cell* **1994**, *6*, 1703–1712.
87. Frueauf, J.; Dolata, M.; Leykam, J.; Lloyd, E.; Gonzales, M.; VandenBosch, K.; Kieliszewski, M. Peptides isolated from cell walls of *Medicago truncatula* nodules and uninfected root.

Phytochemistry **2000**, *55*, 429 438.

88. Ringli, C.; Keller, B.; Ryser, U. Glycine-rich proteins as structural components of plant cell walls. *Cell. Mol. Life Sci.* **2001**, *58*, 1430–1441.
89. Mangeon, A.; Junqueira, R.; Sachetto-Martins, G. Functional diversity of the plant glycine-rich proteins superfamily. *Plant Signal. Behav.* **2010**, *5*, 99–104.
90. Ringli, C.; Hauf, G.; Keller, B. Hydrophobic interactions of the structural protein GRP1.8 in the cell wall of protoxylem elements. *Plant Physiol.* **2001**, *125*, 673–692.
91. Tan, L.; Showalter, A.; Egelund, J.; Hernandez-Sanchez, A.; Doblin, M.; Bacic, A. Arabinogalactan-proteins and the research challenges for these enigmatic plant cell surface proteoglycans. *Front. Plant Sci.* **2012**, *3*, e140.
92. Tan, L.; Eberhard, S.; Pattathil, S.; Warder, C.; Glushka, J.; Yuan, C.; Hao, Z.; Zhu, X.; Avci, U.; Miller, J.; *et al.* An *Arabidopsis* cell wall proteoglycan consists of pectin and arabinoxylan covalently linked to an arabinogalactan protein. *Plant Cell* **2013**, *25*, 270–287.
93. Lamport, D.T.A.; Kieliszewski, M.J.; Showalter, A.M. Salt stress upregulates periplasmic arabinogalactan proteins: using salt stress to analyze AGP function. *New Phytol.* **2006**, *169*, 479–492.
94. Lamport, D. Life behind cell walls: Paradigm lost, paradigm regained. *Cell. Mol. Life Sci.* **2001**, *58*, 1363–1385.
95. San Clemente, H.; Pont-Lezica, R.; Jamet, E. Bioinformatics as a tool for assessing the quality of sub-cellular proteomic strategies and inferring functions of proteins: Plant cell wall proteomics as a test case. *Bioinform. Biol. Insights* **2009**, *3*, 15–28.

13

Proteomic Profiling of Sugar Beet (*Beta vulgaris*) Leaves during Rhizomania Compatible Interactions

Kimberly M. Webb [1,*], Carolyn J. Broccardo [2], Jessica E. Prenni [2] and William M. Wintermantel [3]

1 USDA-ARS-SBRU, Crops Research Laboratory, 1701 Centre Ave., Fort Collins, CO 80526, USA

2 Proteomics and Metabolomics Facility, Colorado State University, C130 Microbiology, 2021 Campus Delivery, Fort Collins, CO 80523, USA; E-Mails: Carolyn.Broccardo@colostate.edu (C.J.B.); Jessica.Prenni@colostate.edu (J.E.P.)

3 USDA-ARS-CIPRU, 1636 E. Alisal St., Salinas, CA 93905, USA; E-Mail: Bill.Wintermantel@ars.usda.gov

* Author to whom correspondence should be addressed; E-Mail: Kimberly.webb@ars.usda.gov

---

Abstract: Rhizomania, caused by *Beet necrotic yellow vein virus* (BNYVV), severely impacts sugar beet (*Beta vulgaris*) production throughout the world, and is widely prevalent in most production regions. Initial efforts to characterize proteome changes focused primarily on identifying putative host factors that elicit resistant interactions with BNYVV, but as resistance breaking strains become more prevalent, effective disease control strategies will require the application of novel methods based on better understanding of disease susceptibility and symptom development. Herein, proteomic profiling was conducted on susceptible sugar beet, infected with two strains of BNYVV, to clarify the types of proteins prevalent during compatible virus-host plant interactions. Total protein was extracted from sugar beet leaf tissue infected with BNYVV, quantified, and analyzed by mass spectrometry. A total of 203 proteins were confidently identified, with a predominance of proteins associated with photosynthesis and energy, metabolism, and response to stimulus. Many proteins identified in this study are typically associated with systemic acquired resistance and general plant defense responses. These results expand on relatively limited proteomic data available for sugar beet and provide the ground work for additional studies focused on understanding the interaction of BNYVV with sugar beet.

Keywords: *Beet necrotic yellow vein virus* (BNYVV); shot-gun proteomics; sugarbeet

1. Introduction

Rhizomania, a disease that reduces root quality and yield in sugar beet (*Beta vulgaris*) is one of the most widely prevalent and economically important diseases affecting sugar beet production throughout the world [1–3]. Rhizomania is caused by *Beet necrotic yellow vein virus* (BNYVV) [4,5] and is transmitted via the plasmodiophorid *Polymyxa betae* [6]. Fields can remain infested with BNYVV indefinitely as *P. betae* cystosori can remain dormant for up to 25 years [4,7]. Therefore, typical remediation approaches such as rotation to non-host crops or lengthening rotations are ineffective at reducing disease incidence, leaving host-plant resistance as the only economically viable means of control [3].

A number of single dominant resistance genes (known as *Rz* genes) have been identified for control of BNYVV beginning with the discovery and introgression of the *Rz1* resistance gene [8,9], which became widely planted throughout all areas where rhizomania threatens sugar beet production. Over the past two decades additional resistance genes have also been identified [10,11]. The different sources of resistance genes appear to have different underlying mechanisms, which are largely undetermined [1,12,13] and several uncharacterized minor genes may contribute to enhanced resistance associated with the primary *Rz* genes [1]. Although the widely used *Rz1* gene prevents symptom development, the virus can still replicate at a low level in resistant plants. This has resulted in emergence of BNYVV variants that can accumulate enough in the presence of the resistance gene and overcome *Rz1* resistance in the field.

Following the introduction of resistant sugar beet varieties containing the single dominant *Rz1* gene, new pathotypes that overcome resistance emerged after only a few cropping seasons [5]. Three major BNYVV types have been reported world-wide; A-type, B-type, and P-type [14–16], with the A-type distributed throughout most sugar beet growing regions of the world [17]. The B-type is predominantly restricted to France and England [16] although limited identifications of the B-type have occurred in Sweden, China, Japan and Iran (summarized in [3]). The A and B types of BNYVV each contain 4 viral RNAs, with RNAs 1 and 2 encoding proteins involved in viral replication, and encapsidation and cell-to-cell movement, respectively. RNA3 encodes a protein that affects pathogenicity and determines the ability of the virus to overcome the most common source of genetic resistance to the virus. RNA4 encodes a single protein involved in transmission by *P. betae* (reviewed in [18]). The P-type is closely related to the A-type, but contains a fifth RNA [17,19]. Isolates containing a fifth RNA usually have increased symptom severity [20]. More importantly, the P-type is able to overcome the *Rz1* resistance gene, which is used to control BNYVV throughout the world. To date the P-type has been restricted to portions of France, Kazakhstan, England, and Iran [15,21–23]. All American isolates identified to date have been A-type, and until a decade ago these were effectively controlled by varieties carrying the *Rz1* gene [8]. In 2002 and 2003, resistant sugar beet in the Imperial Valley of California carrying the *Rz1* resistance gene developed severe rhizomania symptoms due to emergence of a mutant variant of the A-type, now known as the Imperial Strain (BNYVV-IV), which contains two amino acid mutations in the P25 protein encoded by RNA3 that allow the virus to overcome *Rz1* sources of resistance [5]. Subsequent studies have identified similar resistance-breaking variants throughout American sugar beet production regions, although most are limited in distribution [24]. Little is known of the interactions between sugar beet and BNYVV that influence epidemiology, and until the various mechanisms of resistance and infection are understood more fully, alternative disease control methods and additional sources of resistance will be required to control this pathogen.

There have been some proteomic analyses in sugar beet that have focused on abiotic agronomic stress conditions such as drought [25], salt stress [26,27], and nutritional deficiencies [28], and some examined tissue specific differences in protein expression [29,30]. Few published accounts focused primarily on characterizing the interaction of sugar beet with plant pathogens [31,32]. Proteomic approaches to characterize the interaction of BNYVV with sugar beet have been relatively limited. For example, using a subtractive proteomic approach, Larson *et al.* [32] previously identified 50 putative sugar beet proteins that were either up or down regulated in response to infection with a single BNYVV type. These 50 proteins were uniquely expressed in infected plants of a single susceptible genotype at six weeks after planting compared with uninoculated controls. While Larson *et al.* [32] was able to report on proteins that were differentially expressed with a single BNYVV type, due to the lack of a fully sequenced and annotated sugar beet genome and limitations in the ability to detect low abundance proteins in that study, further analysis of the sugar beet proteome was warranted. The goal of this work is to develop methods for exploratory proteomic profiling of susceptible sugar beet leaves. This approach allows for the unbiased evaluation of the detectable proteins in a sample, and lays the groundwork for future quantitative studies between defined treatments. Such an approach provides data for hypothesis generation and serves to better define the hereto poorly annotated sugar beet proteome. We examined the protein expression during compatible (susceptible) interactions with A-type as well as during infection by the Imperial Strain of BNYVV, which overcomes the *Rz1* resistance gene, the most widely used source of resistance to BNYVV in commercial sugar beet germplasm. This was done in order to identify proteins that are expressed during disease development, in the hopes of gaining an understanding of the proteins that may be involved in the underlying plant defense response.

2. Experimental

2.1. Plant Propagation and Inoculation

A proprietary sugar beet variety, R30_rz1, which is susceptible to all forms of BNYVV, was provided by KWS (Einbeck, Germany). To enable detection of a broad range of proteins associated with compatible interactions, two sources of BNYVV were used; standard BNYVV A-type, originally collected from Spence Field at the USDA-ARS in Salinas, CA (BNYVV-A), and a *Rz1* resistance-breaking mutant variant of the A-type originating from the field, Rockwood 158, Imperial County, CA (BNYVV-IV) [5]. Previous studies indicate that these isolates are virtually identical, differing only by two amino acids in the P25 protein, which allow BNYVV-IV to overcome resistance encoded by the *Rz1* gene [2]. Both isolates are well-characterized and represent the broadest known range of variability expected for North American BNYVV isolates in sugar beet with respect to performance against the widely planted *Rz1* source of resistance to BNYVV. Soil containing either BNYVV-A or BNYVV-IV was mixed with equal parts sterile builders sand, placed in new Styrofoam cups, and 100 R30_rz1 sugar beet seeds were planted per cup, with five cups per treatment. Seedlings were grown as described in Larson *et al.* [32] in a Conviron PGC15 Growth Chamber (Winnipeg, MB, Canada) at 24 ℃ with 16 h days and approximately 220 μM $m^{-1}s^{-2}$ light. At 3 weeks post-germination plants were harvested and washed to remove soil. Seedlings from all five cups for each treatment were pooled, and foliar and root portions of the plant were separated at the crown.

Pooled root samples from each treatment were tested by ELISA to confirm BNYVV infection [33]. Only samples with ELISA absorbance readings of at least 2 times the absorbance of healthy controls were subjected to proteomic analysis. Pooled leaf tissue from each treatment was ground in liquid nitrogen and lyophilized. The entire experiment was biologically replicated three times.

2.2. Protein Digestion and Mass Spectrometry

Total protein was extracted from 100 mg (dry weight) of lyophilized leaf tissue for each treatment from each biological replicate, using the Plant Total Protein Extraction Kit (Sigma, St. Louis, MO, USA) according to manufacturer's recommendations. Following extraction, suspended protein was further cleaned for quantification using the ReadyPrep 2-D Cleanup Kit (Bio-Rad, Hercules, CA, USA) in aliquots of 100 μL following manufacturer's recommendations. Protease inhibitors were added to liquid protein extracts (Pierce, Rockford, IL, USA), and samples were stored at −80 ℃ prior to analysis. Protein concentrations were determined via Bradford Assay [34] (Thermo Scientific, Rockford, IL, USA) and 30 μg of each sample underwent in-solution proteolytic digestion as previously described [35].

Briefly, samples were solubilized in 8 M urea, 0.2% Protease Max (Promega, Madison, WI, USA), then reduced with dithiothreitol, alkylated with iodoacetamide, and digested with 1% Protease Max and trypsin at 37 ° Cfor 3 h. Samples were dried in a Speed Vac® vacuum centrifuge, desalted using Pierce PepClean C18 spin columns (Pierce, Rockford, IL, USA), dried and resuspended in 30 μL 3% ACN, 0.1% formic acid. All solvents, water, and acid were LC-MS/MS grade from Sigma (St. Louis, MO, USA). One μLof each sample (from the three biological replicates from each soil type) was analyzed in duplicate injections via LC-MS/MS. Peptides were purified and concentrated using an on-line enrichment column (Thermo Scientific 5 μm, 100 μm ID × 2 cm C18 columns).

Subsequent chromatographic separation was performed on a reverse phase nanospray column (Thermo Scientific EASYnano-LC, 3 μm, 75 μm ID × 100 mm C18 column) using a 90 min linear gradient from 10%–30% buffer B (100% ACN, 0.1% formic acid) at a flow rate of 400 nL/min. Peptides were eluted directly into the mass spectrometer (Thermo Scientific Orbitrap Velos Pro) and spectra collected over a *m/z* range of 400–2,000 Da using a dynamic exclusion limit of 2 MS/MS spectra of a given peptide mass for 30 s (exclusion duration of 90 s). High resolution MS level scans were collected in the FT (resolution of 60,000), and MS/MS spectra were collected in the ion trap (IT). Compound lists of the resulting spectra were generated using Xcalibur 2.2 software (Thermo Scientific, Rockford, IL, USA) with an S/N threshold of 1.5 and 1 scan/group.

2.3. Protein Identification and Data Analysis

MS/MS spectra were searched against a Uniprot *Amaranthaceae* database [36] concatenated to a reverse database and the Uniprot *Mus musculus* (mouse) database [37] using the Mascot database search engine (Matrix Science, version 2.3.2) and SEQUEST (version v.27, rev. 11, Sorcerer, Sage-N Research). Due to the limited size of the *Amaranthaceae* database, the *Mus musculus* sequences were added to ensure adequate database size for statistical scoring and calculation of false discovery rates (FDR) [38]. *Mus musculus* was chosen in order to avoid potential redundant homologous hits with a plant database and any mouse hits were thus ignored in data analysis. While there are EST databases available for *Beta vulgaris* [39,40], we found that there is no straightforward tool to convert the entire database/sequence information to a protein FASTA database. Due to additional complexities associated with assigning the intron/exon sites, start/stop codons, and a lack of meaningful protein annotation from these RNA databases we concluded that conversion would not be an accurate or effective way to identify putative proteins.

The following search parameters were used in Mascot: monoisotopic mass, parent mass tolerance of 20 ppm, fragment ion mass tolerance of 0.8 Da, complete tryptic digestion allowing two missed cleavages, variable modification of methionine oxidation, and a fixed modification of cysteine carbamidomethylation. SEQUEST search parameters were the same except for a fragment ion mass tolerance of 1.0 Da and a parent ion tolerance of 0.0120 Da. Peptide identifications from both of the search engines were combined using probabilistic protein identification algorithms in Scaffold version 4.0.3 (Proteome Software, Portland, OR, USA) [41,42] with protein clustering enabled. All data files for each biological replicate were then combined using the "mudpit" option in Scaffold4 generating a composite listing for all proteins identified. Thresholds were set to 99% protein probability, 95% peptide probability, and a 2 unique peptide minimum was required. Peptides shared across proteins were apportioned between proteins according to a weighting function. The peptide FDR was 0% after manual validation of a subset of proteins identified by 2 unique peptides [43]. Criteria for manual validation included the following: (1) a minimum of at least 5 theoretical y or b ions in consecutive order that are peaks greater than 5% of the maximum intensity; (2) an absence of prominent unassigned peaks greater than 5% of the maximum intensity; and (3) indicative residue specific fragmentation, such as intense ions N-terminal to proline and immediately C-terminal to aspartate and glutamate. GO terms were then mapped by Scaffold4 using Uniprot and confirmed manually by literature review. All mass spectrometry data has been deposited to the ProteomeXchange [44] with identifier PXD000237.

2.4. Statistical Analyses for Differentially Expressed Proteins

Unweighted spectrum counts (SpC) and normalized quantitative value for total spectra and average total ion current (TIC) were exported from Scaffold4. *T*-tests were performed in Excel using the normalized quantitative value (for SpC and Avg TIC). Fold change was analyzed using the normalized quantitative value "plus 1" in Excel. Box plots were visualized in DanteR (Pacific Northwest National Laboratories) to assess the effect of normalization on the median spectral counts of each sample. Venn diagrams were generated in Scaffold4.

3. Results and Discussion

3.1. Protein Characterization

Prior to protein analysis the presence of BNYVV in test samples was confirmed by performing a BNYVV-specific ELISA assay on all pooled material. ELISA confirmed infection of the susceptible sugar beet variety by both BNYVV pathotypes, with a mean O.D. at 405 nm of 0.382 for BNYVV-IV, and 0.298 for BNYVV-A. Positive samples are considered those with O.D. values of at least twice that of healthy sugar beet (mean O.D. 0.145). Protein assignment was completed using Mascot analysis software and the Uniprot *Amaranthaceae* database with the accession number for each identified protein arranged by predicted annotated function (Supplementary Table S1). A total of 203 proteins (Supplementary Table S1) were identified that met our criteria as defined in the methodology, and mass spectrometry data has been deposited to the ProteomeXchange consortium [44] via the PRIDE partner repository [45] with the dataset identifier PXD000237. Normalized spectral abundance factors (NSAF) were calculated [46] (NSAF = (unweighted spectrum count/molecular weight)/sum spectral counts for experiment). The 203 proteins were categorized based on annotated GO terms for those proteins that had terms assigned within the categories biological process, cellular component, and molecular function (Figure 1). Categories with increased representation classified under biological process were predicted to function in metabolic (45%) and cellular processes (40%). The predominant protein classes under cellular components in leaf tissue were associated with cytoplasm (34%) and intracellular organelles (32%), while most proteins assigned under the molecular function category were classified as associated with molecular function (40%) and binding (30%) (Figure 1). The majority of the proteins found in the susceptible sugar variety were present during infection by both BNYVV pathotypes. However, there were unique proteins found in interactions with the standard A-type strain (BNYVV-A) that were not found in the resistance breaking strain (BNYVV-IV) and vice versa (Figure 2, Supplementary Table S1). Additionally, the total number of proteins in each predicted annotated function differed during each interaction. For example, there were more putative photosynthesis/energy production and metabolism proteins identified during the interaction with BNYVV-A than during interaction with the resistance breaking strain (BNYVV-IV). In contrast, more signal transduction and transport proteins were identified during the interaction with BNYVV-IV (Figure 2). This suggests that even in compatible interactions, different biological pathways are being induced to create a susceptible interaction in sugar beet dependent on the virus pathotype in question. In a previous study, Larson *et al.* [32] identified 50 proteins that were differentially up or down regulated in individual susceptible and resistant isogenic lines of sugar beet in response to BNYVV infection. In that study, the authors utilized a subtractive proteomic approach in which proteins were separated using a ProteomeLab PF2D, two-dimensional protein fractionation system (Beckman Coulter, Fullerton, CA, USA) and proteins were identified by MALDI-TOF/TOF mass spectrometry. Such an approach required a large amount of protein (>2 mg) in order to detect differential protein peaks for the subtractive analysis and therefore only those proteins with large differences and significant amounts of protein could be detected. In contrast, an untargeted, global proteomic profiling approach was used in our study, which allowed profiling of all detectable proteins, even those at low concentrations. This approach also allowed for the profiling of proteins expressed during compatible interactions with two pathotypes of BNYVV in a susceptible genotype. This global approach is critical, as it is important to understand all the potential metabolic pathways that may be induced during the response of sugar beet to pathogen infection and perhaps highlight the underlying plant defense response even in susceptible varieties, particularly because so little proteomic data is available for sugar beet. Importantly, this global LC-MS/MS approach is a significant improvement over the use of traditional 2D gels, allowing for easier sample preparation, method development, quantitation, and identification. Despite the difference in methodology, several proteins were identified in the present study that were also found by Larson *et al.* [32] (Supplementary Table S1) including glutamine synthetase, 50S ribosomal proteins, actin 1, profiling, and calmodulin, as well as several putative defense related proteins (superoxide dismutase, ascorbate peroxidase, peroxidases, and chitinase).

Figure 1. Gene Ontology (GO) terms for proteins identified from BNYVV infection of susceptible *Beta vulgaris* leaves. GO terms were collected by Scaffold4 from Uniprot and organized by **(a)** Biological Process **(b)** Cellular Component and **(c)** Molecular Function. Categories are represented as a percent of total identified GO terms.

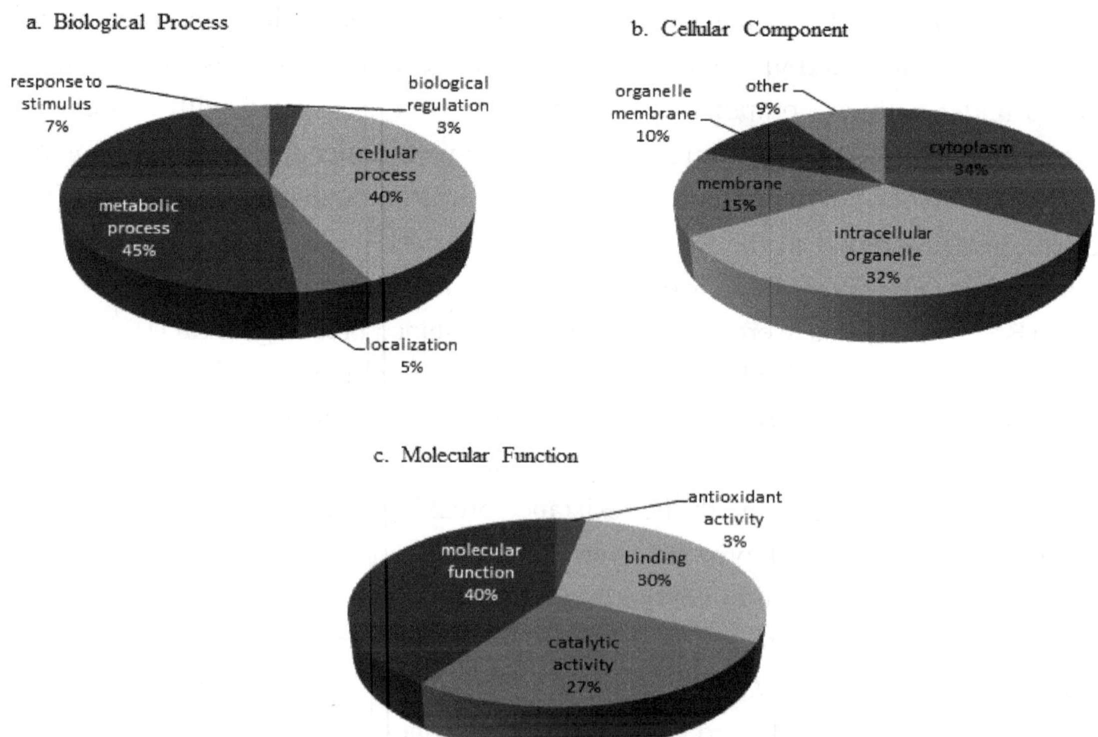

Many photosynthesis and energy related proteins were found to be expressed during compatible BNYVV interactions in sugar beet, including Ribulose bisphosphate(s), Chlorophyll a/b-binding proteins, photosystem reaction center subunits, ATPase(s), *etc.* (Supplementary Table S1). Management of energy resources by plants during periods of stress commonly occurs in order to provide the necessary energy required to mount a defense, and in some compatible virus/host interactions it is predicted that proteins involved in photosynthesis are likely activated by the virus itself, in order to produce the metabolic energy needed for viral replication (reviewed in [47]).

Figure 2. Total number of proteins identified from susceptible (R30_rz1) *Beta vulgaris* leaves during infection with the standard A-type (BNYVV-A) and the *Rz1* resistance-breaking strain (BNYVV-IV). Proteins are assigned to protein classes based on predicted annotated function as assigned by Scaffold4 using Uniprot and confirmed manually by literature review.

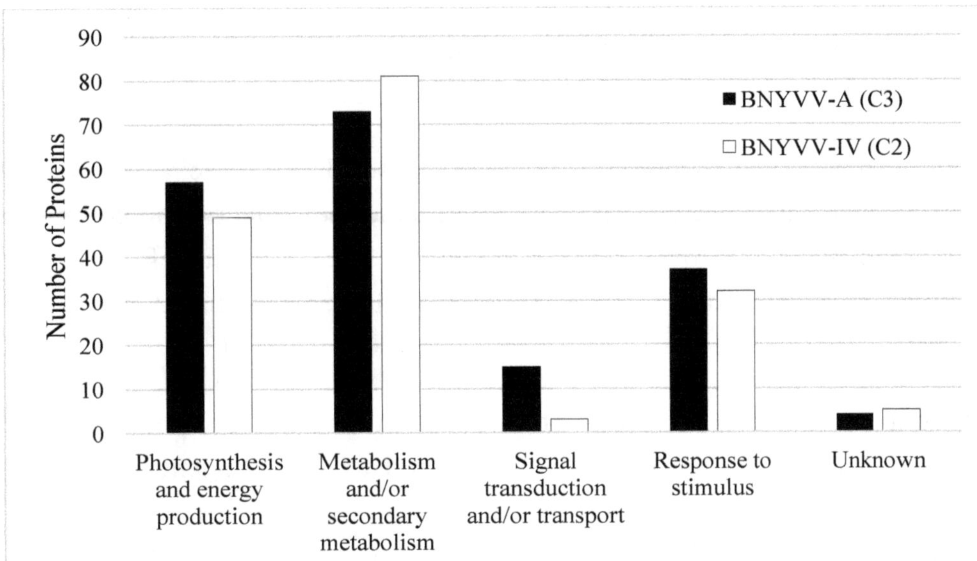

Primary metabolism in plants comprises all metabolic pathways that are essential to the plant's survival, growth, and development. In contrast, secondary metabolites are compounds produced in other metabolic pathways that, although important, are not essential to the functioning of the plant; however, many of these proteins are important in plant defense. Many proteins identified in this study have been associated with primary and secondary metabolic processes in plants (Supplementary Table S1), including glutamine synthetase (B2CZA8) and carbonic anhydrase (P16016).

Signal transduction pathways, which transmit information within individual cells and throughout the plant, are activated whenever biotic and abiotic stresses are recognized at the cellular level, leading to changes in many pathways and cellular processes in plants. Several proteins were expressed during BNYVV infection that are predicted to be associated with signal transduction and transport including three proteins previously reported by Larson *et al.* [32] (Supplementary Table S1). One such protein, calmodulin, has been extensively studied and shown to play crucial role(s) in cellular signaling and regulation of numerous target proteins in plants (reviewed in [48]).

3.2. Response to Stimulus (Including Plant Defense Response) Proteins

Several proteins identified were of primary interest as they had been previously reported to be associated with plant defense responses or responses to other stimuli in plants. Some of these proteins include: glucan endo-1,3-β-D-glucosidase, a 14-3-3 like protein, a pathogenesis related protein, multiple peroxidases, and superoxide dismutase (Supplementary Table S1).

The glucan endo-1,3-β-D-glucosidase (Q9XFW8) is of particular interest. Proteins in this class have been implicated in the hydrolysis of 1,3-β-D-glucosidic linkages in 1,3-β-D-glucans [49] and may contribute to susceptibility of plants to viruses. In plants, callose (a 1,3-β-D-glucan [50,51]) is deposited between the plasma membrane and the cell wall during the hypersensitive response to infection by both fungi and viruses [49,52]. It is believed that callose acts as a physical barrier limiting the spread of pathogens in resistant plants [53]. It was previously shown that down-regulation of another closely related protein β-1,3-glucanase, contributes to restricting the cell-to-cell movement of TMV, reduced plasmodesmatal size exclusion limits, and enhanced callose deposition in plants [49,54]. How glucan endo-1,3-β-D-glucosidase (Q9XFW8) is "behaving" during the susceptible interaction of BNYVV with sugar beet is unknown, but its identification in this study opens up intriguing avenues of future research to characterize compatible BNYVV/sugar beet interactions.

Another protein of interest was identified as a putative 14-3-3-like protein (P29308). 14-3-3 proteins have been reported to bind other proteins in order to regulate their function in complex environmental signaling pathways and have long been thought to play a role in plant defenses against pathogens. A number of studies implicate some 14-3-3 proteins in R-gene mediated plant disease resistance (reviewed in [55,56]).

During pathogen infection various proteins are induced in association with development of systemic acquired resistance (SAR). These are collectively referred to as pathogenesis-related (PR) proteins. PR proteins are defined as host proteins that are induced by the plant after pathogen attack [57]. We only found a single PR protein (Pathogenesis-related protein 1a/B5QTD3) that was expressed during BNYVV compatible interactions, suggesting that although SAR may be induced in the compatible virus-host interactions, the host response may be limited. However, the specific nature and function of putative PR-1 protein family members is unclear and indicates that this generalized defense response should be evaluated further in in subsequent studies.

Three peroxidases were found to be expressed in compatible BNYVV interactions: ascorbate peroxide (Q42459), and two peroxidases (P93552 and P93547). Peroxidases have been reported to play roles in plant defense to pathogens as a means of protection against the increased production of reactive oxygen species during the hypersensitive response and systemic acquired resistance pathways [58–60]. These have been correlated with accumulation of key enzymes important for building plant cell walls in resistant plants [58]. How putative peroxidases are being utilized in sugar beet during compatible interactions is unknown but they could be associated with generalized plant defense responses to BNYVV infection.

Superoxide dismutases are metalloenzymes found in plants and other organisms that protect cells against super-oxide radicals and prevent formation of other active species of oxygen. During the plant defense response superoxide dismutase expression increases, leading to generation of hydrogen peroxide and increased tolerance to pathogens [61,62]. Up-regulation of superoxide dismutase is believed to contribute in preventing oxidative damage in plants under stress [61,62]. A superoxide dismutase (A7WTB6) was expressed in both compatible BNYVV reactions in the susceptible sugar beet variety lending substantial support to the activity of this enzyme during BNYVV infection; how this protein is contributing to plant defense during compatible interactions should be investigated in future experiments.

3.3. Differential Expression of Proteins during Compatible BNYVV Interactions

During initial proteomic profiling we found that, in general, most of the proteins identified were found during interactions with both BNYVV pathotypes (Figure 2). Using statistical analysis of the unweighted spectrum counts (SpC) and average total ion current (Avg TIC) we found that eight proteins were more highly expressed during infection with the standard A-type strain (BNYVV-A) compared with the resistant breaking strain (BNYVV-IV) (Table 1). All eight proteins were identified in both interactions (Supplementary Table S1) but here we show that some proteins were more highly expressed. The proteins found to be differentially expressed during infection by the standard A-type strain generally fell into the categories metabolism/secondary metabolism and photosynthesis and energy production. Additionally, two proteins were identified that were previously found to be associated with stress responses in plants; a superoxide dismutase and a cytosolic heat shock protein.

Table 1. Proteins differentially expressed in a susceptible sugar beet variety by a standard A-type strain (BNYVV-A) compared to infection by the *Rz1* resistance-breaking strain (BNYVV-IV).

Protein Name	Accession Number	Spectral Counting *p*-value ($p < 0.1$) [1]	Average TIC *p*-value ($p < 0.1$) [2]
Photosystem II CP43 chlorophyll apoprotein	PSBC_SPIOL	0.025	0.001
Cytosolic heat shock 70 protein	Q41374_SPIOL	0.053	
Superoxide dismutase [Cu-Zn]	H9BQP7_SUASA	0.100	
Glyceraldehyde-3-phosphate dehydrogenase B, chloroplastic	G3PB_SPIOL		0.010
Choline monooxygenase, chloroplastic	CHMO_BETVU		0.032
Fructose-bisphosphate aldolase, chloroplastic	ALFC_SPIOL		0.033
Dehydroascorbate reductase	Q9FVE4_SPIOL		0.061
Formate—tetrahydrofolate ligase	FTHS_SPIOL		0.061

[1] Spectral counting compares the total number of peptide spectra that match to a protein between two groups.
[2] Average TIC measures the differences in average total ion current off of the mass spectrometer between the two samples.

Pathogen fitness is the ability of the organism to survive and reproduce [63] but can be defined further by using many factors including the level of aggressiveness of the pathogen in susceptible hosts, the ability of the pathogen to survive or persist among a population of other variants or other pathogens, the ability of the pathogen to multiply and spread within the host, as well as having the ability to spread to new hosts ([64], reviewed in [65]). It has been shown that mutations that allow a pathogen to overcome particular resistance genes (*i.e.*, loss of avirulence) also incur a fitness cost to the pathogen during compatible interactions which include decreased aggressiveness (the amount of disease) and reduced symptom expression (reviewed in [65]). Therefore, in susceptible plants, it is possible that the mutation that allows a resistant breaking pathotype to cause disease on resistant hosts, also changes how the pathogen interacts with susceptible hosts and likewise the molecular pathways that are activated in the susceptible host. The finding that there were eight proteins that are more highly expressed in a compatible interaction with the standard A-type BNYVV strain compared with the resistance breaking strain (BNYVV-IV) suggests that a fitness cost may have been incurred in the resistance breaking strain. Characterization of these mechanisms and their role in pathogen fitness should be more fully elucidated in future experiments.

4. Conclusions

Mass spectrometry based proteomic investigations of non-model systems, such as sugar beet, are challenging due to the lack of a well annotated sequenced genome. Larson *et al.* [32] were only able to identify ~42% of the differentially expressed proteins in that study due to reliance on homology with closely related, fully sequenced organisms when performing database searches for protein identification. A complete genomic sequence for sugar beet was only recently released [66] however it has neither been translated into a protein FASTA database, nor fully annotated. This greatly limits its utility for proteomic applications. The study presented herein represents the first large scale shotgun proteomic analysis of sugar beet, which, combined with homology based database searching against the Uniprot *Amaranthaceae* protein sequence database yielded the confident identification of 203 proteins from BNYVV infected sugar beet. Many proteins identified in susceptible sugar beet during infection with BNYVV were expressed with not just one pathotype, but both. These two BNYVV pathotypes represent the known diversity of BNYVV in the United States (with regards to performance against resistant sources), which justifies future downstream studies to determine specific pathways that may be involved in BNYVV pathogenesis. Understanding how sugar beet responds to infection by BNYVV will enable researchers and producers to identify more effective or novel methods for control of BNYVV infection and rhizomania disease through biotechnology, and may result in improved approaches to manage resistant germplasm, and potentially prolong the functional life of the *Rz* set of resistance genes. The results of this study lay the groundwork for expanded research into the mechanisms and pathways of compatible interactions between BNYVV and sugar beet that lead to development of rhizomania disease. Using the methods established herein, we plan future studies to describe the quantitative differences in protein expression between defined treatment groups. Such a large-scale quantitative approach will be the first of its kind in the field of sugar beet proteomics and should provide critical information on interactions between a virus and disease resistance genes.

Acknowledgments

This work was supported by the Beet Sugar Development Foundation, California Beet Growers Association, Sugarbeet Research and Education Board of Minnesota and North Dakota, and Western Sugar Cooperative-Grower Research Committee. We would like to thank Art Cortez, Paul Covey, Laura Hladky, and Addison Reed for assistance in sample preparation and data collection. We also thank the PRIDE team for assistance with deposit of the mass spectrometry data into the ProteomeXchange database. Mention of trade names or commercial products in this publication is solely for the purpose of providing specific information and does not imply recommendation or endorsement by the U.S. Department of Agriculture. USDA is an equal opportunity provider and employer.

Author Contributions

Conception of experimental design and sugar beet/BNYVV treatment structure was performed by K.M.W. and W.M.W. Sugar beet plant preparations and protein extractions were performed by W.M.W. Protein preparation, mass spectrometry, and sample analysis was performed by C.J.B. and J.E.P. K.M.W. compiled experimental results and had primary responsibility for manuscript preparation with C.J.B., J.E.P. and W.M.W. contributing to manuscript preparation.

References

1. Gidner, S.; Lennefors, B.L.; Nilsson, N.O.; Bensefelt, J.; Johansson, E.; Gyllenspetz, U.; Kraft, T. QTL mapping of BNYVV resistance from the WB41 source in sugar beet. *Genome* **2005**, *48*, 279–285.
2. Rush, C.M.; Liu, H.Y.; Lewellen, R.T.; Acosta-Leal, R. The continuing saga of rhizomania of sugar beets in the United States. *Plant Dis.* **2006**, *90*, 4–15.
3. Pavli, O.I.; Stevanato, P.; Biancardi, E.; Skaracis, G.N. Achievements and prospects in breeding for rhizomania resistance in sugar beet. *Field Crops Res.* **2011**, *122*, 165–172.
4. Abe, H.; Tamada, T. Association of *Beet necrotic yellow vein virus* with isolates of *Polymyxa betae* Keskin. *Ann. Phytopathol. Soc. Jpn.* **1986**, *52*, 235–247.
5. Liu, H.; Sears, J.L.; Lewellen, R.T. Occurrence of resistance breaking *Beet necrotic yellow vein virus* of sugar beet. *Plant Dis.* **2005**, *89*, 464–468.
6. Fujisawa, I.; Sugimoto, T. Transmission of *Beet necrotic yellow vein virus* by *Polymyxa betae*. *Ann. Phytopathol. Soc. Jpn.* **1976**, *43*, 583–586.
7. Tamada, T.; Baba, T. *Beet necrotic yellow vein virus* from rhizomania-affected sugar beet in Japan. *Ann. Phytopathol. Soc. Jpn.* **1973**, *39*, 325–332.
8. Lewellen, R.T.; Skoyen, I.O.; Erichsen, A.W. Breeding sugar beet for resistance to rhizomania: Evaluation of host-plant reactions and selection for and inheritance of resistance. In Proceeings of the 50th Congress of the IIRB, Brussels, Belgium, 11–12 February 1987; pp. 139–156.
9. Lewellen, R.T. Selection for resistance to rhizomania in sugar beet. In Proceeings of the 5th International Congress Plant Pathology, Kyoto, Japan, 20–27 August 1988; p. 455.
10. Scholten, O.E.; Lange, W. Breeding for resistance to rhizomania in sugar beet: A review. *Euphytica* **2000**, *112*, 219–231.
11. Biancardi, E.; Lewellen, R.T.; de Biaggi, M.; Erichsen, A.W.; Stevanato, P. The origin of rhizomania resistance to sugar beet. *Euphytica* **2002**, *127*, 383–397.
12. Scholten, O.E.; Klein-Lankhorst, R.M.; Esselink, D.G.; de Bock, T.S.; Lange, W. Identification and mapping of random amplified polymorphic DNA (RAPD) markers linked to resistance against *Beet necrotic yellow vein virus* (BNYVV) in *Beta* accessions. *Theor. Appl. Genet.* **1997**, *94*, 123–130.
13. Scholten, O.E.; de Bock, T.S.; Klein-Lankhorst, R.M.; Lange, W. Inheritance of resistance to *Beet necrotic yellow vein virus* in *Beta vulgaris* conferred by a second gene for resistance. *Theor. Appl. Genet.* **1999**, *99*, 740–746.
14. Koenig, R.; Luddecke, P.; Kaeberle, A.M. Detection of *Beet necrotic yellow vein virus* strains,

variants, and mixed infections by examining single-strand conformation polymorphisms of immunocapture RT-PCR products. *J. Gen. Virol.* **1995**, *76*, 2051–2055.

15. Koenig, R.; Lennefors, B.L. Molecular analyses of European A, B, and P type sources of *Beet necrotic yellow vein virus* and detection of the rare P type in Kazakhstan. *Arch. Virol.* **2000**, *145*, 1561–1570.
16. Kruse, M.; Koenig, R.; Hoffmann, A.; Kaufmann, A.; Commandeur, U.; Solovyev, A.G.; Savenkov, I.; Burgermeister, W. Restriction fragment length polymorphism analysis of reverse transcription PCR products reveals the existence of two mojor strain groups of *Beet necrotic yellow vein virus*. *J. Gen. Virol.* **1994**, *75*, 1835–1842.
17. Schirmer, A.; Link, D.; Cognat, V.; Moury, B.; Bouve, M.; Meunier, A.; Bragard, C.; Gilmer, D.; Lemaire, O. Phylogenetic analysis of isolates of *Beet necrotic yellow vein virus* collected worldwide. *J. Gen. Virol.* **2005**, *86*, 2897–2911.
18. McGrann, G.R.D.; Grimmer, M.K.; Mutasa-Göttgens, E.S.; Stevens, M. Progress towards the understanding and control of sugarbeet rhizomania disease. *Mol. Plant Pathol.* **2009**, *10*, 129–141.
19. Miyanishi, M.; Kusume, T.; Saito, M.; Tamada, T. Evidence for three groups of sequence variants of been necrotic yellow vein virus RNA5. *Arch. Virol.* **1999**, *144*, 879–892.
20. Tamada, T.; Kusume, T.; Uchino, H.; Kiguchi, T.; Saito, M. Evidence that *Beet necrotic yellow vein virus* RNA-5 is involved in symptom development of sugarbeet roots. In Proceedings of the Third Symposium of the International Working Group on Plant Viruses with Fungal Vectors, Dundee, UK, 6–7 August 1996; pp. 49–52.
21. Koenig, R.; Haeberle, A.M.; Commandeur, U. Detection and characterization of resistance-breaking isolates of *Beet necrotic yellow vein virus* in the United States. *Eur. Arch. Virol.* **1997**, *142*, 1499–1504.
22. Ward, L.; Koenig, R.; Budge, G.; Garrido, C.; McGrath, C.; Stubbley, H.; Boonham, N. Occurrence of two different types of RNA-5 containing *Beet necrotic yellow vein virus* in the UK. *Arch. Virol.* **2007**, *152*, 59–73.
23. Mehrvar, M.; Valizadeh, J.; Koenig, R.; Bragard, C. Iranian *Beet necrotic yellow vein virus* (BNYVV): Pronounced diversity of the p25 coding region in A-type BNYVV and identification of P-type BNYVV lacking a fifth RNA species. *Arch. Virol.* **2009**, *154*, 501–506.
24. Liu, H.Y.; Lewellen, R.T. Distribution and molecular characterization of resistance-breaking isolates of *Beet necrotic yellow vein virus* in the United States. *Plant Dis.* **2007**, *91*, 847–851.
25. Hajeidari, M.; Abdollahian-Noghabi, M.; Askari, H.; Heidari, M.; Sadeghian, S.Y.; Ober, E.S.; Salekdeh, G.H. Proteome analysis of sugar beet leaves under drought stress. *Proteomics* **2005**, *5*, 950–960.
26. Wakeel, A.; Asif, A.R.; Pitann, B.; Schubert, S. Proteome analysis of sugar beet (*Beta vulgaris* L.) elucidates constitutive adaptation during the first phase of salt stress. *J. Plant Physiol.* **2011, 168,** 519–526.
27. Yang, L.; Zhang, Y.; Zhu, N.; Koh, J.; Ma, C.; Pan, Y.; Yu, B.; Chen, S.; Li, H. Proteomic analysis of salt tolerance in sugar beet monosomic addition line M14. *J. Proteome Res.* **2013**, *12*, 4931–4920.
28. Rellan-Alvarez, R.; Andaluz, S.; Rodriguez-Celma, J.; Wohlgemuth, G.; Zocchi, G.; Alvarez-Fernandez, A.; Fiehn, O.; Lopez-Millan, A.F.; Abadia, J. Changes in the proteomic and

metabolomic profiles of *Beta vulgaris* root tips in response to iron deficiency and resupply. *BMC Plant Biol.* **2010** *10*, e120.

29. Catusse, J.; Strub, J.; Job, C.; van Dorsselaer, A.; Job, D. Proteome-wide characterization of sugarbeet seed vigor and its tissue specific expression. *Proc. Natl. Acad. Sci. USA* **2008**, *105*, 10262–10267.
30. Catusse, J.; Meinhard, J.; Job, C.; Strub, J.; Fischer, U.; Pestsova, E.; Westhoff, P.; van Dorsselaer, A.; Job, D. Proteomics reveals potential biomarkers of seed vigor in sugarbeet. *Proteomics* **2011**, *11*, 1569–1580.
31. Larson, R.L.; Hill, A.L.; Nunez, A. Characterization of protein changes associated with sugar beet (*Beta vulgaris*) resistance and susceptibility to *Fusarium oxysporum*. *J. Agric. Food Chem.* **2007**, *55*, 7905–7915.
32. Larson, R.L.; Wintermantel, W.M.; Hill, A.L.; Fortis, L.; Nunez, A. Proteome changes in sugar beet in response to *Beet necrotic yellow vein virus*. *Physiol. Mol. Plant Pathol.* **2008**, *72*, 62–72.
33. Wisler, G.C.; Lewellen, R.T.; Sears, J.L.; Liu, H.; Duffus, J.E. Specificity of TAS-ELISA for *Beet necrotic yellow vein virus* and its application for determining rhizomania resistance in field-grown sugar beets. *Plant Dis.* **1999**, *83*, 864–870.
34. Bradford, M.M. A rapid and sensitive method for the quantification of microgram quantities of protein utilizing the principle of protein-dye binding. *Anal. Biochem.* **1976**, *72*, 248–254.
35. Freund, D.M.; Prenni, J.E.; Curthoys, N.P. Response of the mitochondrial proteome of rat renal proximal convoluted tubules to chronic metabolic acidosis. *Am. J. Physiol.* **2013**, *304*, F145–F155.
36. Uniprot *Amaranthaceae* database. Available online: http://www.uniprot.org/taxonomy/3563 (accessed on 6 March 2013).
37. Uniprot *Mus musculus* database. Available online: http://www.uniprot.org/taxonomy/10090 (accessed on 6 March 2013).
38. Giri, P.K.; Kruh, N.A.; Dobos, K.M.; Schorey, J.S. Proteomic analysis identifies highly antigenic proteins in exosomes from *M. tuberculosis*-infected and culture filtrate protein-treated macrophages. *Proteomics* **2010**, *10*, 3190–3202.
39. The Sugarbeet EST database. Available online: http://genomics.msu.edu/cgi-bin/sugarbeet/est_search.cgi (accessed on 5 April 2014).
40. Beta vulgaris Gene Index (BvGI). Available online: http://compbio.dfci.harvard.edu/cgi-bin/tgi/tc_ann.pl?gudb=beet (accessed on 5 April 2014).
41. Keller, A.; Nesvizhskii, A.I.; Kolker, E.; Aebersold, R. Empirical statistical model to estimate the accuracy of peptide identifications made by MS/MS and database search. *Anal. Chem.* **2002**, *74*, 5383–5392.
42. Nesvizhskii, A.I.; Keller, A.; Kolker, E.; Aebersold, R. A statistical model for identifying proteins by tandem mass spectrometry. *Anal. Chem.* **2003**, *75*, 4646–4658.
43. Elias, J.E.; Gygi, S.P. Target-decoy search strategy for increased confidence in large-scale protein identifications by mass spectrometry. *Nat. Methods* **2007**, *4*, 207–214.
44. ProteomeXchange consortium. Available online: http://proteomecentral.proteomexchange.org (accessed on 6 March 2013).
45. Vizcaino, J.A.; Cote, R.G.; Csordas, A.; Dianes, J.A.; Fabregat, A.; Foster, J.M.; Griss, J.; Alpi, E.; Birim, M.; Contell, J.; *et al.* The PRoteomics IDEntifications (PRIDE) database and associated tools: Status in 2013. *Nucleic Acids Res.* **2013**, *4*, D1063–D1069.

46. Paoletti, A.C.; Parmely, T.J.; Tomomori-Sato, C.; Sato, S.; Zhu, D.; Conaway, R.C.; Conaway, J.W.; Florens, L.; Washburn, M.P. Quantitative proteomic analysis of distinct mamalian mediator complexes using normalized spectral abundance factors. *Proc. Natl. Acad. Sci. USA* **2006**, *103*, 18928–18933.
47. Carli, M.D.; Benvenuto, E.; Donini, M. Recent insights into plant-virus interactions through proteomic analysis. *J. Proteome Res.* **2012**, *11*, 4765–4780.
48. Ranty, B.; Alson, D.; Galaud, J. Plant calmodulins and calmodulin-related proteins. *Plant Signal. Behav.* **2006**, *1*, 96–104.
49. Beffa, R.S.; Hofer, R.M.; Thomas, M.; Meins, F., Jr. Decreased susceptibility to viral disease of [beta]-1,3-glucanase-deficient plants generated by antisense transformation. *Plant Cell* **1996**, *8*, 1001–1011.
50. Bell, A.A. Biochemical mechanisms of disease resitance. *Annu. Rev. Plant Physiol.* **1981**, *32*, 21–81.
51. Stone, B.A.; Clarke, A.E. *Chemistry and Biology of (1-3)-beta-glucans*; La Trobe University Press: Melbourne, Austrailia, 1992.
52. Aist, J.R. Papillae and related wounds plugs of plant cells. *Annu. Rev. Phytopathol.* **1976**, *14*, 145–163.
53. Allison, A.V.; Shalla, T.A. The ultrastructure of local lesions induced by potato virus X: A sequence of cytological events in the course of infection. *Phytopathology* **1974**, *64*, 784–793.
54. Iglesias, V.A.; Meins, F., Jr. Movement of plant viruses is delayed in a beta-1,3-glucanase deficient mutant showing a reduced plasmodesmatal size exclusion limit and enhanced callose deposition. *Plant J.* **2000**, *21*, 157–166.
55. Roberts, M.R.; Salinas, J.; Collinge, D.B. 14-3-3 proteins and the response to abiotic and biotic stress. *Plant Mol. Biol.* **2002**, *1031*, 1031–1039.
56. Denison, F.C.; Paul, A.L.; Zupanska, A.K.; Ferl, R.J. 14-3-3 proteins in plant physiology. *Semin. Cell Dev. Biol.* **2011**, *22*, 720–727.
57. Van Loon, L.C.; van Strien, E.A. The families of pathogenesis-related proteins, their activities, and comparative analysis of PR-1 type proteins. *Physiol. Mol. Plant Pathol.* **1999**, *55*, 85–97.
58. Chittoor, J.M.; Leach, J.E.; White, F.F. Induction of peroxidase during defense against pathogens. In *Pathogenesis-Related Proteins in Plants*; Datta, S.K., Muthukrishnan, S., Eds.; CRC Press: Boca Raton, FL, USA, 1999; pp. 171–193.
59. Kawano, T. Roles of the reactive oxygen species-generating peroxidase reactions in plant defese and growth induction. *Plant Cell Rep.* **2003**, *21*, 829–837.
60. Caverzan, A.; Passaia, G.; Barcellos Rosa, S.; Werner Ribeiro, C.; Lazzarotto, F.; Margis-Pinheiro, M. Plant resonses to stresses: Role of ascorbate peroxidase in the antioxidant protection. *Genet. Mol. Biol.* **2012**, *35*, 1011–1019.
61. Gupta, A.S.; Heinen, J.L.; Holaday, A.S.; Burke, J.J.; Allen, R.D. Increased resistance to oxidative stress in transgenic plants that overexpress chloroplastic Cu/Zn superoxide dismutase. *Proc. Natl. Acad. Sci. USA* **1993**, *90*, 1629–1633.
62. Tsang, E.; Bowler, C.; Herouart, D.; van Camp, W.; Villarroel, R.; Genetello, C.; van Montagu, M.; Inze, D. Differential regulation of superoxide dismutases in plants exposed to environmental stress. *Plant Cell* **1991**, *3*, 783–792.

63. Crow, J.F. *Basic Concepts in Population, Quantitative, and Evolutionary Genetics*; Academic Press: New York, NY, USA, 1986; p. 273.

64. Antonovics, J.; Alexander, H.M. The concept of fitness in plant-fungal pathogen systems. In *Plant Disease Epidemiology*; Leonard, K.J., Fry, W.E., Eds.; McGraw-Hill: New York, NY, USA, 1989; pp. 185–214.

65. Leach, J.E.; Vera Cruz, C.M.; Bai, J.; Leung, H. Pathogen fitness penalty as a predictor of durability of disease resistance genes. *Annu. Rev. Phytopathol.* **2001**, *39*, 187–224.

66. Eujayl, I.; Strausbaugh, C. *Beta vulgaris* subsp. Vulgaris. Available online: http://www.ncbi.nlm.nih.gov/Traces/wgs/?val=ARYA01 (accessed on 5 April 2014).

Permissions

All chapters in this book were first published in MDPI; hereby published with permission under the Creative Commons Attribution License or equivalent. Every chapter published in this book has been scrutinized by our experts. Their significance has been extensively debated. The topics covered herein carry significant findings which will fuel the growth of the discipline. They may even be implemented as practical applications or may be referred to as a beginning point for another development.

The contributors of this book come from diverse backgrounds, making this book a truly international effort. This book will bring forth new frontiers with its revolutionizing research information and detailed analysis of the nascent developments around the world.

We would like to thank all the contributing authors for lending their expertise to make the book truly unique. They have played a crucial role in the development of this book. Without their invaluable contributions this book wouldn't have been possible. They have made vital efforts to compile up to date information on the varied aspects of this subject to make this book a valuable addition to the collection of many professionals and students.

This book was conceptualized with the vision of imparting up-to-date information and advanced data in this field. To ensure the same, a matchless editorial board was set up. Every individual on the board went through rigorous rounds of assessment to prove their worth. After which they invested a large part of their time researching and compiling the most relevant data for our readers.

The editorial board has been involved in producing this book since its inception. They have spent rigorous hours researching and exploring the diverse topics which have resulted in the successful publishing of this book. They have passed on their knowledge of decades through this book. To expedite this challenging task, the publisher supported the team at every step. A small team of assistant editors was also appointed to further simplify the editing procedure and attain best results for the readers.

Apart from the editorial board, the designing team has also invested a significant amount of their time in understanding the subject and creating the most relevant covers. They scrutinized every image to scout for the most suitable representation of the subject and create an appropriate cover for the book.

The publishing team has been an ardent support to the editorial, designing and production team. Their endless efforts to recruit the best for this project, has resulted in the accomplishment of this book. They are a veteran in the field of academics and their pool of knowledge is as vast as their experience in printing. Their expertise and guidance has proved useful at every step. Their uncompromising quality standards have made this book an exceptional effort. Their encouragement from time to time has been an inspiration for everyone.

The publisher and the editorial board hope that this book will prove to be a valuable piece of knowledge for researchers, students, practitioners and scholars across the globe.

List of Contributors

Faith M. Robison
Proteomics and Metabolomics Facility, Colorado State University, Fort Collins, CO 80523, USA
Department of Soil and Crop Sciences, Colorado State University, Fort Collins, CO 80523, USA

Mark A Brick
Department of Soil and Crop Sciences, Colorado State University, Fort Collins, CO 80523, USA

Adam L. Heuberger
Department of Horticulture and Landscape Architecture, Colorado State University, Fort Collins, CO 80523, USA

Jessica E. Prenni
Proteomics and Metabolomics Facility, Colorado State University, Fort Collins, CO 80523, USA
Department of Biochemistry, Colorado State University, Fort Collins, CO 80523, USA

Shaojun Dai
College of Life and Environmental Sciences, Shanghai Normal University, Shanghai 200234, China

Juanjuan Yu
College of Life and Environmental Sciences, Shanghai Normal University, Shanghai 200234, China
Sanquan Medical College, Xinxiang Medical University, Xinxiang 453003, He'nan, China

Sixue Chen
Department of Biology, Genetics Institute, Plant Molecular and Cellular Biology Program, Gainesville, FL 32610, USA

Tai Wang
Institute of Botany, Chinese Academy of Sciences, Beijing 100093, China

Guorong Sun
Binzhou Polytechnic College, Binzhou 256603, Shandong, China

Claudia-Nicole Meisrimler and Sabine Lüthj
University of Hamburg, Biocenter Klein Flottbek and Botanical Garden, Oxidative Stress and Plant Proteomics Group, Ohnhorststraße 18, D-22609 Hamburg, Germany

Friedrich Buck
University Hospital Eppendorf, Institute of Clinical Chemistry, Campus Science, Martinistraße 52, D 20246 Hamburg, Germany

Hiromu Suzuki, Yuya Takashima, Futoshi Ishiguri, Nobuo Yoshizawa and Shinso Yokota
Department of Forest Science, Faculty of Agriculture, Utsunomiya University, 350 Mine-machi, Utsunomiya, Tochigi 321-8505, Japan

Jin Yu, Swapan Kumar Roy, Abu Hena Mostafa Kamal, Soo-Jeong Kwon, Yoon-Sup So and Sun Hee Woo
Department of Crop Science, Chungbuk National University, Cheong-ju 361-763, Korea

Kun Cho
Division of Mass Spectrometry Research, Korea Basic Science Institute, Chungbuk 863-883, Korea

Seong-Woo Cho
Lab of Molecular Breeding, Arid land Research Center, Tottori University, Tottori 680-8550, Japan

James B. Holland
USDA-ARS Plant Science Research Unit, Department of Crop Science, Box 7620, North Carolina State University, Raleigh, NC 27695, USA

Zahed Hossain
Plant Stress Biology Lab, Department of Botany, West Bengal State University, Kolkata-700126, India

Setsuko Komatsu
National Institute of Crop Science, National Agriculture and Food Research Organization, Tsukuba 305 - 8518, Japan

Ikenna Okekeogbu, Zhujia Ye, Sasikiran Reddy Sangireddy, Hui Li, Sarabjit Bhatti, Dafeng Hui and Suping Zhou
Department of Agricultural and Environmental Sciences, College of Agriculture, Human and Natural Sciences, Tennessee State University, 3500 John A Merritt Blvd, Nashville, TN 37209, USA

Kevin J. Howe, Tara Fish, Yong Yang and Theodore W. Thannhauser
Plant, Soil and Nutrition Research Unit, USDA-ARS, Tower Rd, Ithaca, NY 14853, USA

Jose T. A. Oliveira, Fredy D. A. Silva, Jose H. Costa and Ilka M. Vasconcelos
Department of Biochemistry and Molecular Biology, Federal University of Ceara, Fortaleza 60451-970, Brazil

Jose H. Araujo-Filho
Department of Biological Sciences, State University of Rio Grande do Norte, Rio Grande do Norte, Mossoro 59610-210, Brazil

Thalles B. Grangeiro and Marina D. P. Lobo
Department of Biology, Federal University of Ceara, Fortaleza 60451-970, Brazil

Darcy M. F. Gondim
Department of Biochemistry and Molecular Biology, Federal University of Ceara, Fortaleza 60451-970, Brazil
University of Fortaleza (UNIFOR), Ceara, Fortaleza 60451-970, Brazil

Jeferson Segalin and Celia R. R. S. Carlini
Department of Biophysics and Center of Biotechnology, Federal University of Rio Grande do Sul, Rio Grande do Sul, Porto Alegre 91501-970, Brazil

Paulo M. Pinto
Institute of Biotechnology, University of Caxias do Sul, Caxias do Sul, RS, Rio Grande do Sul, Caxias do Sul 95070-560, Brazil

Ines Lassowskat, Justin Lee and Dierk Scheel
Leibniz Institute of Plant Biochemistry, Weinberg 3, Halle/Saale, D-06120, Germany

Kai Naumann
Riboxx GmbH, Meissner Str. 191, Radebeul, D-01445, Germany

Ziyang Fu and Pingfang Yang
Key Laboratory of Plant Germplasm Enhancement and Speciality Agriculture, Wuhan Botanical Garden, Chinese Academy of Sciences, Wuhan 430074, China

Emdadul Haque, Fumitaka Abe, Masahiko Mori, Yohei Nanjo, Setsuko Komatsu, Atsushi Oyanagi and Kentaro Kawaguchi
NARO Institute of Crop Science (NICS), National Agriculture and Food Research Organization (NARO), 2-1-18 Kannondai, Tsukuba, Ibaraki 305-8518, Japan

Cécile Albenne, Hervé Canut, Laurent Hoffmann and Elisabeth Jamet
Université de Toulouse; UPS, UMR 5546, Laboratoire de Recherche en Sciences Végétales, BP 42617 Auzeville, F-31326 Castanet-Tolosan, France
CNRS, UMR 5546, BP 42617, F-31326 Castanet-Tolosan, France

Kimberly M. Webb
USDA-ARS-SBRU, Crops Research Laboratory, 1701 Centre Ave., Fort Collins, CO 80526, USA

Carolyn J. Broccardo and Jessica E. Prenni
Proteomics and Metabolomics Facility, Colorado State University, C130 Microbiology, 2021 Campus Delivery, Fort Collins, CO 80523, USA

William M. Wintermantel
USDA-ARS-CIPRU, 1636 E. Alisal St., Salinas, CA 93905, USA

Index

Printed in the USA
CPSIA information can be obtained
at www.ICGtesting.com
LVHW080042211023
761596LV00010BA/1866

9 781639 877546